ADP-Ribosylation Reactions

Guy G. Poirier
Pierre Moreau
Editors

ADP-Ribosylation
Reactions

With 164 Illustrations

Springer-Verlag
New York Berlin Heidelberg London Paris
Tokyo Hong Kong Barcelona Budapest

Guy G. Poirier
Centre de Recherches en
Endocrinologie Moleculaire
Le Centre Hospitalier de
  l'Universite Laval
Quebec, G1V 4G2 Canada

Pierre Moreau
Biochemistry Department
Pavillon Vachon
Universite Laval
Quebec, G1K 7P4 Canada

Library of Congress Cataloging-in-Publication Data
ADP-ribosylation reactions / [edited by] Guy Poirier, Pierre Moreau.
        p.   cm.
    Includes bibliographical references.
    ISBN 0-387-97822-4 — ISBN 3-540-97822-4
    1. ADP-ribosylation.   I. Poirier, Guy, 1948-   .   II. Moreau,
Pierre, 1952-   .
    [DNLM: 1. Adenosine Diphosphate Ribose — metabolism — congresses.
2. DNA Repair — physiology — congresses.   3. NAD, ADP
— Ribosyltransferase — metabolism — congresses.   4. Neoplasms —
metabolism — congresses.   5. Poly Adenosine Diphoshate Ribose —
metabolism — congresses.   QU 58 A2409]
QP625.A29.A359   1992
599'.01925 — dc20
DNLM/DLC
for Library of Congress   92-2197

Printed on acid-free paper.

Production supervised by Henry Krell; manufacturing managed by Jacqui Ashri.
Camera-ready copy prepared by the editors.
Printed and bound by Edwards Brothers, Inc., Ann Arbor, MI.
Printed in the United States of America.

9 8 7 6 5 4 3 2 1

ISBN 0-387-97822-4 Springer-Verlag New York Berlin Heidelberg
ISBN 3-540-97822-4 Springer-Verlag Berlin Heidelberg New York

Preface

This monograph is dedicated to one of the discoverers of poly(ADP-ribose), Professor Paul Mandel, from the Centre de Neurochimie in Strasbourg.

We would like to congratulate him for his distinguished contributions to the field of poly(ADP-ribosyl)ation and express our gratitude for his support in the last years and particularly for his encouragement for the organization of this meeting.

Poly(ADP-ribose) was discovered more than 25 years ago. Since then, excellent progress has been made on the study of the mechanisms of poly(ADP-ribose) reaction. The last five years have been particularly exciting since the development of various molecular biology techniques has revealed the complex nature of this multifunctional enzyme.

Looking at the contributions presented at this meeting, it becomes obvious that more work at the molecular level is needed. Most likely, these experiments will shed some light on the functions of poly(ADP-ribose), but further biophysical studies will still be required to fully understand this complex enzymatic system.

The editors of this monograph are deeply indebted to the work of Dr. Jean Lagueux. Also the work of Mrs. Aline Douville, Mrs. Danièle Poirier and Mrs. Sylvia Kuersteiner is gratefully acknowledged.

Contents

Part 2: Cancer, DNA Repair, and Metabolism

Part 3: Enzymology

Part 4: Mono ADP-Ribosylation

Contributors

ABBONDANDOLO, A.
Lab. of Mutagenesis
Nat. Inst. for Cancer Research
V. le Benedetto XV, 10
I-16132, Genoa
Italy

ABOUL-ELA, Nasreen
Department of Medicine
Texas College of Osteopathic Medicine
University of North Texas
Fort Worth, TX 76107-2690
USA

ADAMIK, Ronald
Lab. Cell. Metabolism
Nat. Heart, Lung and Blood Inst.
National Institutes of Health
Bethesda, MD 20892
USA

AKIYAMA, Tadakazu
Inst. of Basic Medical Sciences
University of Tsukuba
Tsukuba-Shi, Ibaraki 305
Japan

AKTORIES, Klaus
Rudolf-Buchheim-Inst. Pharmakol.
der Justus-Liebig-Univ. Giessen
G-8300 Giessen
Germany

ALTHAUS, Felix R.
Vet. Med. Fakultät (Tierspital)
Inst. für Pharmakologie u Biochemie
Winterthurerstrasse 260
CH-8057 Zurich
Switzerland

ALVAREZ-GONZALEZ, Rafael
Dept. of Microbiology and Immunology
Texas College of Osteopathic Medicine
3500 Camp Bowie Bld.
Fort Worth, Texas 76107-2690
USA

AMI, Yoshihiro
Inst. of Basic Medical Sciences
University of Tsukuba
Tsukuba-Shi, Ibaraki 305
Japan

ARMATO, Ubaldo
Instituto di Anatomia ed Istologia
Strada le Grazie
I-37134 Verona
Italy

AUER, Bernhard
Institut für Biochemie
Universität Innsbruck
A-6020 Innsbruck
Austria

BACHMANN, S.A.
Vet. Med. Fakultät (Tierspital)
Inst. für Pharmakologie u Biochemie
Winterthurerstrasse 260
CH-8057 Zürich
Switzerland

BALDUCCI, Enrico
Dipartimento di Biologia MCA
Universita di Camerino
I-62032 Camerino
Italy

BANASIK, Marek
Dept. of Clinical Science and
Lab. Medicine
Kyoto Univ., Faculty of Medicine
Shogoin, Sakyo-ku
Kyoto 606
Japan

BANSAL, Poonam
Dept. of Biochemistry
All India Institute of Medical Sciences
New Delhi 110029
India

BAUER, Pal I.
Romberg Tiburon Centers
Center for Environmental Studies
San Francisco State Univ.
Tiburon, CA 94920
USA

BERGER, Nathan A.
Hematology/Oncology Division
University Hospitals of Cleveland
Case Western Reserve University
Cleveland, OH 44106
USA

BERGER, Sosamma J.
Hematology/Oncology Division
University Hospitals of Cleveland
Case Western Reserve University
Cleveland, OH 44106
USA

BERGHAMMER, Heinrich
Inst. für Biochemie (nat. Fak.)
University of Innsbruck
Peter-Mayrstr. 1A
A-6020 Innsbruck
Austria

BERTAZZONI, Umberto
Instituto di Genetica Biochimica
Consiglio Nazionale delle Ricerche
Via Abbiategrasso 207
I-27100 Pavia
Italy

BHATIA, Kishor
Basic Science Building, Room 34B
Georgetown Univ. School of Med.
Washington, D.C. 20007
USA

BLIZIOTES, M.M.
NICHD
National Institutes of Health
Bethesda, MD 20892
USA

BOORSTEIN, Robert J.
Dept. of Pathology
NYU School of Medicine
550 First Avenue
New York, NY 10016
USA

BOURASSA, Sylvie
Molecular Endocrinology Lab.
CHUL Research Center
2705, boul. Laurier
Sainte-Foy (Québec) G1V 4G2
Canada

BOYD, Mark
Cell & Molecular Biology Laboratory
University of Sussex, Falmer
Brighton, BN1 9QG
United Kingdom

BRAUN, Stephan
Institute of Pharmacology, Tierspital
Winterthurerstrasse 260
CH-8057 Zurich
Switzerland

BRAZ, Marcus J.
Molecular Medicine Unit
King's College School of Med.
Denmark Hill
London SE5 8RX
England

BUKI, Kalman G.
Romberg Tiburon Centers
Center for Environmental Studies
San Francisco State Univ.
Tiburon, CA 94920
USA

BURKLE, Alexander
German Cancer Research Center
Inst. für Virusforschung
JM Neven Heimer Feld 506
D-6900 Heidelberg
Germany

BUTT, Tauseef
Dept. Biochemistry & Mol. Biology
Georgetown Univ. School of Medicine
Washington, D.C. 20007
USA

CARCERERI DE PRATI, Alessandra
Instituto di Chimica Biologica
Universita di Verona
Strada le Grazie
I-37134 Verona
Italy

CAVANAUGH, Eleanor
Lab. of Cellular Metabolism
Nat. Heart, Lung and Blood Inst.
National Institutes of Health
Bethesda, MD 20892
USA

CESARONE, C. Frederico
Inst. of General Physiology
University of Genoa, Fac. of Sciences
Corso Europa 26, Science Bldg.
16132 Genoa
Italy

CHABERT, Magali G.
Centre de Neurochimie
Centre Nat. de Recherche Scientifique
5, rue Blaise Pascal
F-67084, Strasbourg Cedex
France

CHAN, JOHN R.
Mol. Genetics and Carcinogenesis Lab.
Cross Cancer Institute
11560 University Avenue
Edmonton (Alberta) T6G 1Z2
Canada

CHATTERJEE, Satadal
Hematology/Oncology Division
University Hospitals of Cleveland
Case Western Reserve University
Cleveland, OH 44106
USA

CHAUDHRY, Bushra
Dept. Biochemistry & Mol. Biology
Georgetown Univ. School of Medicine
Washington, D.C. 20007
USA

CHEN, Hao-Chia
NICHD
National Institutes of Health
Bethesda, MD 20892
USA

CHEN, Hai-Ying
Dept. of Biochemistry & Mol. Biol.
Texas College of Osteopathic Medicine
University of North Texas
Fort Worth, Texas 76107-2690
USA

CHENG, Ming-Fang
Hematology/Oncology Division
University Hospitals of Cleveland
Case Western Reserve University
Cleveland, OH 44106
USA

CHERNEY, Barry W.
Dept. of Biochemistry & Mol. Biology
Georgetown University
School of Medicine
Washington, D.C. 20007
USA

CICATIELLO, L.
Istituto di Patologia Generale
ed Oncologia
Universita di Napoli
Napoli
Italy

CLERICI, Libero
CCR Euratom
I-21027 Ispra
Italy

COLLIER, R. John
Microbiology & Molecular Genetics
Harvard Medical School
Boston, MA 02115
USA

COLLINGE, Margaret
Inst. für Pharmakologie u Biochemie
Winterthurerstrasse 260
CH-8057 Zurich
Switzerland

COLLINS, Mary K.L.
Inst. of Cancer Research
Fulliam Road
London SW3 8JB
England

COLUMBANO, A.
Ist. di Farmacologia e Patologia Bioch.
Universita di Cagliari
Italy

COOK, Paul F.
Dept. of Microbiology
Texas College of Osteopathic Medicine
North Texas State University
Fort Worth, TX 76107
USA

COSI, A.
Instituto di Chimica Biologica
Universita di Verona
Strada Le Grazie
I-37134 Verona
Italy

COX, Ray
Cancer Research Lab. no. 151
Veterans Adm. Medical Center
1030 Jefferson Ave.
Memphis, TN 38104
USA

DANIELSEN, Mark
Dept. of Biochemistry & Mol. Biology
School of Medicine
Georgetown University
Washington, D.C. 20007
USA

DARLING, David
Mol. Med. Unit
Kings Coll. School of Med. & Dent.
Denmark Hill
London SE5 8RX
England

DE LUCIA, Filomena
Dip. di Chimica Organica e Biologica
Facolta di Scienze
Universita di Napoli
Napoly
Italy

DE MURCIA, Gilbert
Inst. Biologie Moléculaire et Cellulaire
Centre Nat. de Recherche Scientifique
15 rue Descartes
F-67084 Strasbourg Cedex
France

DE ROSA, Maria
Ist. Biochim. Macrom.
Fac. Medicina
Universita di Napoli
Napoli
Italy

DENDA, Ayumi
Dept. of Oncological Pathology
Cancer Center, Nara Medical College
840 Shijo-cho
Kashihara, Nara 634
Japan

DESMARAIS, Yvan
Molecular Endocrinology lab.
CHUL Research Center
2705, boul. Laurier
Sainte-Foy (Québec) G1V 4G2
Canada

DESNOYERS, Serge
Molecular Endocrinology lab.
CHUL Research Center
2705, boul. Laurier
Sainte-Foy (Québec) G1V 4G2
Canada

DIETRICH, Kevin D.
Mol. Genetics and Carcinogenesis Lab.
Cross Cancer Institute
11560 University Avenue
Edmonton (Alberta) T6G 1Z2
Canada

DURKACZ, Barbara W.
Cancer Research Unit
Medical School
University of Newcastle upon Tyne
Newcastle upon Tyne NE2 4HH
United Kingdom

DUTNALL, R.N.
Dept. of Mol. & Cell Biol.
University of Aberdeen, Marischal Coll.
Aberdeen, Scotland AB9 1AS
United Kingdom

EHRLICH, Wilhelm
Robert Koch-Institute
Abt. Biochemie
Nordufer 20
D-1000 Berlin 65
Germany

EMANUELLI, Monica
Instituto di Biochimica
Facolta di Medicina
Universita di Ancona
I-60131 Ancona
Italy

ESUMI, Hiroyasu
Biochemistry Div.
National Cancer Center Res. Inst
1-1, 5-chome, Tsukiji Chuo-ku
Tokyo 104
Japan

FARAONE-MENNELLA, Maria Rosaria
Dpt. di Chimica Organica e Biol.
Universita di Napoli
Via Mezzocannone 16
I-80134 Napoli
Italy

FARINA, Benedetta
Dpt. di Chimica Organica e Biol.
Universita di Napoli
Via Mezzocannone 16
I-80134 Napoli
Italy

FARZANEH, Farzin
Molecular Medicine Unit
King's College School of Med. & Dent.
Denmark Hill
London SE5 8RX
United Kingdom

FENDRICK, James L.
Dept. of Microbiology and Immunology
University of Arkansas for Med. Sc.
4301 West Markham St., Slot 511
Little Rock, Ar.72205-7199
USA

FISHER, Paul B.
Dept. of Pathology and Urology
Inst. of Cancer Research
Columbia University
New York, NY 10032
USA

FLICK, K.
Institut für Biochemie
Universität Innsbruck
A-6020 Innsbruck
Austria

FRONZA, G.
Lab. of Mutagenesis
Nat. Inst. for Cancer Research
V. le Benedetto XV, 10
I-16132, Genoa
Italy

FUJIMOTO, Shigeyoshi
Dept. of Immunology
Kochi Medical School
Kochi 783
Japan

GALLOWAY, Anne M.
Mol. Genetics and Carcinogenesis Lab.
Cross Cancer Institute
11560 University Avenue
Edmonton (Alberta) T6G 1Z2
Canada

GAMBACORTA, Agata
Ist. M.I.B.
Consiglio Nazionale delle Ricerche
Arco Felice
Napoli
Italy

GARNIER, Jean-Marie
Lab. Gén. Mol. des Eucaryotes
Centre Nat. de Recherche Scientifique
11, rue Humann
67085 Strasbourg Cedex
France

GAULT, W.G.
Dept. of Mol. & Cell Biol.
University of Aberdeen, Marischal Coll.
Aberdeen AB9 1AS
Scotland
United Kingdom

GÄKEN, Joop
Molecular Medicine Unit
King's College School of Medicine
Denmark Hill
London SE5 8RX
England

GHANI, Q. Perveen
Dept. of Stomatology, Box 0650
University of California, SF
San Francisco, CA 94143-0650
USA

GIANNONI, Paolo
Inst. of General Physiology
University of Genoa, Fac. of Sciences
Corso Europa 26, Science Bldg.
16132 Genoa
Italy

GRADWOHL, Gérard
Inst. Biologie Moléculaire et Cellulaire
Centre Nat. de Recherche Scientifique
15, rue Descartes
F-67084 Strasbourg Cedex
France

GRIFFIN, S.
Cancer Research Unit
Medical School
University of Newcastle upon Tyne
Newcastle upon Tyne NE2 4HH
United Kingdom

GRINDLEY, H.
Cancer Research Unit
Medical School
University of Newcastle upon Tyne
Newcastle upon Tyne NE2 4HH
United Kingdom

GRUBE, Karlheinz
Institute of Virus Research
German Cancer Research Center
Im Neuenheimer Feld 506
D-6900 Heidelberg 1
Germany

GUERIN, Sylvain
Molecular Endocrinology Lab.
CHUL Research Center
2705, boul. Laurier
Sainte-Foy (Québec) G1V 4G2
Canada

HAKAM, Alaeddin
Romberg Tiburon Centers
Center for Environmental Studies
San Francisco State Univ.
Tiburon, CA 94920
USA

HALDAR, Joydeep
Dept. of Pathology
New York University Medical Center
New York, NY
USA

HAQUE, Saikh
Dept. of Biochemistry & Mol. Biology
School of Medicine
Georgetown University
Washington, D.C. 20007
USA

HARDAS, Bhushan
Molecular Medicine Unit
King's College School of Med. & Dent.
Denmark Hill
London SE5 8RX
England

HASSAN, M.E.
Dept. of Biochemistry
The University of Texas
Health Science Center
San Antonio, TX 78284-7760
USA

HAUN, Randy S.
Nat. Heart, Lung and Blood Inst.
National Institutes of Health
Bldg. 10, Room 5N307
Bethesda, Maryland 20892
USA

HENGARTNER, Christophe
Molecular Endocrinology Lab.
CHUL Research Center
2705, boul. Laurier
Sainte-Foy (Québec) G1V 4G2
Canada

HERZOG, Herbert
Institut für Biochemie
Universitat Innsbruck
A-6020 Innsbruck
Austria

HIRSCH-KAUFFMANN, M.
Inst. f. Biochemie
Universität Innsbruck
A-6020 Innsbruck
Austria

HITCHMAN, Eva
Dept. of Gastroenterology and Hepat.
University of Vienna
Vienna
Austria

HOG, Fernand
Centre de Neurochimie
Centre Nat. de Recherche Scientifique
5, rue Blaise Pascal
67084 Strasbourg Cedex
France

HORNE, O.
Cancer Research Unit
Medical School
University of Newcastle upon Tyne
Newcastle upon Tyne NE2 4HH
United Kingdom

HÕFFERER, L.
Vet. Med. Fakultat (Tierspital)
Inst. für Pharmakologie u Biochemie
Winterthurerstrasse 260
CH-8057 Zurich
Switzerland

HUNT, Thomas K.
Department of Surgery
Univ. of California
513 Parnassus, HSE-839
San Francisco, CA 94143-0522
USA

HUSER, Hans
Robert Koch-Institut
Abt. Biochemie
Nordufer 20
D-1000 Berlin 65
Germany

HUSSAIN, Zamirul
Dept. of Stomatology
Univ. of California at SF
San Francisco, CA 94143-0650
USA

IGLEWSKI, Wallace J.
Dept. of Microbiology - Immunology
Univ. of Rochester Med. Center
601 Elmwood Ave., Box 8672
Rochester, NY 14642
USA

IKEJIMA, Miyoko
Biochemistry Division
Nat. Cancer Center Res. Inst.
1-1, 5-chome Tsukiji
Chuo-ku, Tokyo 104
Japan

INAGAWA, Jun'ichi
Dept. of Clinial Sc. and Lab. Medicine
Faculty of Medicine
Kyoto University
Kyoto 606
Japan

ITTEL, Marie-Elisabeth
Dept. of Immunology
University of Toronto
399 Bathurst St.
Toronto (Ontario) M5T 2S8
Canada

JACOBSON, Elaine
Department of Medicine
Texas College of Osteopathic Medicine
North Texas State University
Fort Worth, TX 76107-2690
USA

JACOBSON, Myron K.
Department of Biochemistry & Mol. Biol.
Texas College of Osteopathic Medicine
North Texas State University
Fort Worth, TX 76107-2690
USA

JELTSCH, Jean-Marc
Lab. Gén. Mol. des Eucaryotes
Centre Nat. de Recherche Scientifique
11, rue Humann
67085 Strasbourg Cedex
France

JUST, Ingo
Inst. für Pharmakol. und Toxikol.
Univ. des Saarlandes
D-6650 Homburg-Saar
Germany

KAISER, P.
Inst. f. Biochemie
Universität Innsbruck
A-6020 Innsbruck
Insbruck
Austria

KANG, Veronica
Dept. of Biochemistry & Mol. Biology
Georgetown Univ. School of Medicine
Washington, D.C. 20007
USA

KAUFMANN, Scott
Oncology I-127
Johns Hopkins Hospital
600 N. Wolfe Steet
Baltimore, MD 21205
USA

KIDO, Takahiro
Dept. of Clinical Sc. and Lab. Medicine
Faculty of Medicine
Kyoto University
Kyoto 606
Japan

KIER, Peter
Dept. of Hematology
University of Vienna
Vienna
Austria

KIRKLAND, James B.
Dept. of Nutritional Sciences
University of Guelph
Guelph (Ontario) N1G 2W1
Canada

KIRSTEN, Eva S.
Romberg Tiburon Centers
Center for Environmental Studies
San Francisco State Univ.
Tiburon, CA 94920
USA

KOMURA, Hajime
Suntory Inst. for Biorganic Research
Shinanoto-Cho
Mishima-gun, Osaka 618
Japan

KONISHI, Yoichi
Dept. of Oncological Path.,Cancer Center
Nara Medical College
840 Shijo-cho, Kashihara
Nara 634
Japan

KROGER, Hans
Robert Koch-Institut
Des Bundesgesundheitsamtes
Nordufer 20-1000 Berlin 65
Germany

KUN, Ernest
Romberg Tiburon Centers
Center for Environmental Studies
San Francisco State Univ.
Tiburon, CA 94920
USA

KURASHIGE, Takanobu
Dept. of Pediatrics
Kochi Medical School
Kochi 783
Japan

KUSHIDA, Shigeki
Inst. of Basic Medical Sciences
University of Tsukuba
Tsukuba-Shi, Ibaraki 305
Japan

KUYKENDALL, Jim R.
Haskell Laboratory for Toxicology
E.I. du Pont de Nemours and Company
P.O. Box 50, El
Boston, MA 02115
USA

KÜPPER, Jan-Heiner
German Cancer Res. Center
Inst. für Virusforschung
Im Neuenheimer Feld 506
D-6900 Heidelberg
Germany

LAGUEUX, Jean
Molecular Endocrinology Lab.
CHUL Research Center
2705, boul. Laurier
Sainte-Foy (Québec) G1V 4G2
Canada

LAURO, Ludovica
Dipartimento di Biologia MCA
Universita di Camerino
I-62032 Camerino
Italy

LEDDA-COLUMBANO, G.
Inst. di Farmacologia e Patologia Bioch.
Universita di Cagliari
Italy

LEE, Chii-Ming
Lab. of Cellular Metabolism
Nat. Heart, Lung and Blood Inst.
National Institutes of Health
Bethesda, MD 20892
USA

LIBONATI, M.
Inst. of Basic Medical Sciences
University of Tsukuba
Tsukuba-Shi, Ibaraki 305
Japan

LIUZZI, Michel
Mol. Genetics and Carcinogenesis Lab.
Cross Cancer Institute
11560 University Avenue
Edmonton (Alberta) T6G 1Z2
Canada

LUNEC, J.
Cancer Research Unit
Medical School
University of Newcastle upon Tyne
Newcastle upon Tyne NE2 4HH
United Kingdom

LYBAK, Stein
The Wallenberg Laboratory
University of Lund
Box 7031
S-220 07 Lund
Sweden

MAGNI, Giulio
Instituto di Biochimica
Facolta di Medicina
Universita di Ancona
I-60131 Ancona
Italy

MALANGA, Maria.
Vet. Med. Fakultat (Tierspital)
Inst. fur Pharmakologie u Biochemie
Winterthurerstrasse 260
CH-8057 Zurich
Switzerland

MANDEL, Paul
Centre de Neurochimie
Centre Nat. de Recherche Scientifique
5 rue Blaise Pascal
F-67084 Strasbourg Cedex
France

MARIANI, Cristina
Instituto di Genetica Biochimica
Consiglio Nazionale delle Ricerche
Via Abbiategrasso 207
I-27100 Pavia
Italy

MARSISCHKY, Gerald T.
Lab. Microb. & Mol. Gene.
Harvard Medical School
Boston, MA 02115
USA

MARTINEZ, Marcos
Dept. of Microbiol. and Immunol.
Texas College of Osteopathic Medicine
University of North Texas
Fort Worth, Texas 76107-2690
USA

MARTINEZ-CADENA, Ma.-G.
Dept. of Microbiol. & Immunol.
Texas College of Osteopathic Med.
University of North Texas
Fort Worth, TX 76107-2690
USA

MARUTA, Hideharu
Tokyo Institute of Technology
Kanagawa
Japan

MASUTANI, Mitsuko
National Cancer Center Research Inst.
1-1, 5-chome, Tsukiji Chuo-ku
Tokyo 104
Japan

MATSUMURA, Masayuki
Inst. of Basic Medical Sciences
University of Tsukuba
Tsukuba-Shi, Ibaraki 305
Japan

MCDONALD, Lee
National Institutes of Health
Building 10/5N-307
Bethesda, MD 20892
USA

MCMAHON, Kathryn K.
Dept. of Pharmacology
School of Med.
Texas Tech. Univ., Health Sc. Center
Lubbock, TX 79430-0001
USA

MENAPACE, Lia
Instituto di Anatomia ed Istologia
Strada le Grazie
I-37134 Verona
Italy

MENDELEYEV, Jerome
Romberg Tiburon Centers
Center for Environmental Studies
San Francisco State Univ.
Tiburon, CA 94920
USA

MENDOZA-ALVAREZ, Hilda
Biomedical Division
The Noble Foundation
Ardmore, OK 73402
USA

MENEGAZZI, Marta
Instituto di Chimica Biologica
Universita di Verona
Strada le Grazie
I-37134 Verona
Italy

MENISSIER-De MURCIA, Josiane
Inst. Biologie moléculaire et Cellulaire
Centre Nat. de Recherche Scientifique
15 rue Descartes
F-67084 Strasbourg Cedex
France

MÉNARD, Luc
Molecular Endocrinology Lab.
CHUL Research Center
2705, boul. Laurier
Sainte-Foy (Québec) G1V 4G2
Canada

MIDDLESTADT, MICHAEL V.
Mol. Genetics and Carcinogenesis Lab.
Cross Cancer Institute
11560 University Avenue
Edmonton (Alberta) T6G 1Z2
Canada

MISHIMA, Koichi
Laboratory of Cellular Metabolism
Nat. Heart, Lung and Blood Inst.
National Institutes of Health
Bethesda, MD 20852
USA

MIWA, Masanao
Inst. of Basic Medical Sciences
University of Tsukuba
Tsukuba-Shi, Ibaraki 305
Japan

MOEHRING, Joan M.
Dep. of Microbiology & Mol. Genetics
Coll. of Med. & Coll. of Agr.& Life Sc.
University of Vermont
Burlington, Vermont 05405-0068
USA

MOEHRING, Thomas J.
Dep. of Microbiology & Mol. Genetics
Coll. of Med. & Coll. of Agr.& Life Sc.
University of Vermont
Burlington, Vermont 05405-0068
USA

MOLINETE, Miguel
Inst. Biologie Moléculaire et Cellulaire
Centre Nat. de Recherche Scientifique
15 rue Descartes
F-67084 Strasbourg Cedex
France

MONACO, L.
Nat. Heart, Lung and Blood Inst.
National Institutes of Health
Bethesda, MD 20892
USA

MOREAU, Pierre
Département de Biochimie
Faculté des Sciences et Génie
Université Laval
Québec (Québec) G1K 7P4
Canada

MOSS, Joel
Lab. of Cell. Metabolism
Nat. Heart, Lung and Blood Inst.
National Institutes of Health
Bethesda, MD 20892
USA

MURTAGH, J.J.
Nat. Heart, Lung and Blood Inst.
National Institutes of Health
Bethesda, MD 20892
USA

NAEGELI, Hanspeter
Institut of Pharmacology
Tierspital
Winterthurerstrasse 260
CH-8057 Zurich
Switzerland

NAGL, Ulrich
Inst. f. Biochemie
Universität Innsbruck
A-6020 Innsbruck
Austria

NATALINI, Paolo
Dipartimento di Biologia Cellulare
Universita di Camerino
I-62032 Camerino
Italy

NEGRI, Claudia
Istituto di Genetica Biochimica
Consiglio Nazionale delle Ricerche
Via Abbiategrasso 207
I-27100 Pavia
Italy

NEGRONI, Matteo
Istituto di Genetica Biochimica
Consiglio Nazionale delle Ricerche
Via Abbiategrasso 207
I-27100 Pavia
Italy

NICOLAUS, Barbara
Ist. M.I.B.,
Consiglio Nazionale delle Ricerche
Arco Felice
Napoli
Italy

NIEDERGANG, Claude
Centre de Neurochimie
Centre Nat. de Recherche Scientifique
5, rue Blaise Pascal
F-67084 Strasbourg Cedex
France

NIGHTINGALE, M.S.
Nat. Heart, Lung and Blood Inst.
National Institutes of Health
Bethesda, MD 20892
USA

NISHINA, Hiroshi
Tokyo Institute of Technology
Kanagawa
Japan

NOMURA, Ichiro
Dept. of Pediatrics
Kochi Medical School
Nankoku, Kochi 783
Japan

NUNBHAKDI-CRAIG, Viyada
Texas Coll. of Osteopathic Medicine
University of North Texas
Fort Worth, Texas 76107
USA

OBARA, Seiji
Dept. of Biochem.
Shimane Med. Univ.
Izumo 693
Japan

OGURA, Tsutomu
Biochemistry Div.,
National Cancer Center Research Inst.
1-1, 5-chome, Tsukiji Chuo-ku
Tokyo 104
Japan

OLSSON, Hakan
The Wallenberg Laboratory
University of Lund
Box 7031
S-220 07 Lund
Sweden

OPPENHEIMER, N.J.
Dept. of Pharm. Chem.
Univ. of California
San Francisco, CA 94143
USA

ORUNESU, Mauro
Instituto Fisiologia Generale
Facolta di Science
Corso Europa 26
I-16132 Genova
Italy

OZAWA, Youichi
Inst. of Basic Medical Sciences
University of Tsukuba
Tsukuba-Shi, Ibaraki 305
Japan

PANZETER, Phyllis
University of Zurich-Tierspital
Winterthurerstrasse 260
CH-8057 Zurich
Switzerland

PARTISANI, Mariagrazia
Centre de Neurochimie
Centre Nat. de Recherche Scientifique
5, rue Blaise Pascal
67084 Strasbourg Cedex
France

PATERSON, Malcolm C.
Mol. Genetics and Carcinogenesis Lab.
Cross Cancer Institute
11560 University Avenue
Edmonton (Alberta) T6G 1Z2
Canada

PEARSON, C.E.
Dept. of Oncology, Faculty of Medicine
McGill Cancer Center
3655 Drummond Street
Montréal (Québec) H3G 1Y6
Canada

PEARSON, Colin K.
Dept. of Mol. et Cell Biology
Marischmal College, Univ. of Aberdeen
Aberdeen, AB9 1AS
Scotland
United Kingdom

PEDRAZA-REYES, M.
Dept. of Microbiol. & Immunol.
Texas College or Osteopathic Med.
University of North Texas
Fort Worth, TX 76107-2690
USA

PENNER, Edward
Dept. of Gastroenterology and Hepat.
University of Vienna
Vienna
Austria

PERO, Ronald W.
Molecular Ecogenetics
The Wallenberg Laboratory
University of Lund, Box 7031
S-220 07 Lund
Sweden

PERTUSI, Raymond M.
Texas Coll. of Osteopathic Medicine
Univ. of North Texas
Fort Worth, Texas 76107
USA

PETZOLD, Shirley J.
Hematology/Oncology Division
University Hospitals of Cleveland
Case Western Reserve University
Cleveland, OH 44106
USA

PHILIPPE, Michel
Lab. de Biol. et Gén. du Développement
URA Centre Nat. de Recherche Scientifique
Université de Rennes
35042 Rennes Cedex
France

PIRON, Kristen J.
Dept. of Pharmacology
Health Sciences Center
Texas Tech. Univ.
Lubbock 79430
USA

POIRIER, Guy G.
Molecular Endocrinology Lab.
CHUL Research Center
2705, boul. Laurier
Sainte-Foy (Québec) G1V 4G2
Canada

POLTRONIERI, P.
Instituto di Chimica Biologica
Universita di Verona
Strada Le Grazie
I-37134 Verona
Italy

POTVIN, Frédéric
Molecular Endocrinology Lab.
CHUL Research Center
2705, boul. Laurier
Sainte-Foy (Québec) G1V 4G2
Canada

PRICE, Gerald B.
Dept. of Oncology, Faculty of Medicine
McGill Cancer Center
3655 Drummond Street
Montréal (Québec) H3G 1Y6
Canada

PRICE, S. Russ
Nat. Heart, Lung and Blood Inst.
National Institutes of Health
Bethesda, Maryland 20892
USA

PUTNAM, Donna
Dept. of Pathology
New York University Medical Center
New York, NY
USA

QUESADA, Piera
Dipartimento di Chimica Organica
e Biol.
Universita di Napoli
Via Mezzocannone 16
I-80134 Napoli
Italy

RAFFAELLI, Nadia
Instituto di Biochimica
Facolta di Medicina
Universita di Ancona
I-60131 Ancona
Italy

REALINI, Claudio
Institut of Pharmacology
Tierspital
Winterthurerstrasse 260
CH-8057 Zurich
Switzerland

RIBECCO, Maria
Instituto di Anatomia ed Istologia
Strada le Grazie
I-37134 Verona
Italy

RICHARD, Marie-Claude
Institute fur Pharmakologie
und Biochemie
Tierspital
Winterthurerstrasse 260
CH-8057 Zurich
Switzerland

RINGER, David P.
Biomedical Division
The Noble Foundation
Ardmore, OK 73402
USA

RUBIN, Bernard R.
Texas Coll. of Osteopathic Medicine
University of North Texas
Fort Worth, Texas 76107
USA

RUDRA, Nandini
Dept. of Biochemistry
All India Inst. of Medical Sciences
New Delhi 110029
India

RUGGIERI, Silverio
Dipartimento di Biologia MCA
Universita di Camerino
I-62032 Camerino
Italy

SABIR, Jamal
Genetics and Development Group
School of Biological Sciences
University of Sussex, Falmer
Brighton, BN1 9QG
England

SAITO, Isao
Dept. of Clinical Sc. and Lab. Medicine
Kyoto University Faculty of Medicine
Kyoto 606
Japan

SALFORD, Leif
Molecular Ecogenetics
The Wallenberg Laboratory
University of Lund, Box 7031
S-220 07 Lund
Sweden

SAUERMANN, Georg
Inst. of Tumorbiology Cancer Research
University of Vienna
Borschkeg 8a
A-1090 Vienna
Austria

SAULIER, Bénédicte
Lab. de Biologie Gén. du Développement
Université de Rennes I
Centre Nat. de Recherche Scientifique
F-35042 Rennes
France

SCARABELLI, Linda
Instituto di Fisiologia Generale
Facolta di Scienze
Corso Europa 26
I-16132 Genova
Italy

SCHALLEHN, Gisela
Inst. für Med. Mikrobiologie und Imm.
Universität Bonn
D-5300 Bonn
Germany

SCHNEIDER, Rainer
Institut für Biochemie
Universität Innsbruck
A-6020 Innsbruck
Austria

SCHREIBER, Valérie
Laboratoire de Biochemie II
Inst. de Biologie Mol. et Cell.
Centre Nat. de Recherche Scientifique
F-67084 Strasbourg Cedex
France

SCHWEIGER, Manfred
Institut für Biochemie
Universität Innsbruck
A-6020 Innsbruck
Austria

SCOVASSI, Anna Ivana
Istituto di Genetica Biochimica
Consiglio Nazionale delle Ricerche
Via Abbiategrasso 207
I-27100 Pavia
Italy

SERVENTI, Inez M.
Lab. of Cell. Metabolism
Nat. Heart, Lung and Blood Inst.
National Institutes of Health
Bethesda, MD 20892
USA

SHALL, Sydney
Cell & Molecular Biology Laboratory
University of Sussex, Falmer
Brighton, BN1 9QG
United Kingdom

SHIMOYAMA, Makoto
Department of Biochemistry
Shimane Medical University
Izumo 693
Japan

SIMM, A.
Cancer Research Unit
Medical School
University of Newcastle upon Tyne
Newcastle upon Tyne NE2 4HH
United Kingdom

SIMMONS, Anne M.
Dept. of Biochemistry
The University of Texas
Health Science Center
San Antonio, TX 78284-7760
USA

SIMONIN, Frédéric
Inst. Biologie Moléculaire et Cellulaire
Centre Nat. de Recherche Scientifique
15 rue René Descartes
F-67084 Strasbourg Cedex
France

SINGH, Neeta
Dept. of Biochemistry
All India Institute of Medical Sciences
Ansari Nagar
New Delhi 11029
India

SLAMA, James T.
Dept. of Biochemistry
University of Texas,
7703 Floyd Cercle Drive
San Antonio, TX 78284-7760
USA

SMITH, Debra G.
Texas Coll. of Osteopathic Medicine
University of North Texas
Fort Worth, Texas 76107
USA

SMULSON, Mark E.
Dept. Biochemistry & Mol. Biology
Georgetown Univ.
School of Medicine
Washington, D.C. 20007
USA

STANLEY, Sally J.
Nat. Heart, Lung and Blood Inst.
National Institutes of Health
Bethesda, MD 20892
USA

SU, Zao-Zong
Dept. of Pathology and Urology
Inst. of Cancer Research
Columbia University
New York, NY 10032
USA

SUGIMURA, Takashi
National Cancer Center Research Inst.
1-1, 5-chome, Tsukiji Chuo-ku
Tokyo 104
Japan

SUZUKI, Hisanori
Instituto di Chimica Biologica
Universita di Verona
Strada Le Grazie
I-37134 Verona
Italy

TAKENOUCHI, Nobuko
National Cancer Center Research Inst.
1-1, 5-chome, Tsukiji Chuo-ku
Tokyo 104
Japan

TALBOT, Brian
Dép. de Biologie
Fac. des Sciences
Université de Sherbrooke
Sherbrooke (Québec) J1K 2R1
Canada

TANIGUCHI, Taketoshi
Medical Research Laboratory
Kochi Medical School
Nankoku, Kochi 783
Japan

TANUMA, Sei-ichi
Dept. of Physiological Chemistry
Faculty of Pharmaceutical Sciences
Sagamiko, Kanagawa 199-01
Japan

TAVASSOLI, Manoochehr
Genetics and Development Group
University of Sussex
Falmer, Brighton, BN1 9QG
United Kingdom

TERASHIMA, Masaharu
Dept. of Biochemistry
Shimane Medical University
Izumo 693
Japan

TESTOLIN, Lucia
Instituto di Anatomia ed Istologia
Strada le Grazie
I-37134 Verona
Italy

THIBODEAU, Jacques
Immunologie
Institut de Recherche Clinique
 de Montréal
110 Pine Avenue West
Montréal H2W 1R7
Canada

THOMASSIN, Hélène
Endocrinologie Moléculaire
et du Développement
Centre Nat. de Recherche Scientifique
92190 MEUDON
France

TOMODA, Takashi
Dept. of Pediatrics
Kochi Medical School
Nankoku, Kochi 783
Japan

TORNESE BUONAMASSA, D.
Sclavo Research Institute
Siena
Italy

TROLL, Walter
Molecular Ecogenetics
The Wallenberg Laboratory
University of Lund, Box 7031
S-220 07 Lund
Sweden

TSAI, Su-chen
Lab. of Cell. Metabolism
Nat. Heart, Lung and Blood Inst.
National Institutes of Health
Bethesda, MD 20892
USA

TSUCHIYA, Mikado
Dept. of Biochemistry
Shimane Medical University
Izumo 693
Japan

TSUJIUCHI, Toshifumi
Dept. of Oncological Path.
Cancer Center, Nara Medical College
840 Shijo-cho, Kashihara
Nara 634
Japan

TSUTSUMI, Masahiro
Dept. of Oncological Pathology
Cancer Center, Nara Medical College
840 Shijo-cho
Kashihara, Nara 634
Japan

UCHIDA, Kazuhiko
Inst. of Basic Medical Sciences
University of Tsukuba
Tsukuba-shi, Ibaraki 305
Japan

UCHIDA, Masako
Inst. of Basic Medical Sciences
University of Tsukuba
Tsukuba-Shi, Ibaraki 305
Japan

UEDA, Kunihiro
Dept. of Clinical Sc. and Lab. Medicine
Kyoto Univ. , Faculty of Medicine
Shogoin, Sakyo-ku
Kyoto 606
Japan

VAUGHAN, Martha
Lab. of Cell. Metabolism
Nat. Heart, Lung and Blood Inst.
National Institutes of Health
Bethesda, MD 20892
USA

VENANZI, Franco M.
Dipartimento di Biologia MCA
Universita di Camerino
I-62032 Camerino
Italy

WAINSCHEL, L.
Dept. of Pharm. Chem.
Univ. of California at SF
San Francisco, CA 94143
USA

WASER, S.
Vet. Med. Fakultat (Tierspital)
Inst. fur Pharmakologie u Biochemie
Winterthurerstrasse 260
CH-8057 Zurich
Switzerland

WASSON, Bryan L.
Texas Coll. of Osteopathic Medicine
University of North Texas
Fort Worth, Texas 76107
USA

WEINFELD, Michael
Mol. Genetics and Carcinogenesis Lab.
Cross Cancer Institute
11560 University Avenue
Edmonton (Alberta) T6G 1Z2
Canada

WEISZ, Alessandra
Istituto di Patologia Generale
ed Oncologia
Universita di Napoli
Napoli
Italy

WELSH, C.F.
Lab. of Cell. Metabolism
Nat. Heart, Lung and Blood Inst.
National Institutes of Health
Bethesda, MD 20892
USA

WESIERSKA-GADEK, Jozefa
Institute of Tumorbiology
Cancer Research
University of Vienna
Borschkegasse 8a
A-1090 Vienna
Austria

WILLMORE, Elaine
Cancer Research Unit
Medical School
University of Newcastle upon Tyne
Newcastle upon Tyne NE2 4HH
United Kingdom

WILSON, S.B.
Dept. of Mol. & Cell Biol.
University of Aberdeen, Marischal Coll.
Aberdeen, Scotland AB9 1AS
United Kingdom

YAMADA, Kazuo
Dept. of Biochem.
Shimane Med. Univ.
Izumo 693
Japan

YAMAMOTO, Hiroshi
Dept. of Biochemistry
Kanazawa Univ. School of Medicine
13-1 Takara-machi
Kanazawa 920
Japan

ZANNIS-HADJOPOULOS, M.
McGill Cancer Centre
McGill University
3655 Drummond Street
Montréal (Québec) H3G 1Y6
Canada

ZHANG, Jinguan
Cancer Research Laboratory
University of Western Ontario
London (Ontario) N6A 5B7
Canada

ZWEIFEL, Barbara
Univ. of Zurich-Tierspital
Winterthurerstr. 260
Ch.-8057
Zurich
Switzerland

ZUCCONI-GRASSI, G.
Instituto di Chimica Biologica
Universita di Verona
Strada Le Grazie
I-37134 Verona
Italy

Part 1
Molecular Biology

Structure and function of the human poly(ADP-ribose) polymerase

Miguel Molinete, Valérie Schreiber, Frédéric Simonin, Gérard Gradwohl, Josiane Ménissier-de Murcia and Gilbert de Murcia

Institut de Biologie Moléculaire et Cellulaire du CNRS, 15 rue René Descartes, 67084 Strasbourg cedex, France.

Introduction

A number of roles have been ascribed to poly(ADP-ribose) polymerase, (PARP; EC. 2.4.2.30), including involvement in DNA repair, cell proliferation, differentiation and transformation (1-4). One of our major goals is to understand the molecular basis of the complex mechanism leading to the PARP activation in response to DNA strand breaks. Cloning of the gene has allowed the development of molecular biological tools to elucidate the structure and the function(s) of this highly conserved enzyme. This paper describes the recent results obtained in our laboratory using these new approaches.

Results and Discussion

PARP: a multifunctional highly conserved enzyme. PARP is a multifunctional enzyme, following limited proteolysis three functional domains have been identified in the enzyme molecule (5) : a 46 kDa fragment including the DNA binding domain, located in the N-terminal region, a central 22 kDa polypeptide fragment containing the automodification sites and a C-terminal fragment of 54 kDa bearing the NAD binding domain (figure 1).

The human PARP cDNA has been cloned in several laboratories (6-9). The nucleotide sequence contained in a single open reading frame of 3,042 nucleotides, encodes a 1,014 aa residues polypeptide with a calculated molecular weight of 113,153 . A comparison of the whole PARP sequence with the National Biomedical Research Foundation's protein data bases revealed no extensive identities. However several protein sequence motifs could be detected within the different domains, indicating putative functions: (i) the N-terminal DNA binding domain contains a potential zinc finger like motif of the form: C - Xaa 2 - C - Xaa 28, 30 -

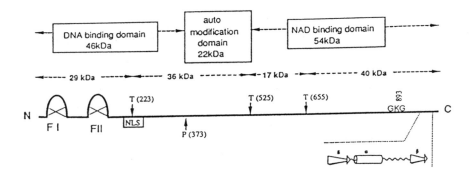

figure 1 : Schematic representation of the different functional domains of the PARP molecule obtained after mild trypsin (T) or papain (P) digestion, according to Kameshita et al. (5). The two zinc fingers, FI and FII, are represented as loops with X. The nuclear location signal, NLS, is located at position 221-226. The motif G K G located at position 893 is involved in the catalytic activity. The putative β – α –β fold supposed to interact with the ADP moiety of NAD, lies from residue 978 to residue 993.

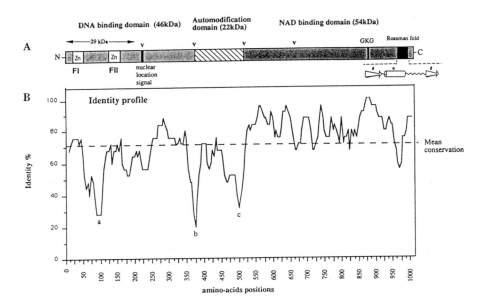

figure 2 : Identity profile obtained after alignment of the human (6-9), murine (11), bovine (12) and chicken (13) PARP sequences. Each position of the alignment is scored as +1 when the corresponding aa residue is strictly conserved. The average percentage of identity was calculated every 5th residue using a window of 20 aa.

H - Xaa 2 - C; (ii) a nuclear location signal (221- K K K S K K 226) is also present in this N-terminal part of the molecule; (iii) the C-terminal region including the NAD binding domain contains the motif G xxx G K G (residues 888 to 894) recalling the NTP binding motif (10) .

More recently, the sequence of the mouse (11), bovine (12), chicken (13) and a part of the rat (14) cDNAs have been determined revealing an extensive conservation at the amino acid level. Using an identity profile, a comparison of these four sequences together with the human PARP sequence, is displayed in figure 2. The mean conservation of the entire amino acid sequence is 72 %. Interestingly, two variable regions of low conservation (indicated b and c in figure 2) coincide with the boundaries of the automodification domain and the NAD binding domain respectively, as defined by limited proteolysis (figure 1). This observation is in agreement with the concept of concatenated functional domains (15). The best amino acid conservation is observed within the NAD binding domain for which 83% identity is found. The DNA binding domain appears to be interrupted by a variable region (numbered a in figure 2) corresponding exactly to the boundary between the two homologous sequences containing the zinc-finger like motifs .

The DNA binding domain. Since the work of Zahradka and Ebisuzaki (16) we know, that PARP needs zinc for its activity. We have recently determined that each enzyme molecule contains two $Zn(II)$ ions. Using the Zinc-blot assay (figure 3) the metal binding sites were located in the 29kDa N-terminal fragment included in the DNA binding domain (see figure 1). By means of the South-Western blot technique (figure 3) we have demonstrated that zinc is essential for the binding of this fragment to DNA (17). DNase I protection studies indicated that PARP specifically binds to a single-stranded break in DNA by its metal binding domain, in a zinc dependent manner. The 29 kDa domain is sufficient to protect symmetrically 7 ± 1 nucleotides on either side of a single-stranded break (figure 4) (18). Using the site directed mutagenesis approach, we have identified the amino acids involved in metal coordination (figure 5) and analyzed the consequences of altering the proposed zinc-finger structure on DNA binding : disruption of the metal binding ability of the second zinc finger (FII) dramatically reduces target DNA binding. In contrast, when the postulated $Zn(II)$ ligands of FI were mutated the DNA binding activity was only slightly affected (19). Taken together, these results demonstrate that PARP contains a novel type of zinc fingers that differs from previously recognized classes in terms of both structure and function.

6

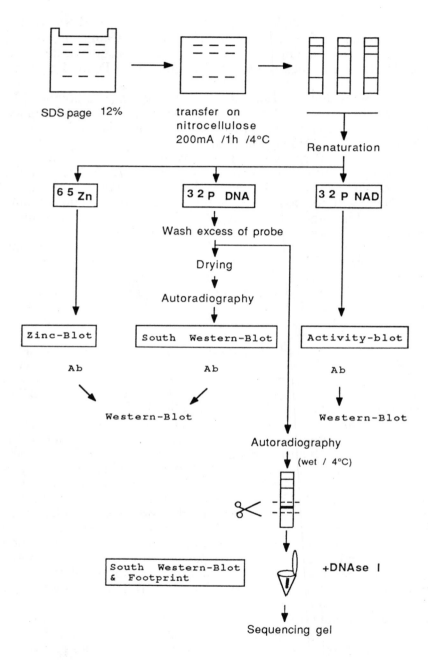

figure 3 : Methodology used to study the functional domains of purified poly (ADP-ribose) polymerase

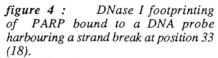

figure 4 : *DNase I footprinting of PARP bound to a DNA probe harbouring a strand break at position 33 (18).*
G : degradation products of the DNA probe by the G sequencing reaction.
lane 1: control of DNA degradation without PARP.
lane 2: degradation of the PARP - DNA complex.
lane 3: zinc-depleted PARP-DNA complex degradation
lane 4: 29 kDa N-terminal peptide - DNA complex degradation.
lane 5:. bacterially expressed N-terminal peptide-DNA complex degradation.

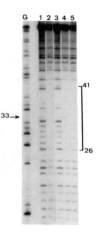

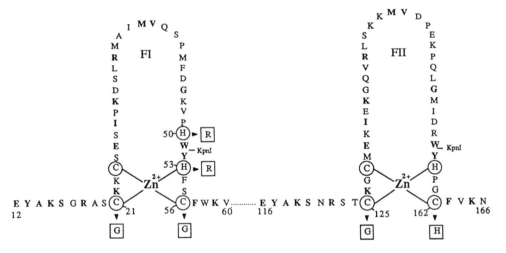

figure 5 : Structure of the human PARP DNA binding region showing the two zinc fingers like subdomains. The symbols for the zinc-coordinating aminoacids are circled . Amino acid substitution performed by site directed mutagenesis are indicated by arrows heads pointing to the mutant residues. Duplicated aminoacids are shown in boldface according to ref. (6) .

The nuclear location signal The nuclear location signal, (NLS), is a short amino acid sequence that specifies accumulation of large proteins into the cell nucleus (reviewed in ref. 20). To test the ability of the motif K K K S K K (aa residues 221 to 226) to play a role in the nuclear transport of PARP, the cDNA encoding the N-terminal fragment (residues 1 to 233) has been cloned in the

pCH110   pCH SV NLS   pCH N-PARP   pCH N-PARPΔNLS

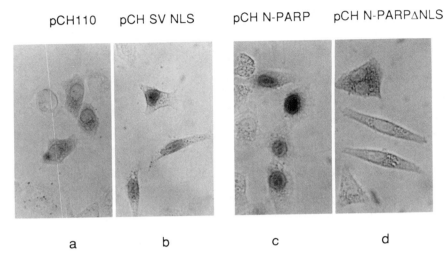

a   b   c   d

figure 6 : *Histochemical staining of β-galactosidase expressed in HeLa cells using the pCH 110 eucaryotic vector.*

    a) HeLa cells transfected by the vector alone (negative control)

    b) HeLa cells transfected by the pCH NLS SV construct expressing the nuclear location signal of SV 40 T-Ag (residues 125 to 135) in fusion with β - galactosidase used here as a reporter - protein.

    c) HeLa cells transfected by the pCH N PARP construct expressing the 29 kDa N-terminal domain of the human PARP fused with β – galactosidase.

    d) HeLa cells transfected by the pCH PARP Δ NLS construct expressing the 29 kDa N-terminal domain of the human PARP deleted of the sequence K K K S K K (residues 221 to 226).

eukaryotic expression vector pCH 110 designed for high level expression in mammalian cells. This vector bears the β-galactosidase encoded lac Z gene under the control of the SV 40 early promoter. As a positive control, we have cloned the SV 40 T-antigen NLS sequence (aa residues 125 - P P K K K R K V E D P - 135) (21) using a set of appropriate oligonucleotides in frame with the lac Z gene, resulting in the pCH NLS SV vector. The location of the β-galactosidase fusion protein was revealed by its activity using a histochemical method with X-gal . The results displayed in figure 6 indicate that the SV 40 NLS as well as the N-terminal domain of PARP (figure 6b and 6c respectively) are necessary to target the reporter protein (β-galactosidase) to the nucleus. In contrast, omission of the SV 40 T-antigen NLS motif or deletion of the PARP NLS sequence (residues 221 to 226) (figure 6a and 6c respectively)

9

maintains the β-galactosidase activity exclusively in the cytoplasm, thus demonstrating that this sequence is necessary to target the PARP molecule to the nucleus (22). Although nothing is known about the three-dimensional structure of this motif, it is interesting to note that in the case of PARP, the NLS sequence must reside in an exposed region since it is preferentially cleaved by proteases (see figure 1).

The catalytic domain. Since the work of Kameshita *et al.* (5) the C-terminal region of PARP has been identified as the NAD binding domain (figure 1). With respect to the NAD binding function, PARP is of special interest since in this case, the NAD molecule is used as a substrate rather than a coenzyme involved in electron transfer mechanisms. To study this highly conserved part of the PARP molecule, we have developed a novel assay, the "activity blot" (figure 3), which allows the detection of transferred and renatured polypeptides involved in poly(ADP-ribose) synthesis (23-24). Deletion experiments indicate that the C-terminal region of the molecule contains the catalytic domain exhibiting both the polymerizing and branching activities. The minimum size of this catalytic domain is 40 kDa (figure 7). The basal activity associated

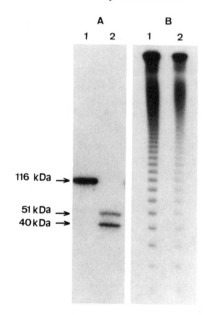

figure 7 :   *Activity blot and analysis of the reaction product using the whole PARP enzyme or its catalytic domain.*
panel A :Activity blot using :lane 1: purified calf thymus PARP; lane 2 :the catalytic region expressed in E.coli (the 40 kDa peptide represents the minimal active domain) .
panel B :poly(ADP-ribose synthesized in experiment A by:lane 1: purified calf thymus PARP ;lane 2:40 kDa catalytic domain

116 kDa →
51 kDa →
40 kDa →

with this domain can also be detected in solution but at a higher protein concentration compared to the activity of the entire enzyme stimulated by DNA breaks. Since under these conditions the 40 kDa domain is automodified by poly(ADP-ribosyl)ation, one can conclude that acceptor sites previously located exclusively in the automodification domain (5), may also be present in the catalytic domain . These experiments indicate that the C-terminal part is functionally autonomous like the DNA binding domain. However, the N-terminal part of the enzyme molecule is clearly essential for the stimulation of the enzymatic activity by DNA strand breaks (23, 24).

Recently an alignment between the 40 kDa catalytic domain and NAD(P) amino acid dehydrogenases (AADH), was found by our group (23). According to the sequence similarities obtained, the G rich segment (G X X X G K G, aa residues 888 to 894) may not be related to a potential NTP binding site as previously proposed (7, 8) for the following reasons: (i) in the PARP sequence, the canonical NTP binding motif exhibiting all the reported invariant aminoacid residues and a correct spacing between them is absent, (ii) the polymerizing activity is not dependent upon the presence of Mg^{2+} which is of critical importance in NTP binding proteins for their hydrolytic activity. The results from the mutagenesis experiments performed on this domain pointed to the crucial importance of Lys-893 for the catalytic activity. This same invariant Lys was reported in *E.coli* glutamate dehydrogenase to be intimately involved in either direct or indirect interaction with two substrates: NADPH and glutamic acid (25). Thus, although in PARP and AADH the catalytic processes are probably highly divergent. It is attractive to speculate that the loss of PARP activity obtained upon substitution of the Lys-893 by an aliphatic residue (Ile), may eventually reflect the loss of a critical residue which interacts with the two substrates of PARP (NAD and a glutamic residue) allowing the proper positioning of the two molecules. It is worth mentioning that Lys-893 is included in the most conserved region (100 % identity) of the PARP catalytic domain (see figure 2).

An interesting structural possibility arising from the alignment of PARP with the AADH sequences is the preservation within the extreme C-terminal part of the PARP protein of a segment which may be involved in a particular β–α–β (Rossman fold) which binds the ADP moiety of the dinucleotide in a variety of NAD(P) binding proteins (figure 2).

Prospects Cloning of the PARP gene has allowed the identification of a number of sequence motifs which mediate nuclear targeting of the protein and DNA strand break recognition to which the catalytic activity is linked. A detailed structural information from the crystallized protein will eventually allow, for example, to understand the complex mechanism of PARP activation by DNA strand breaks. However, the major challenge for the investigator still remains the determination of the *in-vivo* role of this protein modification. In the absence of any mutant cell line known to be defective in the PARP activity, the possibility to generate now mouse models by homologous recombination should help to address this question in the future.

the abbreviations used are : PARP, poly(ADP-ribose) polymerase (EC 2.4.2.30); aa, amino acid; bp, base pair; AADH, amino acid dehydrogenase; NLS, nuclear location signal; NTP, nucleoside triphosphate; FI, finger I; FII, finger II.

Acknowledgements

We thank Dr. O. Poch for stimulating discussions and Dr. C. Niedergang for communicating the chicken PARP cDNA sequence. The work described in this paper was supported by grants from Association pour la Recherche contre le Cancer, ARC, and from La Ligue Nationale contre le Cancer.

References

1. Ueda, K. and Hayaishi, O. (1985). ADP-ribosylation, *Ann. Rev. Biochem.* **54**, 73-100.
2. Althaus, F.R. and Richter, C. (1987). ADP-ribosylation of proteins. Enzymology and biological significance. *Mol. Biol. Biochem. Biophys.* **37**, 1-126.
3. G.de Murcia, G., Huletsky, A. and Poirier, G.G. (1988). Review : Modulation of chromatin structure by poly(ADP-ribosyl)ation, *Biochem. Cell Biol.* **66**, 626-635
4. Shall, S. (1984). ADP-ribose in DNA repair: a new component of DNA excision repair *Adv. Rad. Biol.* **11**, 1-69.
5. Kameshita, I. Matsuda, Z., Tanigushi, T. and Shizuta, Y. (1984). Poly(ADP-ribose)synthetase. Separation and identification of three proteolytic fragments as the substrate-binding domain, the DNA binding domain, and the automodification domain. *J. Biol. Chem.* **259**, 4770-4776.
6. Uchida, K., Morita, T., Sato, T., Ogura, T., Yamashita, R., Nogushi, S., Suzuki, H., Nyunoya, H., Miwa, M. and Sugimura, T. (1987). Nucleotide sequence of a full-length cDNA for human fibroblast poly(ADP-ribose) polymerase. *Biochem. Biophys. Res. Commun.* **148**, 617-622.
7. Kurosaki, T., Ushiro, H., Mitsuchi, Y., Suzuki, S., Matsuda, M., Matsuda, Y., Katunuma, N., Kangawa, K., Matsuo, H., Hirose, T., Inayama, S. and Shizuta, Y. (1987). Primary structure of human poly(ADP-ribose) synthetase as deduced from cDNA sequence. *J. Biol. Chem.* **262**, 15990-15997.
8. Cherney, B. W., Mc Bride, O.W., Chen, D., Alkhatib, H., Bhatia, K., Hensley ,P. and Smulson, M.E. (1987). cDNA sequence, protein structure and chromosomal location of the human gene for poly(ADP-ribose) polymerase. *Proc. Natl. Acad. Sci. USA.* **84**, 8370-8374.

9. Herzog, H., Zabel, B.U., Schneider, R., Auer, B., Hirsh-Kauffmann, M. and Schweiger, M. (1989). Human nuclear NAD$^+$ ADP-ribosyltransferase: Localization of the gene on chromosome 1q41-q42 and expression of an active human enzyme in Escherichia coli. *Proc. Natl. Acad. Sci. USA.* **86**, 3514-3518.

10. Walker, E. J., Saraste, M. , Runwick, M. J., and Gay, N.J. (1982). Distantly related sequences in the α− and β− subunits of ATP synthase, myosin, kinases and other ATP-requiring enzymes and a common nucleotide binding fold. *EMBO J.* **1**, 945-951.

11. Huppi, K., Bhatia, K., Siwarski, D., Klinman, D., Cherney, B. and Smulson, M. (1989). Sequence and organization of the mouse poly (ADP-ribose) polymerase gene. *Nucleic Acids Res.* **17**, 3387-3401.

12. Saito, I., Hatakeyama, K., Kido, T., Ohkubo, H., Nakanishi S. and Ueda, K. (1990). Cloning of full-length cDNA encoding bovine thymus poly(ADP-ribose) synthetase: evolutionarily conserved segments and their potential functions. *Gene* **90**, 249-254.

13. Ittel, M. E., Garnier, J.M., Jeltsh, J.M., and Niedergang, C.P. (1991). Chicken poly(ADP-ribose) polymerase: complete sequence deduced from cDNA, comparison with mammalian enzyme sequences. *Gene* (in press)

14. Thibodeau, J., Gradwohl, G., Dumas, C., Clairoux-Moreau, S., Brunet, G., Penning, C., Poirier, G.G. and Moreau, P. (1989). Cloning of rodent cDNA encoding the poly(ADP-ribose) polymerase catalytic domain and analysis of mRNA levels during the cell cycle. *Biochem. Cell Biol.* **67**, 653-660.

15. Feldhaus, A.L. and Lesnaw, J.A. (1988). Nucleotide sequence of the gene of vesicular somatis virus (New Jersey): identification of conserved domains in the New Jersey and Indiana L proteins. *Virology* **163**, 359-368.

16. Zahradka, P. and Ebisuzaki, K. (1984). Poly(ADP-ribose)polymerase is a zinc metalloenzyme. *Eur. J. Biochem.* **142**, 503-509.

17. Mazen, A., Ménissier-de Murcia, J., Molinete, M., Simonin, F., Gradwohl, G., Poirier, G.G. and de Murcia, G. (1989). Poly(ADP-ribosose) polymerase : A novel Finger Protein. *Nucleic Acids Res.* **17** , 4689-4698.

18. Ménissier-de Murcia, J., Molinete, M., Gradwohl, G., Simonin, F. et de Murcia, G. (1989). Zinc binding domain of poly(ADP-ribose)polymerase participates in the recognition of single strand breaks on DNA. *J. Mol. Biol.* **210**, 229-233.

19. Gradwohl, G., Ménissier-de Murcia, J., Molinete, M., Simonin, F., Koken, M., Hoeijmakers, J.H.J. and de Murcia, G. (1990). The second zinc-finger domain of poly(ADP-ribose) polymerase determines specificity for single-stranded breaks in DNA. *Proc. Natl. Acad. Sci. USA* **87**, 2990-2994.

20. Garcia-Bustos, J., Heitman, J. and Hall, M.N. (1991). Nuclear protein localization. *Biochim. Biophys. Acta.* **1071**, 83-101

21. Kalderon, D., Richardson, W.D., Markham, A.F., and Smith, A. (1984). Sequence requirements for nuclear location of simian virus 40 large-T antigen *Nature (London)* **311**, 33-38.

22. Schreiber, V., Boeuf, H., Molinete, M., de Murcia, G. and Ménissier-de Murcia, J. (in preparation).

23. Simonin, F., Ménissier-de Murcia, J., Poch, O., Muller, S., Gradwohl, G., Molinete, M., Penning, C., Keith, G. and de Murcia, G. (1990). Expression and site directed mutagenesis of the catalytic domain of human poly(ADP-ribose) polymerase in Escherichia coli. *J. Biol. Chem.* **265**, 19249-19256.

24. Simonin, F. , Briand, J.P., Muller, S. and de Murcia, G. *Anal. Biochem* (in-press)

25. McPherson, M.J., Baron, A.J., Jones, K.M. Price, G.J., and Wootton, J.C. (1988). Multiple interactions of Lys-128 of Escherichia coli glutamate dehydrogenase revealed by site-directed mutagenesis. *Protein Engineering* **2**, 147-152

Cloning of Poly(ADP-ribose) Polymerase cDNA from Lower Eukaryotes

Masanao Miwa, Youichi Ozawa, Masako Uchida, Shigeki Kushida,
Yoshihiro Ami, Masayuki Matsumura, and Kazuhiko Uchida

Introduciton

Poly(ADP-ribose) polymerase (EC 2.4.4.30) catalyzes
DNA-dependent ADP-ribosylation of nuclear proteins with NAD as
a substrate (1-8). Poly(ADP-ribose) polymerase activity was
reported from various eukaryotic cells but not from prokaryotic
cells (1-8).

Although frunction of poly(ADP-ribose) polymerase reaction
is suggested to be involved in regulation of DNA repair, cell
growth, differentiation and transformation, there have not been
the direct evidence to prove the above function. It would be
useful to isolate the gene for poly(ADP-ribose) polymerase from
lower eukaryotes like Drosophila melanogaster and slime mold
and to search for the phenotypes with mutation at the same
chromosomal localization or after desruption of gene for
poly(ADP-ribose) polymerase.

For the first step to attain this goal, we tried to
isolate cDNA from Xenopus laevis and cherry trout.

Results and Discussion

1. Southern blot hybridization analysis of poly(ADP-ribose)
   polymerase genes in various organisms using human
   poly(ADP-ribose) polymerase cDNA as a probe.

When cDNA for human poly(ADP-ribose) polymerase was used as a
probe, it hybridized to EcoRI DNA fragments of human, pig,
mouse and rat, but it did not significantly hybridize to those
of Xenopus laevis, Drosophila melanogaster, Cynops
pyrrhogaster, turtle, C. plecoglossus, C. salvelinum, cherry
trout, salmon, squid, lamprey, D. disscoidium, and S.
cerevisiae. Although most of the eukaryotes except C.

recent publication showed that there were significant
conservation of nucleotide sequence in various regions among
human, mouse, rat and bovine poly(ADP-ribose) polymerase cDNAs
(9-14). Recently chicken poly(ADP-ribose) polymerase cDNA
sequence became available (15), and there is still conservation
of various regions.

2. Identification of DNA fragments by PCR homologous to
   poly(ADP-ribose) polymerase.

Comparing the nucleic acid and putative amino acid sequences of
cDNAs from human, mouse, rat and bovine poly(ADP-ribose)
polymerases, we synthesized mixed 20-mer oligonucleotide primer
sets corresponding to sequences which showed high homology
among the above organisms (Fig. 1). We could amplify DNA
fragments with expected size by polymerase chain reaction (PCR)
from cDNA libraries of Xenopus laevis and cherry trout. The
DNA fragment with expected size of about 450 bp was amplified
with a set of primer 5 and primer 8, and that about 400 bp was
amplified with a set of primer 5 and primer 9 from cDNA of
Xenopus laevis and cherry trout.

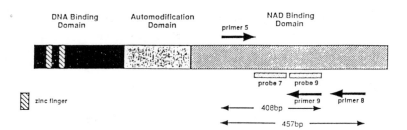

Fig. 1.   Strategy for amplification of DNA fragments
          homologous to poly(ADP-ribose) polymerase cDNA
          sequence by polymerase chain reaction.

Southern blot analysis revealed a single major band around
the size of 457 bp when primer 5 and primer 8 were used to
amplify DNA fragments from cherry trout, Xenopus laevis and
human cDNA libraries and hybridized with [^{32}P] labeled probe 9
at 25°C. The hybridization signal from cherry trout almost

disappeared at 37°C (Fig. 2) and that from Xenopus laevis
disappeared at 42°C, but that of human cDNA remained. Similar
results were obtained when primer 5 and primer 9 were used to
amplify about 400 bp DNA fragments and prove 7 was hybridized
to DNA fragments from cherry trout, Xenopus laevis and human
cDNAs. The results strongly suggest that the amplified DNA
fragments are cDNA for poly(ADP-ribose) polymerase from these
species.

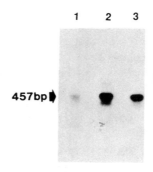

Fig. 2.   Southern blot hybridization of amplified DNA
           fragments with primer 5 and primer 8 using probe 9.
           1; cherry trout.   2; Xenopus laevis.   3; human.

3. Cloning and sequencing of DNA fragments from Xenopus laevis
   and cherry trout homologous to poly(ADP-ribose) polymerase
   cDNA.

The DNA fragments of about 400 bp which was amplified from
Xenopus laevis or cherry trout cDNA libraries using primer 5
and primer 9 were subcloned. The DNA sequence analysis of
cloned DNA fragment from Xenopus laevis showed that it
consisted of 402 bp and had 74% homology in nucleotide level to
that of human cDNA for poly(ADP-ribose) polymerase, while the
DNA sequence amplified from cherry trout had 405 bp and had 73%
homology. When putative amino acid sequences were compared,
Xenopus laevis and cherry trout had 84% and 77% homology to
human poly(ADP-ribose) polymerase, respectively. Fig. 3 showed
the genomic sothern blot hybridization analysis when the DNA
fragment cloned from the cherry trout was used as a probe.
Strong hybridization signal were detected in cherry trout and
sweetfish.

17

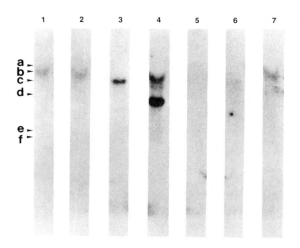

Fig. 3.   Southern blot hybridization of DNA from various
species using a cloned cDNA fragment from cherry
trout as a probe. 1; White salmon. 2; char.
3; sweetfish. 4; cherry trout. 5; Xenopus laevis.
6; Cynops pyrrhogaster. 7; human. The molecular
weitht markers are Hind III digested DNA fragments
of lambda phage. a; 23 kb. b; 6.6 kb. c; 4.4 kb.
d; 2.3 kb. e; 2.0 kb. f; 0.6 kb.

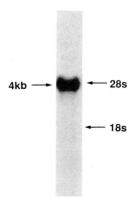

Fig. 4.   Northern blot hybridization of the total RNA of
oocytes of Xenopus laevis using ^{32}p-labeled subcloned
frgment from Xenopus laevis as a probe.

18

4. Northern blot analysis of mRNA of poly(ADP-ribose) polymerase.

A single band around 4 kb was observed when total RNA from oocytes (Fig. 4), brian and liver from Xenopus oocytes were hybridized to [^{32}P] labeled cloned DNA fragment. Similarly a single band with about 4 kb was observed when total RNA from oocytes, brain and liver of cherry trout was hybridized to [^{32}P] labeled cloned DNA fragment.

We are now trying to clone the full length cDNA from both Xenopus laevis and cherry trout. Next step might be cloning cDNA for poly(ADP-ribose) polymerase from even lower eukaryotes.

Acknowledgment
This work was supported in part by Grants-in-aid for Cancer Research from the Ministry of Education, Science and Culture, from Sagawa Foundation and for University of Tsukuba Project Research, Japan.

References
1. Miwa, M.; Sugimura, T. ADP-ribosylation and carcinogenesis. In: Moss, J.; Vaughan, M., eds. ADP-Ribosylating toxins and G proteins; insights into signal transduction. Washington, DC: American Society for Microbiology; 1990:543-560.
2. Ueda, K. Nonredox reactions of pyridine nucleotides. In: Dolphin, D.; Rozanne, P.; Arramovic, O., eds. Pyridine nucleotide coenzymes; chemical, biochemical and medical aspects. New York, John Wiley & Sons, Inc.; 1987:549-597 (vol. 2B).
3. De Murcia, G.; Huletskey, A.; Poirier, G.G. Modulation of chromatin structure by poly(ADP-ribosyl)ation. Biochem. Cell Biol. 66:625-635; 1988.
4. Shall, S. ADP-ribose in DNA repair; a new component of DNA excision repair. Adv. Radiat. Biol. 11:1-69;1984.
5. Gaal, J.C.; Smith, K.R.; Pearson, C.K. Cellular euthanasia mediated by a nuclear enzyme; a central role for nuclear ADP-ribosylation in cellular metabolism. Trends Biochem. Sci. 12:129-130; 1987.
6. Althaus, F.R. Poly-ADP-Ribosylation reactions. In: Althaus, F.R.; Richter, C., eds. ADP-Ribosylation of proteins; enzymology and biological significance. Berlin,

Springer-Verlag, K.G.; 1987:1-126.

7. Mandel, P.; Okazaki, H.; Niedergang, C. Poly(adenosine diphosphate ribose). Prog. Nucleic Acid Res. Mol. Biol. 27:1-51;1982.

8. Sugimura, T. Poly(adenosine diphosphate ribose). Prog. Nucleic Acid Res. Mol. Biol. 13:127-151;1973.

9. Uchida, K.; Morita, T.; Sato, T.; Ogura, T.; Yamashita, R.; Noguchi S.; Suzuki, H.; Nyunoya, H.; Miwa, M.; Sugimura, T. Nucleotide sequence of a full-lungth cDNA for human fibroblast poly(ADP-ribose) polymerase. Biochem. Biophys. Res. Commun. 148:617-622;1987.

10. Kurosaki, T.; Ushiro, H.; Mitsuuchi, Y.; Suzuki, S.; Matsuda, M.; Matsuda, Y.; Katunuma N.; Kangawa, K.; Matsuo, H.; Hirose, T.; Inayama, S.; Shizuta, Y. Primary structure of human poly(ADP-ribose) synthetase as deduced from cDNA sequence. J. Biol. Chem. 262:15990-15997;1987.

11. Cherney, B.W.; McBride, O.W.; Chen, D.; Alkhatib, H.; Bhatia, K.; Hensley, P.; Smulson, M.E. cDNA sequence, protein structure, and chromosomal location of the human gene for poly(ADP-ribose) polymerase. Proc. Natl. Acad. Sci. USA 84:8370-8374;1987.

12. Huppi, K.; Bhatia, K.; Siwarski, D.; Klinman, D.; Cherney, B.; Smulson, M. Sequence and organization of the mouse poly(ADP-ribose) polymerase gene. Nucleic Acids Res 17:3387-3401;1989.

13. Thibodeau, J.; Gradwohl, G.; Dumas, C.; Clairoux-Moreau, S.; Brunet, G.; Penning, C.; Poirier, G.G.; Moreau, P. Cloning of rodent cDNA coding the poly(ADP-ribose) polymerase catalytic domain and analysis of mRNA levels during the cell cycle. Biochem. Cell. Biol. 67:653-660;1989.

14. Saito, I.; Hatakeyama, K.; Kido, T.; Ohkubo, H.; Nakanishi, S.; Ueda, K. Cloning of a full-length cDNA encoding bovine thymus' poly(ADP-ribose) synthetase; evolutionarily conserved segments and their potential functions. Gene 90:249-259;1990.

15. Ittel, M.-E.; Garnier, J.-M.; Jeltsch, J.M.; Niedergang, C.P. Chicken poly(ADP-ribose) polymerase; complete deduced amino acid sequence and comparison with mammalion enzyme sequences. Gene in press; 1991.

Molecular Biology of Human Nuclear NAD+: ADP-Ribosyl-transferase (polymerizing)

M.Schweiger, B.Auer, H.Herzog, M.Hirsch-Kauffmann, P.Kaiser, K.Flick, U.Nagl, R.Schneider

Institut für Biochemie (Nat.Fak.), Universität Innsbruck, Peter-Mayr-Straße 1 a, A-6020 Innsbruck, Tel. 0043-512-507-3200, Telefax 0043-512-507-2544

Introduction

Nuclear NAD+: protein ADP-ribosyltransferase (polymerizing) [pADPRT; EC 2.4.2.30] is a most interesting enzyme. It transfers ADP-ribosyl moieties from nicotinamide adenine dinucleotide (NAD+) to proteins and forms a protein bound branched ADP-ribosyl-polymer and seems to be ubiquitous in eukaryotic cells (although its presence in Saccharomyces cereviseae is not confirmed yet). The first row of target proteins of ADP-ribosylation are, besides pADPRT itself, histones, high mobility group proteins (HMG), DNA ligase, topoisomerase I and II and several others. The ADP-ribosyl polymer has a very short half life, in the range of a minute, implying an involvement in control mechanisms. Indeed, the enzyme is involved in central processes of the cell such as DNA repair, differentiation, tumorigenic cell transformation and others (for a review see 1).

Single strand DNA breaks, which act as coenzyme, play a fundamental role in the action of pADPRT. During DNA-repair, the appearance of DNA single-strand breaks are a necessary step to activate the enzyme, after which the cellular NAD+ is consumed and the NAD+ concentration within the cell is dramatically lowered and nuclear proteins are ADP-ribosylated. If the pADPRT activity is blocked by specific inhibitors, the lethality of noxious chemicals is potentiated, the NAD+ concentration is not reduced and ADP-ribosylation of proteins does not take place. Therefore, it is believed that pADPRT is involved in DNA repair. However, the specific function of ADP-ribosylation in DNA repair is as yet unknown. In accordance with its central biochemical role, the amino acid sequence and structure of this enzyme is highly conserved between distant species as seen by the crossreaction of antibodies against the human enzyme with purfied trout and snail (2, 3) pADPRT.

Inhibitors of pADPRT trigger differentiation of teratocarcinoma cells. This fits well with the observation that differentiation in these cells, which may be induced by retinoic acid, is accompanied by diminishment of pADPRT activity (4). Thus, the enzyme seems to play a regulating role in differentiation and proliferation. This notion is confirmed by the dramatic stimulation of activity of pADPRT during the induction of antibody production in B lymphocytes, which is accompanied by cellular proliferation (5). Similarly inhibitors of pADPRT prevent tumorigenic cell transformation (6, 7). Thus, the ADP-ribosyl transferase seems to act as the central switch board between proliferation and differentiation. High activity triggers cell proliferation and low activity differentiation.

The involvement of pADPRT in various cellular processes together with the high turnover of poly ADP-ribose and single stranded gaps of DNA acting as cosubstrate of the enzyme has led to the discussion of two lines for its action, which appear to be compatible with each other. The major functions of pADPRT could be the control of the number of incisions to prevent the DNA from disintegrating by the accumulation of too many gaps as well as providing a mechanism to remove histones and other proteins from the DNA. For this later function, a higher affinity of the histones for poly ADP-ribose than for DNA would displace these proteins from the DNA (8). This double function of pADPRT is supported by several observations. The number of pADPRT molecules per cell is remarkably constant. Reduction of enzyme activity by inhibitors blocks DNA repair without causing an excess of single stranded gaps in the DNA.

Results and Discussion

Our interest arose from the observation that cells derived from patients suffering from Fanconi´s anemia do not respond to DNA damage with diminishment of the NAD^+ concentrations, indicating that NAD^+ metabolism might be affected in this disease (9, 10). Since the reaction of pADPRT is the main degradative pathway of NAD^+ this enzyme seemed to be a good candidate for the defect of this hereditary disease.

The access to this most interesting enzyme was opened by the development of an efficient affinity chromatography (11). cDNA for pADPRT was isolated and characterized in several laboratories (12-16). The deduced amino acid sequence and protein structure prediction revealed two zinc fingers for DNA binding. Interestingly, mRNAs for pADPRT possess differential polyadenylation. In several untransformed cells the poly (dA) tail starts 186 nucleotides downstream from a polyadenylation site active in Hela cells and SV40-transformed fibroblasts (17). Expression of pADPRT cDNA in *E.coli* resulted in several products since within the coding sequence there are 3 "ATGs" in combination with perfect "Shine and Dalgarno" ribosome binding sequences (17). It is not clear what sense it makes that in a higher eukaryote the coding sequence for a gene, in this case pADPRT, contains prokaryotic start signals. This does not seem to be random, since the probability of the combination ATG/ribosome binding site sequences occuring is far higher than that predicted by a chance distribution. One rational explanation for these cryptic sequences could be that the coding sequence of pADPRT was adapted from prokaryotes.

Expression of pADPRT in yeast

Since expression of complete pADPRT protein appeared to be difficult in *E.coli* we tried to use *Saccharomyces cerevisiae* as a lower eukaryote for the purpose of expression of pADPRT. This experiment was extremely interesting also from another point of view: *S. cerevisiae* does not show ADP-ribosylation in detectable amounts, thereby raising the question if this organism would react to the action of human pADPRT expressed in this system. Expression plasmids, containing the pADPRT cDNA and portions thereof in various orientations,were constructed as depicted in Fig. 1.

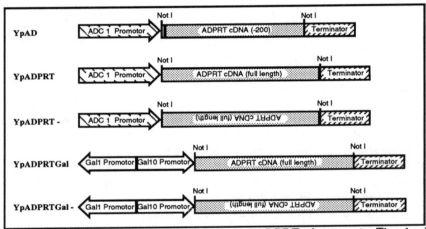

Fig. 1. **Constructs for expression of pADPRT in yeast.** The basic plasmid contains a 2μ segment and the *leu 2* gene from yeast as well as an *E.coli* origin of replication and the ampicillin resistance gene.

YpAD expressing a truncated pADPRT polypeptide lacking the first 36 amino acids including the first "zinc finger"under constitutive control of the alcohol dehydrogenase (ADH) promoter, could be transferred into *S. cerevisiae* cells with normal efficiency. On the other hand,YpADPRT, expressing the entire pADPRT protein, showed almost zero transformation rate. As a control YpADPRT(-), where pADPRT cDNA is inserted in the opposite direction to the ADH promotor and will not be expressed, showed transformation rates equal to YpAD (Fig. 2). Under conditions where pADPRT activity is inhibited by 3-methoxybenzamide the transformation rate of all three constructs goes up to the control level (Fig. 2).

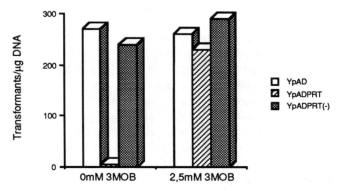

Fig. 2. **Transformation rates of indicated plasmids in yeast with and without inhibition of pADPRT**

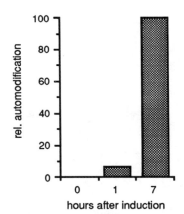

Fig. 3. **Induction of pADPRT activity in YpADPRTGal transformants by galactose**

Expression of full length pADPRT cDNA under control of the regulated galactose epimerase promotor (Gal 1/10) showed a strong increase of pADPRT activity after induction, as measured by activity gel analysis (Fig.3). The appearance of pADPRT protein could also be visualised by immunoblot analyses with affinity purified pADPRT antibodies as a single protein with M_r 116000 (not shown). This means that the eukaryotic translation machinery of S. *cerevisiae* has not recognized the internal prokaryotic translation start signals and produced a single polypeptide from pADPRT mRNA. The cells underwent dramatic morphological changes after induction of pADPRT expression in that the budding index decreased, cell size increased and odd-shaped aggregates were formed.

These results indicate that *S. cerevisiae*, despite not containing measurable amounts of endogenous pADPRT activity, recognizes the signal of ADP-ribosylation and reacts by stopping cell growth. This reaction is caused by enzymatic activity of pADPRT, since it is abolished by inhibition of ADP-ribosylation by NAD^+ analogues and a truncated pADPRT peptide, which does not possess the full enzyme activity, did not show any effect (Fig. 1). DNA-binding as the reason for growth inhibition seems to be unlikely since it cannot be inhibited by NAD^+ analogues and in addition the truncated polypeptide is still able to bind DNA. Depletion of the NAD^+ pool by increased ADP-ribosylation does not take place in the transformed and induced cells as shown in Fig. 4. Most likely, poly ADP-ribose itself or ADP-ribosylation of proteins involved in growth regulation are the reason for inhibition of growth in *S. cerevisiae* by expression of human pADPRT.

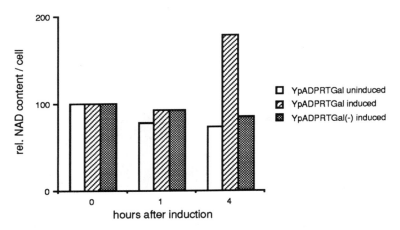

Fig. 4. NAD^+ **content of indicated yeast transformants**

The human gene for the ADP-ribosyltransferase maps on the chromosome 1, at the region q41-q42. There is another site with hybridization capacity to the pADPRT cDNA on chromosome 14 (17). This might be a pseudogene. Given the role of pADPRT and tumorigenic transformation, it is both remarkable and perhaps not a coincidence that the genes for protooncogene trk and for the transforming growth factor b also map to region q41-42 of chromosome 1.

The pADPRT gene consists of 23 exons ranging in size from 62-553 bp spread over a region of 43 kb, which was found within three overlapping cosmid clones from a human genomic library. All the exons contain canonical splice consensus sequences. In contrast to the "zinc finger" structures of steroid receptors, which are encoded by one exon each, two exons form one of each of the "zinc fingers" in the DNA binding domain of pADPRT. One of the cosmid clones contained several kb adjacent to the mRNA start site (18).

Analysis of the human pADPRT promoter

The central role of poly ADP-ribosyltransferase, the stimulation of its activity by a factor of up to 20 during the early phase of B-lymphocyte stimulation as well as a number of other arguments implies that the synthesis of this enzyme is regulated. The remarkably constant number of enzyme molecules in cells could be most easily explained by autoregulation. This was also suggested to us by the earlier finding that T7 protein kinase (19, 20), which is also involved in regulation of gene expression and is regulated by automodification like pADPRT, is autoregulated in its synthesis. Consequently, we focused on the characterization of the 5´ upstream sequence of pADPRT.

A DNA clone of 5kb derived from human liver genomic DNA was isolated. This fragment contains 2 kbp 5´ flanking region, 287 bp exon 1 and 2,7kbp from intron 1. The major transcription start site was determined by primer extension to be 160 nucleotides upstream from the translational start ATG. This coincides with the sequence of the full length cDNA.This experiment also revealed other transcription initiation sites.

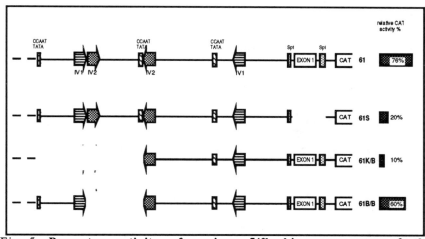

Fig. 5. Promoter activity of various 5´flanking sequences of the human pADPRT gene linked to the CAT reporter gene. Location and direction of the inverted repeats is given by differently filled arrows (IV1, IV2), SP1 binding site clusters, CCAAT-TATA boxes and open reading frames are depicted by boxes.

A cluster of two SP1 binding sites is located in front of the transcription start (38 and 55 nucleotides upstream) as well as a cluster of four SP1 binding sites 3´of the first exon. However, there are no CCAAT/TATA regions in the immediate vinicity (the first one is located more than 580 bp upstream).

Interestingly there are several short direct and inverted repeats which could form two types of DNA loops-cruciform structures. One type of DNA loop would not change the distance between the most proximal SP1 site and the next upstream CCAAT/TATA sequence. However, the second type would shorten this distance considerably and would bring thereby the CCAAT/TATA sequence into the vicinity of the SP1 site. This would imply that the later DNA loops are the active promoter structure (Fig. 6) To prove this, several deletions in the relevant region were constructed and attached to the chloramphenicol acetyltransferase, (CAT) reporter gene (Fig 5). After transfection the activity of the expressed CAT was determined as a measure of the promoter strength. The high expression SV40 promoter was taken as control (100 percent). The complete 2 kbp upstream sequence (together with exon I and 700 bp from intron I) had significant promoter activity (75 % of the SV40 promoter activity). Deletion of the proximal parts of the inverted repeats preventing formation of both types of DNA loops reduced the promoter activity considerably (to 10 %), whereas eliminating the first type of DNA loop formation, which still allowed the DNA looping to bring the CCAAT/TATA regions into the vicinity of the transcription start site, had little, if any, consequences to the promoter activity (60 % of SV 40 promoter). This together with several controls confirms the notion that the promoter activity of the pADPRT gene is dependant on DNA loop formation (Fig.6). The ADP-ribosyltransferase binds specifically to cruciform structures (21). Since its own promoter can form cruciform structures it is a straightforward approach to ask whether this enzyme regulates its own expression via binding to the specific DNA loop in its promoter.

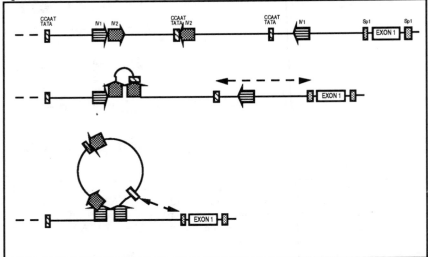

Fig. 6. Schematic representation of the possible secondary structures of the pADPRT promoter. Location and direction of the inverted repeats is given by differently filled arrows (IV1, IV2), SP1 binding site clusters, CCAAT-TATA boxes and open reading frames are depicted by boxes.

To elucidate this problem, experiments were conducted where pADPRT promoter driven CAT activity was measured with and without coexpression of pADPRT. By cotransfection of a construct containing the pADPRT promotor with the CAT gene and a plasmid carrying pADPRT cDNA behind the MMTV or SV 40 promotor, the activity of the pADPRT promoter in the presence of an excess of poly ADP-ribosyl transferase was studied. Induction of pADPRT cDNA (controlled by the MMTV promoter) by dexamethasone diminished the pADPRT promoter dependent expression of CAT. There was no influence of pADPRT on the SV 40 and MMTV promoters. The diminishment of activity of pADPRT promoter is thus specific. This specific repression of the pADPRT promotor was also observed with overexpression of the DNA binding domain of pADPRT (aminoacids 1-534) by itself (Fig.7).

Inhibitors of pADPRT which interact with the NAD$^+$ binding domain, do not interfere with the autorepression of the pADPRT promoter. Since pADPRT binds to the stem of cruciform DNA structures the most direct mode of action would be that pADPRT binds to the "inactive" DNA loop thereby stabilizing it and/or to the base of the "active" cruciform structure covering the CCAAT/TATA sequence thereby repressing the promoter activity. This mechanism of autoregulation of gene expression would lead to a constant concentration of this enzyme in the cell, since any excess would repress further synthesis.

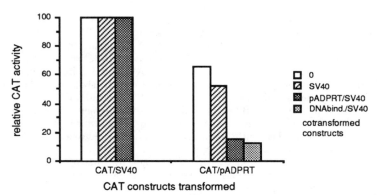

Fig. 7. **CAT activity of the SV40 promoter and the pADPRT promoter** after cotransfection with pADPRT cDNA and the cDNA for the DNA binding domain of pADPRT (aa 1-534) both under the SV40 promoter.

In conclusion, our knowledge on pADPRT has increased tremendously in recent years. The structure of its cDNA and its gene are known, expression in *E.coli* and yeast has provided interesting insight and made large scale preparations of the enzyme feasible. The elucidation of the promotor structure resulted in the discovery of the cruciform structures which appear to be involved in the regulation of activity of its promotor and suggested a mechanism for the autorepression of expression of this gene. Although the knowledge on the enzyme is profound, the most fundamental problem on the mode of exerting its central role in cellular processes is as yet unsolved.

References

1. Althaus, F. R., Richter, C. (1987). ADP-ribosylation of proteins. Molecular Biology Biochemistry and biophysics 37, 1-122

2. Burtscher,H.J., Klocker,H., Schneider,R., Auer,B., Hirsch-Kauffmann,M., Schweiger,M.: ADP-ribosyltransferase from Helix pomatia. Purification and characterization. Biochem.J. 248, 859-864 (1987).

3. Burtscher, H.J., Schneider, R., Klocker, H., Auer, B., Hirsch-Kauffmann, M., Schweiger,M.: ADP-ribosyltransferase is highly conserved: purification and characterization of ADP-ribosyltransferase from a fish and its comparison with the human enzyme. J.Comp.Physiol.B. 157, 567-572 (1987).

4. Ohashi,Y., Ueda,K., Hayashi,O., Ikai,K., Niwa,O.: (Induction of murine teratocarcinoma cell differentiation by suppression of poly (ADP-ribose) synthesis. Proc.Natl.Acad.Sci.USA 81,7132-7136 (1984).

5. Johnstone,A.P.:Requirement for ADP-ribosyltransferase activity and rejoining of DNA strand breaks during lymphocyte stimulation, in ADP-Ribosylation of Proteins (Althaus,F.R.,Hilz,H Shall,S.eds.) pp.424-428, Springer Verlag, Berlin (1985).

6. Kun,E.,Kirsten,E.,Milo,G.E.,Kurian,P.&Kumari,H.L.: Cell cycle-dependent intervention by benzamide of carcinogen-induced neoplastic transformation and in vitro poly(ADP-ribosyl)ation of nuclear proteins in human fibroblasts. Proc.Natl.Acad.Sci.USA 80,7219-7223 (1983).

7. Borek,C.,Morgan,W.F.,Ong,A.,Cleaver,J.E.: Inhibition of malignant transformation in vitro by inhibitors of poly(ADP-ribose) synthesis. Proc.Natl.Acad.Sci.USA 81, 243-247 (1984).

8. Althaus, F.R., Bachmann, S., Braun, S.A., Collinge, M.A., Höfferer, L., Malanga, M., Panzeter, P.L., Realini, C., Richard, M.C., Waser, S., Zweifel, B. (1991). The poly(ADP-ribose)-protein shuttle of chromatin. In ADP-Ribosylation reactions (G.G.Poirier and P. Moreau, eds.) Springer New York.

9. Klocker, H., Auer, B., Hirsch-Kauffmann, M., Altmann, H., Burtscher, H.J., Schweiger, M. (1983). DNA repair dependent NAD+ metabolismis impaired in cells from patients with Fanconi´s anemia. EMBO J. 2, 303-307

10. Schweiger, M., Auer, B., Burtscher, H.J., Hirsch-Kauffmann, M., Klocker, H., Schneider, R. (1987). DNA repair in human cells: Biochemistry of the hereditary diseases Fanconi´s anaemia and Cockayne syndrome. Eur.J.Biochem. 165, 235-242

11. Burtscher, H.J., Auer, B., Klocker, H., Schweiger, M.,Hirsch-Kauffman, M. (1986). Isolation of ADP-ribosyltransferase by affinity chromatography.. Anal. Biochem. 152, 285-290

12. Schneider, R., Auer, B., Kühne, C., Herzog, H., Klocker, H., Burtscher, H.J., Hirsch-Kauffmann, M., Wintersberger, U., Sch (1987). Isolation of a cDNA clone for human NAD+:protein ADP-ribosyltransferase. Eur. J. Cell Biol. 44, 302-307

13. Alkahatib, H. M., Chen, D., Cherney, B., Bhatia, K., Notario, V., Giri, C., Stein, G., Slattery, E., Roeder, R. G., Smulson, (1987). Cloning and expression of cDNA for human poly(ADP-ribose) polymerase. Proc. Natl. Acad. Sci. USA 84, 1224-1228

14. Kurosaki, T., Ushiro, H., Mitsuuchi, Y., Suzuki, S., Matsuda, M., Katanuma, N., Kangawa, K., Matsuo, H., Hirose, T., Inayama (1987). Primary structure of

human poly(ADP-ribose)synthetaseas deduced from cDNA sequence.. J.Biol.Chem. 262, 15990-15997

15. Uchida, K., Morita, T., Sato, T., Ogura, T., Yamashita, R., Noguchi, S., Suzuki, H., Nyunoya, H., Miwa, M., Sugimura, T. (1987). Nucleotide sequence of a full-length cDNA for human fibroblast poly(ADP-ribose)-polymerase.. Biochem. Biophys. Res. Commun. 148, 617-622

16. Huppi, K., Bhatia, K., Siwarski, D., Klinman, D., Cherney, B., Smulson, M., (1989). Sequence and organization of the mouse poly(ADP-ribose) polymerase gene. Nucleic Acids Res. 17, 3387-3401

17. Herzog,H.,Zabel,B.U.,Schneider,R.,Auer,B.,Hirsch-Kauffmann,M.,Schweiger,M. (1989). Human nuclear NAD+ ADP-ribosyltransferase: Localization of the gene on chromosome 1q41-q42 and expression of an active human enzyme in Eschrichia coli. Proc. Natl. Acad.Sci. USA 86, 3514-3518

18. Auer, B., Nagl, U., Herzog, H., Schneider, R., Schweiger, M. (1989). Human nuclear NAD+ ADP-ribosyltransferase(polymerizing): Organization of the gene. DNA 8, 575-580

19. Mailhammer,R., Reiness,G., Ponta,H., Yang,H.Ll, Schweiger,M., Zillig,M., Zubay,G.: RNA Polymerase modifications after T-Phage infections of E.coli. In metabolic interconversion of enzymes.Edit. S.Shaltiel, Springer Verlag Berlin, p.161 (1975).

20. Pai,S.H., Rahmsdorf,H.J., Ponta,H., Hirsch-Kauffmann,M., Herrlich,P., Schweiger,M.: Protein kinase of bacteriophage T7. 2.Properties, enzyme synthesis in vitro and regulation of enzyme synthesis and activity in vivo. Eur.J.Biochem. 55,305-314 (1975).

21. Sastry,S.S., Kun,E.: The interaction of adenosine diphosphoribosyl transferase (ADPRT) with a cruciform DNA. Biochem. Biophys. Res. Commun. 167, 842-847 (1990)

Strategies for Studying the Functions of PADPRP Genes on Human
Chromosomes 1 and 13
Mark E. Smulson
Department of Biochemistry & Molecular Biology
Georgetown University School of Medicine
Washington, D.C. 20007 USA

Introduction

Our laboratory has focused in part on utilizing molecular aspects of the
PADPRP gene sequences located on chromosome 1q as well as on 13 to
help clarify the biological role of these genes. With respect to
chromosome 1q stably transfected cell lines have been established in
HeLa cells which express a reverse complementary transcription
product *in vivo* to the exons coded for on this chromosome locus. Also, in
the case of the PADPRP-like gene on chromosome 13q, we have fine
mapped this sequence and have observed a linkage of deletion of this
region with predisposition to various forms of cancer, especially in the
African-American population.

Results

In the current study of PADPRP deregulation, we established a stably
transfected cell line of human Hela-S3 cells (PADPRP-as(7)) that
express PADPRP antisense transcripts driven by a mouse mammary
tumor virus LTR promoter (MMTV) in response to glucocorticoid
hormone. In spite of the fact that the biological half-life of PADPRP
in cells is relatively long (15 hours), we found that induction of the
PADPRP antisense transcripts could be adjusted to cause a progressive
decrease in both nuclear PADPRP activity as well as content.

Expression of Antisense PADPRP RNA in Transfected Cells. Since
PADPRP-as(7) contained multicopy, integrated inverse PADPRP
sequences, and enzyme activity appeared to be significantly reduced
after hormone induction, we questioned whether antisense PADPRP
transcripts accumulated in PADPRP-as(7) in response to Dex. PADPRP-
as(7) and vector alone control cells were incubated in the presence of
$1\mu M$ Dex and total RNA was isolated from 0 to 72 hours and RNA was
analyzed by Northern hybridization (Fig. 1).

No antisense transcripts could be detected, in the absence of Dex (A lane
1) indicating that in cell line PADPRP-as(7) the MMTV promoter is
under tight control. Antisense mRNA was initiated by two hours of Dex
treatment (Fig. 1A) and its level remained relatively constant for at

least 48 hr. Partial or total degradation of antisense transcripts after 48 hr was observed (Fig. 1A, lane 7).

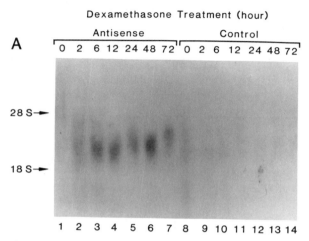

Fig. 1. Expression and Stability of Antisense Transcripts of PADPRP in PADPRP-as(7) and Control Cell Lines after Dex Induction. HeLa cell line PADPRP-as(7) and control cells (transfected with vector empty) (5×10^7 cells) were treated in the presence or absence (time 0) of 1 μM Dex. Total RNA was isolated and analyzed by Northern transfer and hybridization to a probe which detects anti-PADPRP mRNA. (Taken from ref. 1).

Time course of lowering of PADPRP activity and nuclear content. The specific activity of PADPRP was observed to remain constant in the control cells during 72 hr of culture; Dex *per se* had no effect on the PADPRP activity of the control cells transfected with the MMTV vector alone (Table 1). More significantly, no marked change in PADPRP activity in antisense cells incubated in the absence of Dex for the 72 hr culture period was noted. In contrast, Dex incubation caused a significant inhibition of PADPRP specific activity in the antisense cell culture. The specific activity in PADPRP-as(7) was reduced by 55% and 83% after 48 and 72 h induction of Dex, respectively.

Depletion in nuclear PADPRP content with antisense induction. The data in Fig. 2, a Coomassie blue stain of a duplicate of the gel used for the Western blot, shows that the amount of protein loading in all samples was essentially the same. In control cells (Fig. 2B, lanes 4, 5) no apparent changes in PADPRP content was observed for as long as 72 hr due to incubation with Dex, additionally, the PADPRP concentration, in the absence of induction, was approximately the same in antisense cells compared to control cells (cf. lanes 1 and 4). In contrast to these data, the enzyme content was markedly reduced by induction of complementary PADPRP mRNA. After 48 hr of induction (lane 2) only a minute amount of immunologically reactive PADPRP could be detected , and by 72 hr essentially no PADPRP band was visible (lane 3), consistent with the 83% inhibitory level of activity noted in Table 1.

Table 1. *Kinetics of antisense induced loss of PADPRP activity in HeLa incubated with dexamethasone*

Incubation time h	PADPRP Activity pmol (ADP-Rib)/min/mg protein		Inhibition %
	(-)Dex	(+)Dex	
A. *Control cells*			
48	442 ± 50	429 ± 26	–
72	461 ± 38	446 ± 23	–
B. *pInverted PADPRP cells*			
48	429 ± 43	186 ± 9 (p<0.01)	57
72	390 ± 30	67 ± 10 (p<0.01)	83

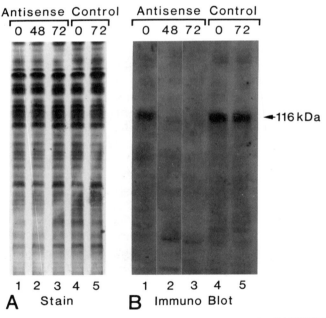

Dexamethasone Treatment (hour)

Fig. 2. Dex Induction of Antisense Results in Depletion of Cellular PADPRP Content. **A.** represents Coomassie blue staining; **B.** represents enzyme-linked immunoblotting using anti-PADPRP antibody. Taken from ref. 1.

Phenotypic characteristics of cells expressing PADPRP antisense RNA.
These results were confirmed in intact cells by the experiment
demonstrated by immunohistological staining utilizing anti-PADPRP
antibody. It was noted that the cells had altered morphology. In
general the induced antisense cells appeared to be more spindle shaped
and displayed a mosaic structure. The most prominent morphological
change was a tendency in some cells to exist as multinuclear aggregates.

Altered chromatin structure in antisense cells. We have earlier
proposed a structural, as well as catalytic, role for PADPRP in
chromatin (2). A marked difference in DNase I sensitivity was
observed when comparing nuclear DNA derived from cells with
depleted PADPRP content. During the linear portion of digestion (2
min) 30% of DNA was digested from nuclei derived antisense compared
to 10% in control nuclei. These observations suggest that chromatin of
the two cell lines may be organized differently; future studies will be
directed at exploiting the ability to experimentally alter nuclear
content of this enzyme to better clarify whether PADPRP represents a
significant structural component of chromatin fibers.

*Antisense-induced reduction in poly ADP-ribosylation and DNA
repair.* At 2.0 mM MMS, DNA damage representing 600 rad-
equivalents was observed of which over 90% was repaired by 5 hr. No
difference in the *extent* of SSB-repair in the antisense cells, whether
induced or uninduced for antisense transcription was observed. Thus,
only minute nuclear amounts of PADPRP seem sufficient for SSB repair.

Based on the above results initial repair periods were examined (Fig.
3). In agreement with the results of others, when endogenous PADPRP
activity was completely inhibited by benzamide (5 mM) no DNA
repair was observed. As noted, SSB-repair under the conditions in
these experiments progresses rapidly thus in PADPRP-as(7) cells, in
the absence of hormone, SSB-repair was nearly 60% complete within 20
min. During the same time period it was of significance that *no repair
was observed* in the cells expressing antisense mRNA to PADPRP.
DNA repair was initiated later (i.e. 30-40 min) in cells with reduced
PADPRP.

Thus, the initiation of DNA repair in antisense cells, as occasioned by
incubation of cells with MMS, was almost completely reduced in cells
with limiting PADPRP. Later, DNA repair capacity was
reestablished indicating that the enzyme concentration normally may
not be limiting for these reactions.

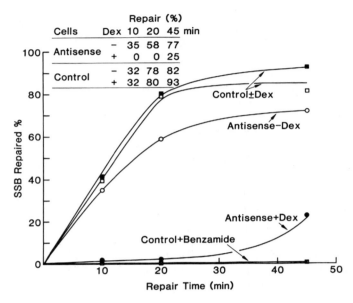

Fig. 3. The Short Time Repair of MMS-Induced SSB in Control and Antisense Cells. Control and antisense cells were treated with MMS following Dex induction. The treated cells were allowed to repair for 10, 20 and 45 min. MMS-induced SSB was measured by alkaline elution. Taken from ref. 1.

II.   Strategies to Investigate the PADPRP-like Sequences on Chromosome 13q. Our studies on the PADPRP-like exon sequences on chromosome 13q have centered mainly on correlations with a deletion in this region (i.e. 13 33qter) and predisposition to cancer (3). The initial analysis has revealed the following:

Table 2 *Increased frequency of B allele in Burkitt DNA from different population groups*

Distribution of genotypes based upon RFLP banding patterns obtained following restriction of DNA with *Hind*III. Taken from ref. 3.

Population	DNA from normal cells with genotype distribution of				B allele frequency	BB genotypes over noncancer control (fold increase)
	AA	AB	BB	Total		
Black						
American	15	18	4	37	0.35	
African	23	18	9	50	0.36	
Caucasian						
American	45	12	2	59	0.14	
Chinese	18	3	0	21	0.07	
		Tumor DNA				
Black	0	11	8	19	0.70	2.8
Caucasian	12	9	2	23	0.28	2.5

Table 3 *Allelic zygosity of the chromosome 13 PADPRP sequences in genomic DNA from matched normal and tumor samples*
Taken from ref. 3.

DNA	Genotype AA	AB	BB	Total	B-allele frequency	p1a	p2
Tumor							
B-cell lymphoma	12	7	4	23	0.33	0.0001	
Lung carcinoma	11	9	4	24[b]	0.35	0.001	
Breast carcinoma	34	16	3	53	0.21	0.003[c]	
Myeloid leukemia	11	2	0	13	0.08	0.45	
Colorectal carcinoma	41	15	7	63	0.22	0.0006[c]	
From normal tissue of above individuals with cancer							
Lung carcinoma	4	4	1	9	0.33	0.012	0.88
Breast carcinoma	13	16	2	31	0.32	0.0006	0.395
Colorectal carcinoma	38	20	4	62	0.23	0.023	0.70
From normal tissue of black individuals with various cancers[d]	4	15	9	28	0.59		

(1) As shown by the composite data of Table 2 and 3, the frequency of the deletion was high in multiple types of cancers, and also in the normal tissue of the same individuals.

(2) Instances of loss of heterozygosity were observed (Fig. 4 below), suggesting a linkage to a new tumor suppressor (3).

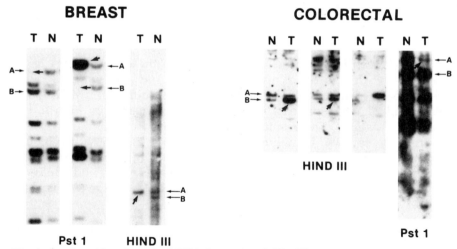

BREAST   **COLORECTAL**

Fig. 4. Southern blot of matched DNA digested with *Hind*III or *Pst*I and hybridized with full-length cDNA PADPRP probe. The blot shows representative normal heterozygous samples (N) with tumor-specific alteration (T) from a matched study of 102 samples. Taken from ref. 3.

(3) The frequency of the deletion, detected in normal tissue, is 3-fold higher in non-cancerous African-Americans compared to Caucasians as demonstrated by the data in Table 2, above.

More recent studies in the laboratory on this PADPRP gene have focused on the following:

A high frequency of this deletion in black cancer patients was noted using the PADPRP probe as a marker specifically focusing on the high risk cancers, multiple myeloma, prostate and lung. Shown below (Table 4) is the frequency of the deleted chromosome 13q allele in DNA from *normal* tissue of African American cancer patients.

Table 4 *Frequency of Chromosome 13q33-qter Deletion on Average is 0.56 in Normal Tissue of Black Cancer Patients with Variety of Tumors*

Cancer	AA	Genotype AB	BB	B Allele Frequency
Breast	3	3	1	0.4
Gallbladder			1	1.0
Colon		1	2	0.5
Glossal	1			0.0
Prostate		3	2	0.7
Cervix		1		0.5
Ovary		1		0.5
Lung		3	2	0.7
Kaposi Sarcoma		1		0.5
Rhabdoyo Sarcoma		1		0.5
Thyroid		1		1.0
Lymphoma		1		0.5
Gastric		1		0.5
			Average	0.56

We have recently analyzed the frequency of these two alleles in 32 patients with multiple myeloma (MM) and 11 patients with monoclonal gammopathy of undetermined significance (MGUS) (22 black and 21 white patients) since these cancers are very high in the black population. The results demonstrate the frequency of B allelic expression is 48% in whites (normal=14%) and 60% in blacks (normal=35%). B allelic frequency was similar in patients with MM and MGUS. Matched germline and tumor DNA showed identical patterns. These results suggest that the B allelic genotype predisposes to MM and MGUS (4).

Discussion

With the establishment of the tightly controlled PADPRP-as(7) antisense cell line available a number of new experimental approaches are now available towards elucidating a better understanding of the role of PADPRP and its mechanism in chromatin repair/replication/recombination reactions (1).

Secondly, we initially investigated whether structural differences associated with B-cell tumorigenesis could be detected in human genomic DNA by hybridization with PADPRP cDNA (3, 4). There was no difference in the frequency of polymorphic loci harboring the active gene on chromosome 1 or the pseudogene on chromosome 14. Unexpectedly, a polymorphism on chromosome 13 showed a significant increase in DNA from individuals with specific cancer versus individuals without cancer. The decreased frequency of the *A* allele in specific subpopulations of those with cancer suggests that individuals with a loss of the *A* allele are predisposed to specific cancers. These data are reminiscent of the hypothesis regarding candidate antioncogenes as discussed by Murphree and Benedict.

Acknowledgements

This work was supported by grant CA25344 from the National Cancer Institute, and a contract (DAMD17-90-C-0053) from the U.S. Army Medical Research and Development Command.

References

1. Ding, R., Pommier, Y., Kang, V. and Smulson, M., Intranuclear Depletion of Poly(ADP-Ribose) Polymerase by Induced Antisense Expression in Human Cells Causes Delay in DNA Repair. Manuscript submitted 1991.

2. Butt, T. R., Jump, D.E. and Smulson, M.E., The Nucleosome Periodicity in HeLa Cells Chromatin as Probed by Microccocal Nuclease Proc. Natl. Acad. Sci., 76, 1628-1632, (1979).

3. Bhatia, K., Cherney, B.W., Huppi, K., Magrath, I.T., Cossman, J., Sausville, E., Barriga, F., Johnson, B., Gause, B., Bonney, G., Neequayi, J., DeBernadi, M. and Smulson, M.E., A Deletion Linked to a Poly (ADP-Ribose) polymerase Gene on Chromosome 13q33-qter occurs Frequently in the Normal Black Population as well as in Multiple Tumor DNA Cancer Research, 50, 5406-5413 (1990).

4. Cao, J., Smulson, M., Waldburger, K., Lichtenstein, A.K. and Berenson, J.R., Deletion of Genetic Material from a Poly(ADP-Ribose) Polymerase-like Gene on Chromosome 13 occurs frequently in Patients with Monoclonal Gammopathies Proceedings of the American Association for Cancer Research, 32, 310 (1991)

Expression of the DNA-Binding Domain of Human Poly(ADP-Ribose) Polymerase as a *Trans*-Dominant Inhibitor of Poly(ADP-Ribosyl)ation in Transfected Eucaryotic Cell Lines

Jan-Heiner Küpper and Alexander Bürkle

Introduction

Most of the studies concerning the role of poly(ADP-ribosyl)ation in cellular physiology relied on NAD analogs, e.g. benzamide and derivatives, as competitive PARP[*] inhibitors. Since such drugs have side-effects on other cellular functions we decided to overexpress selectively the PARP DNA-binding domain as a dominant-negative mutant of for inhibiting this enzyme activity in a highly specific manner (Küpper et al., 1990). Here we review the construction of eucaryotic expression plasmids carrying the PARP full-length open reading frame and a truncated cDNA coding for the DNA-binding domain, respectively. Transfection of these constructs into eucarytic cell lines and monitoring PARP activity in transfected cells clearly showed enhanced enzyme activity in the case of overexpression of the full-length open reading frame. By contrast, transfection of plasmids coding for the DNA-binding domain resulted in a drastic inhibition of poly(ADP-ribosyl)ation, as predicted (Küpper et al., 1990).

Experimental procedures

cDNA cloning and construction of expression plasmids—Oligo-dT- and randomly primed HEF cDNA libraries were screened with oligonucleotides synthesized according to published PARP cDNA sequences (Cherney et al., 1987; Kurosaki et al., 1987; Uchida et al., 1987). Using oligonucleotide 11 (nucleotide position #985-1024, antisense orientation) several phage clones of about 1kb insert length covering the 5'-end of the open reading frame were

---

[*]The abbreviations used are: FITC, fluorescein isothiocyanate; HEF, human embryonic fibroblasts; MNNG, N-methyl-N'-nitro-N-nitrosoguanidine; PARP, poly(ADP-ribose) polymerase; PBS, phosphate-buffered saline; PCR, polymerase chain reaction; SV40, Simian virus 40; TCA, trichloro acetic acid; TRITC, tetramethylrhodamine isothiocyanate.

isolated. The insert of one of these clones was subcloned into Bluescript (Stratagene), giving rise to a plasmid designated pH86.2. Library screening, isolation and preparation of phage clones, and subcloning procedures into plasmids were performed using standard techniques (Maniatis et al., 1982). Using oligonucleotide 1 (#1321-1360, antisense) several clones of about 2kb insert length spanning the 3'-portion of the open reading frame were isolated. Subcloning of these inserts into Bluescript gave rise to pH3.1. These 5' and 3' PARP sequences were not overlapping, however. Therefore, the missing central part of the cDNA was generated by enzymatic amplification of PARP cDNA using a modification of the method of Harbarth and Vosberg (1988): Total RNA of HEF cells was prepared according to the guanidinium thiocyanate protocol of Chomczynski and Sacchi (1986). One µg of total RNA was denatured for 5 min at 95°C in 9.5 µl of a buffer containing 50 mM Tris/HCl, pH 8.3, 50 mM KCl, 6 mM $MgCl_2$, 10 mM dithiothreitol, 10% dimethylsulfoxid, 0.5 mM of each deoxynucleotide and 500 ng oligonucleotide 13 (#1399-1418, antisense) as a specific primer. Stringent annealing of the primer to the RNA was performed by incubating the reaction mixture for 2 min at 42°C. Elongation with 12.5 units AMV reverse transcriptase (Boehringer, Mannheim, Germany) was then performed for 15 min at 42°C. cDNA and primer 13 were ethanol precipitated and dissolved in 29 µl of 67 mM Tris/HCl, pH 8.8, 6.7 mM $MgCl_2$, 17 mM $(NH_4)_2SO_4$, 10 mM ß-mercaptoethanol, 7 µM EDTA, 170 µg/ml BSA, 250 µM of each deoxynucleotide and 500 ng of oligonucleotide 12 (#997-1016, sense). Using 2.5 units Taq DNA polymerase (BioLabs, Schwalbach, Germany) per reaction volume (30 µl), the cDNA was amplified in 40 PCR cycles (Saiki et al., 1988). The 421-bp amplification product was purified from a low-melting-agarose gel and cloned into the SmaI site of Bluescript vector by standard techniques (Maniatis et al., 1982). Because of the reported low fidelity of Taq DNA polymerase (Dunning et al., 1988) the PCR inserts of several Bluescript clones were sequenced using standard techniques. In one of these clones (pPCR7) we found no sequence alteration except for one silent base substitution (#1131, T→A), as compared to published sequences.

A PARP cDNA clone (pPARP23, spanning the complete open reading frame (#-29 through the EcoRI site at #3086) being referred to as "full-length cDNA" was then constructed by ligation of pH86.2, pPCR7 and pH3.1 inserts using the BstEII and AatII sites at #999 and #1381, respectively. This full-length cDNA and, in parallel, a truncated version coding for the DNA-binding domain (#-29 to the NlaIV site at #1127) were subcloned into a SV40 late-replacement vector, giving rise to pPARP62 and pPARP51, respectively (Fig. 1). In addition, the DNA-binding domain coding region was subcloned into a pUC19-based eucaryotic expression vector which carries the human cytomegalovirus promoter/enhancer and the polyadenylation signal of the herpes simplex virus thymidine kinase gene, giving rise to pPARP6 (Fig. 1).

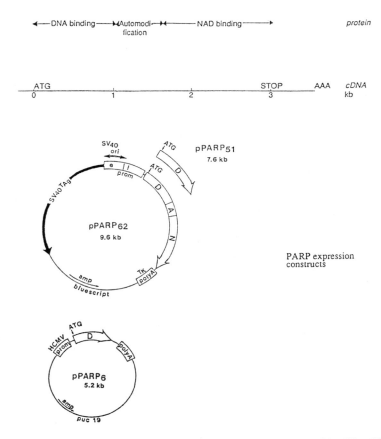

Fig. 1: Construction of eucaryotic PARP expression plasmids. The 3042-bp open reading frame coding for the DNA-binding- (D) automodification- (A) and NAD-binding domain (N) of PARP was cloned as described in the text. Full-length open reading frame of pPARP62 (# -29 up to 3086 of PARP cDNA) is under the control of SV40 late promotor (l), whereas SV40 Large TAg is under the control of the early promotor of SV40 (e). The cDNA is processed by the polyadenylation signal of the herpes simplex virus thymidine kinase gene (TK polyA). pPARP51 comprises the information for the DNA-binding domain only (#-29 up to 1127 of cDNA). Expression construct pPARP6 comprises the information for the DNA-binding domain (#-29 up to 1127 of cDNA) which is under the control of the human cytomegalovirus promoter/enhancer (HCMV). Translation initiation codons are indicated with ATG.

Stop codons for DNA-binding domain expression plasmids were contributed by a *Xba*I site within the polylinker sequences. For transfections, plasmid DNA was purified with Qiagen-pack 500 columns (Diagen GmbH, Düsseldorf, Germany).

Transfection procedure—CV-1 monkey cells were transfected as described (Küpper et al., 1990).

Immunoblot procedure—Crude extracts were prepared exactly as described (Scovassi et al., 1984). Extracted proteins from transfected CV-1 cells were separated on 10% SDS-polyacrylamide gels. Proteins were blotted onto nitrocellulose. The blot was incubated overnight at 4°C with a polyclonal rabbit antiserum raised against the second PARP "zinc finger" (anti-FII serum) diluted 1:2000 in TBS-Tween/dry milk. The blot was developed with peroxidase-conjugated goat anti-rabbit immunoglobulins.

Indirect immunofluorescence—Fixation of transfected cells, NAD incubation and subsequent indirect immunoflourescence were done exactly as described (Küpper et al., 1990).

Results and Discussion

Cloning and analysis of expression constructs with immunoblotting
A cDNA spanning the complete PARP open reading frame (#-29 through the *Eco*RI site at #3086) was cloned as described in Experimental Procedures. This cDNA, being referred to as "full-length cDNA", as well as a truncated version coding for the DNA-binding domain (#-29 to the *Nla*IV site at #1127) were subcloned into a SV40 late-replacement vector, giving rise to pPARP62 and pPARP51, respectively (Fig. 1). The coding capacity of the truncated version corresponds to the DNA-binding domain as defined by the aminoterminal papain cleavage fragment of the enzyme (Kameshita et al., 1984), the cleavage site being mapped at amino acid position 372 (Kurosaki et al., 1988). This fragment comprises all structural features which were predicted, or already shown, to be involved in nuclear location and DNA binding (Cherney et al., 1987; Kurosaki et al., 1987; Uchida et al., 1987; Gradwohl et al., 1990).

Additionally, the DNA-binding domain coding region was subcloned under the control of the very strong human cytomegalovirus promoter/enhancer in a vector lacking SV40 sequences, giving rise to pPARP6 (Fig. 1).

Expression constructs were transfected into CV-1 cells by electroporation, and one or two days post-transfection cells were processed for immunoblotting or immunofluorescence. For immunoblotting we used anti-FII serum. Transfection of pPARP62 which carries a full-length PARP cDNA led to overexpression of the 116-kDa PARP protein as revealed by immunoblot (Fig. 2, lane 2). Under the same conditions the resident enzyme in control cells transfected with vector carrying no PARP sequences was barely detectable (Fig. 2, lane 1). Transfection of pPARP51, on the other hand, led to overexpression of a 43-kDa protein (Fig. 2, lane 3), as expected from the length of the truncated open reading frame. Again, the resident 116-kDa PARP is barely detectable under these conditions. Faster migrating material of about 25 kDa is likely to be due to proteolytic degradation for the following reasons: (a) Material of the same size appeared also in transfections of pPARP6, where the same open reading frame is expressed from a different vector (data not shown). Thus the 25-kDa band is not construct-specific. (b) This material was of quite variable abundance in separate experiments, sometimes being present only in minute amounts, whereas the 43-kDa protein was always very prominent (data not shown). Presumably, proteolysis may have been also a consequence of the marked cytotoxicity associated with electroporation.

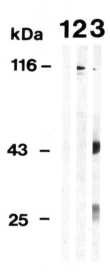

Fig. 2: Immunoblot of crude extracts from transfected CV-1 cells
Transfection of PARP expression plasmids and immunoblotting was performed as described in Experimental Procedures. Extracts from 1×10^5 cells were loaded. Lane 1, transfection of SV40 vector carrying no PARP cDNA sequences; lane 2, transfection of pPARP62; lane 3, transfection of pPARP51.

Analysis of PARP activity in ethanol-fixed cells
To see whether the overexpressed proteins indeed are biologically active, we used an "*in vitro*" assay. Cells grown on coverslips were ethanol-fixed and PARP activity was reconstituted by postincubation with NAD, leading to the *in situ* production of poly(ADP-ribose), as described (Ikai et al., 1980). Poly(ADP-

ribose) synthesis was then detected by indirect immunofluorescence, using the
monoclonal antibody H10 (Kawamitsu et al., 1984) against poly(ADP-ribose).

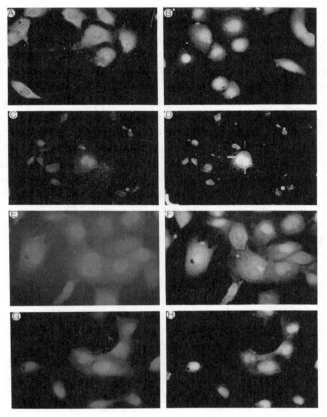

Fig. 3: Indirect double immunofluorescence of *in vitro* stimulated PARP.
Transfection and immunofluorescence of CV-1 cells were done as described
(Küpper et al., 1990). After ethanol fixation, cells were postincubated with
NAD (except in panel A); the second zinc finger of PARP is identified with
TRITC-conjugated goat anti-rabbit immunoglobulins (panels C, E, G), and
poly(ADP-ribose) is identified with FITC-conjugated goat anti-mouse
immmunglobulins (panels A, B, D, F, H); panels A and B, transfection of SV40
late-replacement vector lacking PARP sequences, no NAD postincubation in
panel A; panels C and D, transfection with pPARP62; panel E and F,
transfection with pPARP51; panels G and H, transfection with pPARP51,
ethanol-fixed cells were γ-irradiated (240 gray) before NAD postincubation.

Without NAD postincubation a very faint nuclear staining appeared as compared to the nonspecific cytoplasmic background (Fig. 3A). By contrast, NAD postincubation led to a distinct nuclear staining (note the much shorter photographic exposure time in Fig. 3B).

Since transient transfection procedures hit only a fraction of cells (about 50% in our experiments), it was important to identify transfected cells *in situ*. Therefore, we mixed monoclonal antibody H10 with the anti-FII serum and performed double immunofluorescence studies. Transfection of the full-length expression construct pPARP62 led to a strong increase in the nuclear anti-FII fluorescence (Fig. 3C), whereas cells transfected with vector carrying no insert gave a weak and homogenous staining (data not shown). Overexpressing cells showed a drastic increase in immunoreactivity with H10 against poly(ADP-ribose) (Fig. 3D), as compared to neighboring nontransfected cells. This confirms that the full-length PARP cDNA expression construct was biologically active. On the other hand, transfection of pPARP51 coding for the DNA-binding domain led to strong protein signals in nuclei of transfected cells, too (Fig. 3E), whereas polymer staining and hence PARP activity in the same nuclei was strongly reduced (Fig. 3F). The same result was seen with transfection of pPARP6 in CV-1 and CO60 Chinese hamster cells (data not shown).

If the overexpressed PARP DNA-binding domain exerts this *trans*-dominant inhibitory effect on resident CV-1 enzyme molecules by competition for DNA strand breaks this effect should be competed out by an excess of breaks. Indeed, inhibition was not found when saturating amounts of DNA strand breaks were introduced by γ-irradiation of ethanol-fixed cells before NAD postincubation (Fig. 3G and H). Alternatively, using saturating amounts of double-stranded oligonucleotides which can substitute for DNA strand breaks and very efficiently stimulate PARP in permeabilized cells (Grube et al., 1991, Bürkle et al., this volume), we were also able to titrate out the inhibitory effect (data not shown).

To see whether this *trans*-dominant inhibition occurs in the same way in living cells, we developed a TCA/ethanol fixation method for the fixation of cells stimulated with MNNG to synthesize poly(ADP-ribose) *in vivo*. After transfection of the DNA-binding domain coding region (pPARP51), treatment of cells with 50 μM MNNG, TCA/ethanol fixation, and subsequent analysis with double immunofluorescence we could show the same inhibitory effect of resident PARP enzyme by the overexpressed DNA-binding domain in living cells (data not shown, *cf.* Küpper et al., 1990), as was obtained with the NAD postincubation assay (Fig. 3E and F).

Our results show that a selectively overexpressed PARP DNA-binding domain acts as an efficient inhibitor of poly(ADP-ribosyl)ation in a *trans*-dominant fashion.

Most likely, this inhibition is due to a competition for DNA strand breaks necessary to activate the enzyme, since it can be overcome by excessive amounts of breaks either induced by high-dose carcinogen treatment, γ-irradiation, or high concentrations of double-stranded oligonucleotides. Thus, it appears feasible to study the role(s) of poly(ADP-ribosyl)ation in living cells without application of chemical inhibitors of poly(ADP-ribosyl)ation which have been widely used so far, but unfortunately are not free of side effects. Other molecular genetic approaches to inhibit poly(ADP-ribosyl)ation in living cells are currently investigated, e.g. anti-sense RNA and DNA strategies or gene targeting by homologous recombination, which should block enzyme synthesis. That situation, however, would be different from the one reported here, since not only catalytic activity but also DNA-binding activity of PARP would be lost, which might have independent functions. Thus, overexpression of the DNA-binding domain mimicks the inhibitory effect of competitive NAD analogs like 3-aminobenzamide.

We are currently establishing stably transfected cell lines which express the PARP DNA-binding domain under the control of constitutive or inducible promoters. Such cell lines should prove to be powerful tools for investigating the roles of poly(ADP-ribosyl)ation in various cellular phenomena including DNA replication and repair, cell cycle, differentiation as well as gene amplification and other events in the multistep process of carcinogenesis.

Acknowledgement- We wish to thank Prof. Harald zur Hausen for continuous support, Dr. Valerie Bosch for the SV40 late-replacement vector, Dr. Ansbert Schneider-Gädicke for the human embryonic fibroblast cDNA library, Dr. Hisae Kawamitsu for monoclonal antibody H10, and Dr. Gilbert de Murcia for anti-FII serum.

References

1. Cherney, B. W.; McBride, O. W.; Chen, D.; Alkhatib, H.; Bhatia, K.; Hensley, P.; Smulson, M. E. cDNA sequence, protein stucture, and chromosomal location of the human gene for poly(ADP-ribose) polymerase. Proc. Natl. Acad. Sci. USA 84: 8370-8374; 1987
2. Chomczynski, P.; Sacchi, N. Single step method of RNA isolation by acid guanidinium thiocyanate-phenol-chloroform extraction. Anal. Biochem. 162: 156-159; 1986
3. Dunning, A. M.; Talmud, P.; Humphries, S. E. Errors in the polymerase chain reaction. Nucleic Acids Res. 16: 10393; 1988
4. Gradwohl, G.; Ménissier de Murcia, J.; Molinete, M.; Simonin, F.; Koken, M.; Hoeijmakers, J. H. J.; de Murcia, G. The second zinc-finger domain of poly(ADP-ribose) polymerase determines specificity for single-stranded breaks in DNA. Proc. Natl. Acad. Sci. USA 87: 2990-2994; 1990

5. Grube, K.; Küpper, J. H.; Bürkle, A. Direct stimulation of poly(ADP-ribose) polymerase in permeabilized cells by double-stranded DNA oligomers. Anal. Biochem. 193: 236-239; 1991

6. Harbarth, P.; Vosberg, H.-P. Enzymatic amplification of myosin heavy-chain mRNA sequences *in vitro*. DNA 7: 297-306; 1988

7. Ikai, K.; Ueda, K.; Hayaishi, O. Immunohistochemical demonstration of poly(adenosine diphosphate-ribose) in nuclei of various rat tissues. J. Histochem. Cytochem. 28: 670-676; 1980

8. Kameshita, I.; Matsuda, Z.; Taniguchi, T.; Shizuta, Y. Poly(ADP-ribose) synthetase. Separation and identification of three proteolytic fragments as the substrate-binding domain, the DNA-binding domain, and the automodification domain. J. Biol. Chem. 259: 4770-4776; 1984

9. Kawamitsu, H.; Hoshino, H.; Okada, H.; Miwa, M.; Momoi, H.; Sugimura, T. Monoclonal antibodies to poly(adenosine diphosphate ribose) recognize different structures. Biochemistry 23: 3771-3777; 1984

10. Küpper, J. H.; de Murcia, G.; Bürkle, A. Inhibition of poly(ADP-ribosyl)ation by overexpressing the poly(ADP-ribose) polymerase DNA-binding domain in mammalian cells. J. Biol. Chem. 265: 18721-18724; 1990

11. Kurosaki, T.; Ushiro, H.; Mitsuuchi, Y.; Suzuki, S.; Matsuda, M.; Matsuda, Y.; Katunuma, N.; Kangawa, N.; Matsuo, H.; Hirose, T.; Inayama, S.; Shizuta, Y. Primary structure of human poly(ADP-ribose) synthetase as deduced from cDNA sequence. J. Biol. Chem. 262: 15990-15997; 1987

12. Maniatis, T.; Fritsch, E. F.; Sambrook, J. Molecular cloning. A laboratory manual. Cold Spring Harbor Laboratory; 1982

13. Saiki, R. K.; Gelfand, D. H.; Stoffel, S.; Scharf, S. J.; Higuchi, R.; Horn, G. T.; Mullis, K. B.; Erlich, H. A. Primer-directed enzymatic amplification of DNA with a thermostable DNA polymerase. Science 239: 487-491; 1988

14. Scovassi, A. I.; Stefanini, M.; Bertazzoni, U. Catalytic activities of human poly(ADP-ribose) polymerase from normal and mutagenized cells detected after sodium dodecyl sulfate-polyacrylamide gel electrophoresis. J. Biol. Chem. 259: 10973-10977; 1984

15. Uchida, K.; Morita, T.; Sato, T.; Ogura, T.; Yamashita, R.; Noguchi, S.; Suzuki, H.; Nyunoya, H.; Miwa, M.; Sugimura, T. Nucleotide sequence of a full-length cDNA for human fibroblast poly(ADP-ribose) polymerase. Biochem. Biophys. Res. Commun. 148: 617-622; 1987

Directed Mutagenesis of Glutamic Acid 988 of Poly(ADP-ribose) Polymerase

Marsischky, G.T.[1], Ikejima, M[2]., Suzuki, H.[3], Sugimura, T.[2], Esumi, H.[2], Miwa, M.[4], and R.J. Collier[5]

Tufts University School of Medicine, Boston MA 02111 (USA)[1], Biochemistry Division, National Cancer Research Institute, Tokyo 104 (Japan)[2], Institute of Biological Chemistry, University of Verona, Verona (Italy)[3], Dept. of Biochemistry, Institute of Basic Medical Sciences, University of Tsukuba, Tsukuba-shi, Ibaraki 305 (Japan)[4], and Shipley Institute of Medicine, Harvard Medical School, Boston MA 02115 (USA)[5]

Introduction

Although poly(ADP-ribose) polymerase and ADP-ribosylating bacterial toxins catalyze similar reactions, no structural relationship has yet been shown between the two classes of enzymes. Recently, evidence of such a relationship was suggested by Domenighini *et al.* 1991 (1), who found weak sequence similarity between the COOH-terminus of the polymerase and certain toxins. This sequence similarity includes a glutamic acid residue that has been photolabeled in several bacterial toxins, e.g. Glu-148 of diphtheria toxin (DT), in the presence of labeled NAD (2), and which is crucial to the ADP-ribosyl transferase (ADPRT) activity of these toxins. Additionally, the determination of the structure of the native form of *Pseudomonas aeruginosa* Exotoxin A (ETA) by X-ray crystallography reveals that the corresponding glutamic acid residue, ETA 553, resides in the putative active cleft of the catalytic domain (3). This putative active site glutamic acid residue corresponds to Glu-988 of the poly(ADP-ribose) polymerase, and lies within the COOH-terminal catalytic domain of the enzyme (4). Here we report experiments to explore the possible role of Glu-988 in polymer synthesis using site directed mutagenesis.

Results

Site-directed mutagenesis of Glu-988. The putative active site glutamic acid residues in DT and ETA have been changed using site directed mutagenesis. It was found that even conservative substitutions, such as Asp and Gln of DT Glu-148, result in a large (>100-fold) decrease in toxin ADPRT activity (6). Less conservative have much more drastic affects on activity (7). Despite their affect on ADPRT activity, however, these mutations have no affect on the affinity of the toxin for NAD, or on toxin NAD glycohydrolase activity, indicating that

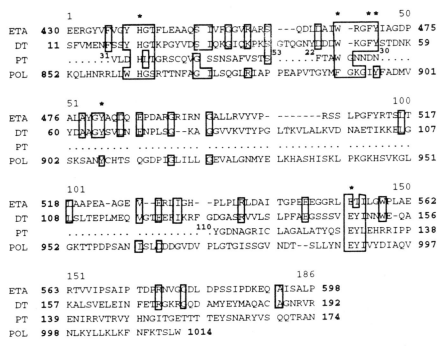

```
         1                *                                *    ** 50
ETA  430 EERGYVFVCY HGTFLEAAQS IVFGGVRARS ---QDLDAIW -RGFYIAGDP 475
DT    11 SFVMENFSSY HGTKPGYVDS IQKGIQKPKS GTQGNYDDIW -KGFYSTDNK 59
                         31            53    22  30
PT       .....:.VLD HIIGRSCQVG SSNSAFVSTS ......FTAW GNNDN..... 
POL  852 KQLHNRRLLW HGSRTTNFAG ILSQGLRIAP PEAPVTGYMF GKGIYFADMV 901

         51      *                                         100
ETA  476 ALAMGYAQLQ EPDARGRIRN GALLRVYVP- -------RSS LPGFYRTSLT 517
DT    60 YDAAGYSVDN ENPLSG--KA GGVVKVTYPG LTKVLALKVD NAETIKKELG 107
PT       .......... .......... .......... .......... .......... 
POL  902 SKSANMCHTS QGDPIGLILL GEVALGNMYE LKHASHISKL PKGKHSVKGL 951

         101                                 *            150
ETA  518 IAAPEA-AGE M--ERIIGH- -PLPLRLDAI TGPEEEGGRL EIILGWPLAE 562
DT   108 LSLTEPLMEQ VGTEERIKRF GDGASRVVLS LPFAEGSSSV EYINNWE-QA 156
                                110
PT       .......... ..........   YGDNAGRIC LAGALATYQS EYLEHRRIPP 138
POL  952 GKTTPDPSAN ISLQDDGVDV PLGTGISSGV NDT--SLLYN EYIVYDIAQV 997

         151                         186
ETA  563 RTVVIPSAIP TDFRNVGQDL DPSSIPDKEQ AISALP 598
DT   157 KALSVELEIN FETRGKRQQD AMYEYMAQAC AGNRVR 192
PT   139 ENIRRVTRVY HNGITGETTT TEYSNARYVS QQTRAN 174
POL  998 NLKYLLKLKF NFKTSLW 1014
```

Figure 1. Alignment of Exotoxin A (ETA), diphtheria toxin (DT), pertussis toxin (PT) and poly(ADP-ribose) polymerase (POL) was adapted from Carroll and Collier, 1988 (5), and Domenighini *et al.* 1991 (1). *, active site residues.

this glutamic acid residue has a specific and important role in the covalent addition of ADP-ribose to an amino acid side chain (6). We therefore decided to make the same conservative Asp and Gln mutations of Glu-988 of the polymerase (Figure 1). Mutagenesis was accomplished using the Amersham oligonucleotide-directed *in vitro* mutagenesis kit, version 2, and synthetic oligonucleotide primers on an M13 mp19 derivative carrying an Eco RI fragment of the poly(ADP-ribose) polymerase cDNA which codes for the entire COOH terminus. Mutants were confirmed by DNA sequencing and subcloned into the *E. coli* expression vector pTP as an Nco I Dra III fragment. Subclones were verified by double-strand DNA sequencing.

Residual activity of Glu-988 mutants. Substitutions at Glu-988 have a profound affect on polymerase activity. Both the Asp-988 (E988D) and Gln-988 (E988Q) mutant extracts showed a greater than 100-fold reduction in polymer synthesis compared to wild type extracts (Figure 2). Western blots indicated that equivalent amounts of wild-type and mutant polymerase were present in the extracts (data not shown). Activity of both mutants is inhibited

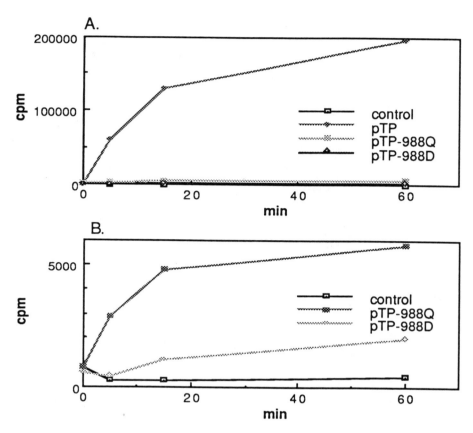

Figure 2. Time course of ADP-ribose incorporation. Extracts were made from E. coli expressing the mutant polymerase according to Ikejima *et al.* 1990 (8). Reaction: 10 µl crude extract was mixed with 100 µl assay mixture and incubated at 25° C. Assay mix contained 100mM Tris-Cl pH 8.0, 10 mM MgCl$_2$, 100 µM [32P]NAD (ca. 10 Ci/mmol), 20 µg/ml activated DNA, and 20 µg/ml calf thymus histone fraction II. At time points, 25 µl aliquots were spotted to TCA saturated 3MM paper, washed twice with 5% TCA, once with methanol, dried and counted in a liquid scintillation counter (Cerenkov cpm). Control lysate: *E. coli lon* strain ME8417 (▣), ME8417 wild type polymerase lysate: pTP (❖), ME8417 988-Gln polymerase lysate: pTP-988Q (▨), ME8417 988-Asp polymerase lysate: pTP-988D (◈).

by 3-aminobenzamide and is DNA dependent when assayed using an activity gel (data not shown), indicating that these mutations do not grossly alter either NAD binding or DNA activation.

The first and primary ADP-ribosylated species is the full-length enzyme itself for both mutants, as has been previously shown for the wild-type enzyme (9), although a number of protein species are found to be ADP-ribosylated in crude extracts of both mutants (Figure 3). Not all of these species correspond to polymerase fragments as determined by western blotting. Although it is

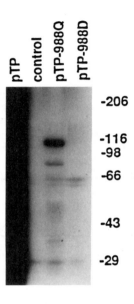

Figure 3. Proteins ADP-ribosylated by Glu-988 mutants. Reaction: 10 µl crude extract was mixed with 100 µl assay mixture and incubated for 30 minutes at 25° C. Reactions were stopped by the addition of 900 µl -20° C EtOH and allowed to precipitate at -20° C for 2 hours. Samples were then resuspended in 10 µl 10 mM Tris-Cl pH 8.0, 1 mM EDTA and 1 mM 3-aminobenzamide (3ab), after which 10 µl 2x Solubilizing Buffer was added and the samples incubated at 56° C for 2 minutes. Samples were then loaded on a 7% polyacrylamide minigel (10) and subjected to electrophoresis at 125 V for 2 hours. The gel was then stained for 30 minutes and destained overnight, dried and subjected to autoradiography.

possible that at least some of these are derived from the nonimmunogenic COOH terminus, it seems likely that some are transADP-ribosylated E. coli proteins. Also, addition of histone to the reaction mixture results in a new ADP-ribosylated species (data not shown).

When the reaction products of the mutant enzymes were analyzed, it was found that the E988D mutant makes short polymer and also monomer. The E988Q mutant makes only monomer (Figure 4). The attachment of these products to each mutant polymerase was found to be highly sensitive to alkali (Figure 4) or neutral hydroxylamine (data not shown) indicating that they are attached by the carboxylate ester linkage characteristic of the wild-type enzyme.

Discussion

The results presented suggest that Glu-988 is important for elongation of polymer, but not for initiation of polymer synthesis. Mutations which affect only one of these processes are perhaps to be expected since the two reactions arechemically distinct, i.e. initiation of polymer synthesis results from the addition of an ADP-ribose to an acidic amino acid side chain to form an ester linkage, while elongation results from the addition of an ADP-ribose to the 2'OH of the ribose moiety of another ADP-ribose to form a glycosidic bond. It had, however, been expected that mutations affecting an aspect of the polymerase which resembles a bacterial toxin should affect addition of ADP-ribose to an

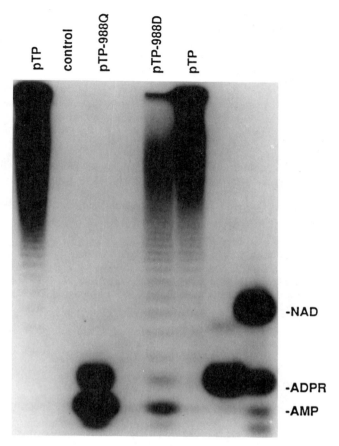

Figure 4. ADP-ribosylation products of Glu-988 mutants. Reaction: 15 µl crude extract and 150 µl assay mixture were mixed and incubated for 1 hour at 25° C. The reaction was terminated by running the sample through a G50 spun column (12) and protein-bound ADP-ribosylation products were released by treating the samples with 0.1 N NaOH, 5mM EDTA for 30 minutes at 37° C. Samples were then neutralized with HCl, and run on a 20% polyacrylamide Tris Borate EDTA gel (12). The gel was then dried and subjected to autoradiography. Some hydrolysis of ADP-ribose to form AMP occured during the treatment with alkali. The identification of the lower band as AMP was confirmed by 2-dimensional thin layer chromatography (GTM unpublished results).

amino acid side chain. On the other hand, it should be considered that bacterial toxins catalyze the addition of ADP-ribose to a number of different side chains, so a toxin-like function of the polymerase which adds ADP-ribose to a 2'OH group is perhaps not too surprising.

It remains to be seen whether another active site is involved in polymer initiation, or whether the site already identified is sufficient for both processes.

In the latter case, Glu-988 would be a substituent of the active site which is required only for an aspect of ADP-ribose addition which is specific to polymer elongation, e.g. enhancing the nucleophilicity of the substrate. Site directed mutagenesis of other putative active site residues is being done in order to explore this possibility.

References

1. Domenighini, M., Montecucco, C., Ripka, W.C., and Rappouli, R. (1991) Molecular Microbiology 5: 23-31.

2. Carrol, S.F., and Collier, R.J. (1984) Proc. Natl. Acad. Sci. U.S.A. 81: 3307-3311.

3. Allured, V.S., Collier, R.J., Carrol, S.F., and McKay, D.B. (1986) Proc. Natl. Acad. Sci. U.S.A. 83: 1320-1324.

4. Simonin, F., Menissier-de Murcia, J., Poch, O., Muller, S., Gradwohl, G., Molinette, M., Penning, C., Keith, G., and de Murcia, G. (1990) J. Biol. Chem. 265: 19249-19256.

5. Carrol, S.F., and Collier, R.J. (1988) Molecular Microbiology 2: 293-296.

6. Wilson, B.A., Reich, K.A., Weinstein, B.R., and Collier, R.J. (1990) Biochemistry 29: 8643-8651.

7. Lukac, M., and Collier, R.J. (1988) J. Biol. Chem. 263: 6146-6149.

8. Ikejima, M., Noguchi, S., Yamashita, R., Ogura,T., Sugimura, T., Gill, D.M., Miwa, M. (1990) J.Biol. Chem. 265: 21907-21913.

9. Ikejima, M., Marsischky, G.T., and Gill, D.M. (1987) J. Biol Chem. 262: 17641-17650.

10. Laemmli, U.K. (1970) Nature 227: 680-685.

11. Payne, D.M., Jacobson, E.L., Moss, J., and Jacobson, M.K. (1985) Biochemistry 24: 7540-7549.

12. Maniatis, T., Fritsch, E.F., and Sambrook, J. (1982) Molecular Cloning: A Laboratory Manual, Cold Spring Harbor Laboratory, Cold Spring Harbor, N.Y.

Chicken poly(ADP-ribose) polymerase. Complete protein sequence deduced from cDNA, comparison with mammalian enzyme sequences.

Claude P. Niedergang, Jean-Marie Garnier, Jean-Marc Jeltsch and Marie-Elisabeth Ittel.

Introduction

In the past few years human poly(ADP-ribose)polymerase (pADPRP) cDNA (Uchida et al., 1987, Kurosaki et al., 1987) as well as the bovine (Saito et al., 1990), mouse (Huppi et al., 1989) and part of the rat (Thibodeau et al., 1989) cDNAs have been cloned revealing an extensive aa sequence conservation in mammals. To further study the interspecies conservation, we isolated a 1.8 kb human pADPRP cDNA clone from an oligo (dT) primed lgt10 cDNA library (Walter et al., 1985). This clone encodes the entire automodification domain of the enzyme and a great part of the NAD binding domain. Cross hybridization of chicken genomic DNA and mRNA with this human probe has been obtained in low-stringency conditions (Ittel et al., 1989). Using this 3' probe and another one covering the 5' region of the pADPRP human cDNA from nucleotide 1 to 697 (kindly provided by Dr G. Gradwohl), and called 5' probe, we cloned and sequenced the chicken pADPRP cDNA. Its deduced aa sequence has been compared with the known mammalian aa sequences.

Results and discussion

cDNA cloning of chicken pADPRP. The human pADPRP cDNA 3' probe was used to screen an oligo (dT)-primed lgt10 cDNA library constructed from hen oviduct poly(A)$^{+}$ RNA (Krust et al., 1986). Overlapping clones, BS6 and BS136 (Fig.1) were isolated at low stringency conditions and subcloned into the EcoRI site of pBluescribe. Sequencing of both strands revealed that these two clones correspond to the NAD-binding domain and part of the automodification domain.

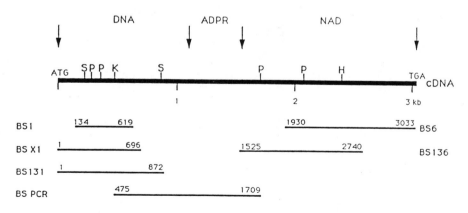

Fig.1 Restriction map of chicken pADPRP cDNA. H, HincII; K, kpnI; P,PstI; S, SacI.

To obtain the cDNA N-terminal part, the same library was rescreened with the human 5' probe. The positive clone BS1 was isolated. This chicken cDNA clone, used as probe, allowed us to isolate at high-stringency conditions two other clones, BSX1 and BS131(Fig.1) By characterization and comparison with human cDNA it appears that these two clones overlapping the initial BS1 clone, are localized in the DNA binding domain.

The midle part of the chicken pADPRP, not found in the library, was cloned by PCR using synthetic oligonucleotides flanking the missing part. The PCR product was analysed by gel electrophoresis showing the major reaction product at the expected size. After reamplification of this fragment the final PCR product was gel-purified (Koenen 1989) and cloned between the KpnI and PstI sites of pBluescribe to give the BS PCR recombinant plasmid. Two positive clones were sequenced on both strand.

Comparison with the full-length human cDNA sequence (Uchida et al.,1987) reveals that the combined overlapping chicken sequences (BS131; BSPCR; BS6 and BS136) extend from nt 1 of the start codon (Met) to the stop codon TGA. The specificity of these different chicken cDNA clones was confirmed by Northern blotting of total RNA isolated from chicken astrocytes or chicken neurones cultures. A clear specific band estimated to be about 3.7 kb, was observed when the insert of clone BS1 was used as probe to hybridize homologous chicken mRNA. The same pattern was obtained with the insert of the five other isolated clones (data not shown).

Comparison between the chicken and mammalian coding regions of pADPRP. The size of nt sequence of chicken pADPRP coding regions is 3033 bp, whereas in mouse it is 3039 bp (Huppi et al., 1989), in human 3042 bp (Uchida et al., 1987) and in bovine 3048 bp (Saito et al., 1990). The nt sequence similarity between chicken and mammalian cDNAs is about 70 to 75%. At the protein level the identity is slightly higher. In terms of conservative aa residues the complete sequences are near 90% homologous (Table 1).

Specific regions reveal greater aa identity than others (Fig. 2), particularly in motifs which are implied in functional properties. Table 1 shows the % of aa identical or conservatively maintained between chicken, human bovine and mouse in the three proteolytically separable domains. The results suggest a higher homology among mammalian pADPRP than between mammalian and chicken pADPRP. The two putative zinc finger domains have been pointed out and it appears that the greatest identity and conservation is obtained within these two specific structures. As shown in Fig. 2, Cys^{21}, Cys^{24}, His^{53} and Cys^{56} as well as Cys^{125}, Cys^{128}, His^{159} and Cys^{162}, which are presumed to fold into a finger-like structure (Gradwohl et al., 1990), are well conserved. Moreover, the sequence duplication between the fingers (Uchida et al., 1987) can also be observed in chicken (data not shown).

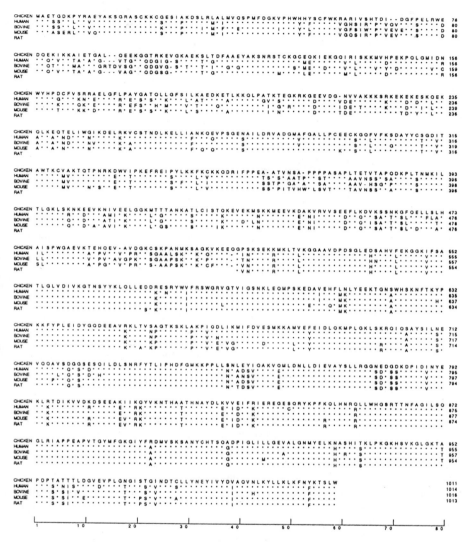

Fig. 2 Comparison of the cDNA-derived aa sequences of chicken vs. human, bovine, mouse, and part of rat pADPRP. The dashes within the sequences represent gaps introduced to allow maximum sequence alignment.

The putative automodification domain has only 65% identical aa between mammals and chicken. Nevertheless, it is noteworthy that 12 glutamic acid residues which might be the sites for automodification by poly(ADPR) are conserved within these different species. Additional common automodification sites can be defined when Glu residues correspond to Asp residues functionally equivalent.

Species	pADPRP		DNA-binding domain				ADPR-binding domain	NAD-binding domain
	aa 1-1011	aa 1-20	aa 21-56 Zn-finger 1	aa 57-124	aa 125-162 Zn-finger 2	aa 163-370	aa 371-522	aa 523-1011
Human	79 (90.8)	80 (90)	82.9 (97.2)	57.4 (82.4)	81.6 (89.5)	74.5 (89.4)	65.1 (81.6)	87.5 (95.7)
Bovine	76.6 (89.5)	80 (90)	75 (94.4)	48.5 (73.5)	68.4 (89.5)	71.6 (88)	63.8 (81.6)	87.1 (95.5)
Mouse	77 (89.7)	65 (90)	82.9 (97.2)	57.4 (79.4)	81.6 (89.5)	73 (84.1)	63.8 (81.6)	85.7 (95.9)
Rat	-	-	-	-	-	-	-	87.1 (96.1)

Table 1. Percent aa identity -() plus conservative changes - in mammalian pADPRP compared to chicken pADPRP.

The best aa conservation is observed within the NAD-binding domain where the identity is near 86% (Table 1). The identity is still increased in the regions flanking the two Gly-rich sequences in the C-terminal domain of pADPRP (aa 885-891 and aa 947-951 of chicken sequence)(Fig.2). These consensus aa sequences are presumed to be characteristic of proteins interacting with adenine nucleotides (Kurosaki et al., 1987; Walker et al., 1982). Within these structures Lys[890] in the chicken sequence can be identified to Lys [893] in the human sequence determined to be essential for enzyme activity (Simonin et al., 1990). Finally, it should be mentioned that the C-terminal chicken pADPRP sequence portion (aa 840-1011) exhibits extensive sequence similarity with NAD(P)[+] aa dehydrogenases known sequences as it has already been shown for the human enzyme by Simonin et al., (1990). As a matter of fact, using the same criteria of conservative residue definition, we found that this chicken pADPRP sequence region only differs from the same human region by 8 aa on a total of 172 aa. Another interesting feature is the extensive similarity of the chicken aa residues involved in a b-a-b-fold which has been implicated in the binding of the ADP moiety of NAD(P)[+] molecules.

It may be concluded that in mammals and birds the specific features of each pAPRP domain are well conserved. In lower vertebrates and invertebrates the enzyme is still present but some of its physical properties were modified indicating that the cDNA identity will be decreased. Thus chicken cDNA may be an useful tool in the cloning of pADPRP-like sequences in lower vertebrates and invertebrates.

Acknowledgements

J.L. Toussaint is gratefullly acknowledged for his aid in screening the data by computer analysis.The nt sequence has been submitted to the EMBL DNA Sequence Library under accession n°X52690.

References

Gradwohl, G., Menissier De Murcia, J., Molinete, M., Simonin, F., Koken, M., Hoeijmakers, J.H.J. and De Murcia, G. The second zinc-finger domain of poly(ADP-ribose) polymerase determines specificity for single-stranded breaks in DNA. Proc.Natl.Acad.Sci. USA 87 (1990) 2990-2994.

Koenen, M., Recovery of DNA from agarose gels using liquid nitrogen. Trends Genet. 5 (1989) 137.

Krust, A., Green S., Argos, P., Kumar, V., Walter, P., Bornert, J.M. and Chambon, P. The chicken oestrogen receptor sequence: homology with u-erbA and the human oestrogen and glucocorticoid receptors. EMBO J. 5 (1986) 891-897.

Kurosaki, T., Ushiro, H., Mitsuuchi, Y., Susuki, S., Matsuda, M., Matsuda, Y., Katunuma, N., Kangawa, K., Matsuo, H., Hirose, T., Inayama,S. and Shizuta, Y. Primary structure of human poly(ADP-ribose) synthetase as deduced from cDNA sequence. J. Biol. Chem. 262 (1987) 15990-15997.

Saito, I., Hatakeyama, K., Kido, T., Ohkubo, H. Nakanishi, S. and Ueda, K. Cloning of a full-length cDNA encoding bovine thymus poly(ADP-ribose)synthetase:evolutionary conserved segments and their potential functions. Gene 90 (1990) 249-254.

Simonin, F., Menissier De Murcia, J., Poch, O., Muller, S., Gradwohl, G., Molinete, M., Penning, C., Keith, G. and De Murcia, G. Expression and site-directed mutagenesis of the catalytic domain of human poly(ADP-ribose) polymerase in *E.coli* . lysine-893 is critical for activity. J. Biol. Chem. 265 (1990) 19249-19256.

Thibodeau, J., Gradwohl, G., Dumas, C., Clairoux-Moreau, S., Brunet, G., Penning, C., Poirier, G.G. and Moreau, P. Cloning of rodent cDNA encoding the poly(ADP-ribose) polymerase catalytic domain and analysis of mRNA levels the cell cycle. Biochem. Cell. Biol. 67 (1989) 653-660.

Uchida, K., Morita, T., Sato, T., Ogura, T., Yamashita, R., Nogushi, S., Suzuki, H., Nyunoya, H., Miwa, M., and Sugimura, T. Nucleotide sequence of a full-length cDNA for human fibroblast poly(ADP-ribose) polymerase. Biochem. Biophys. Res. Commun 148 (1987) 617-622.

Walter, P., Green, S., Greene, G.L., Krust, A., Bornert, J.M., Jeltsch, J.M., Staub, A., Jensen, E., Scrace, G., Waterfield, M. and Chambon, P. Cloning of the human oestrogen receptor cDNA. Proc. Natl.Acad. Sci. USA 82 (1985) 7889-7893.

Walker, J.E., Saraste, M., Runswick, M.J. and Gay, N.J. Distantly related sequences in the a- and b-subunits of ATP synthase, myosin, kinases and other ATP-requiring enzymes and a common nucleotide binding fold. EMBO J. 1 (1982) 945-951.

Nuclear poly(ADPR)Polymerase expression and activity in rat astrocytes culture: effects of bFGF.

Magali G. Chabert, Claude P. Niedergang, Fernand Hog, Mariagrazia Partisani and Paul Mandel.

Introduction

Due to the availability of rat poly(ADPR)P cDNA (Thibodeau et al.,1989), the investigation of the poly(ADPR)P gene expression, in various experimental conditions, became possible. Long term primary cultures of rat astrocytes were used to investigate the poly(ADPR)P gene expression during the proliferation and the differentiation. In addition, the effects of basic Fibroblast Growth Factor (bFGF) have been investigated.

Results and discussion

In long term rat astrocytes cultures, the proliferation period with a parallel increase in DNA occurs between 8 and 21 days of culture and is followed by a plateau which corresponds to a period of cell differentiation (data not shown). In presence of bFGF (5ng/ml), striking changes appear in cell morphology: a great number of expansions is observed (Fig.1); however, the DNA amount profile remains quite similar.

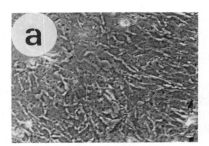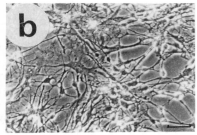

Figure 1: Phase contrast micrographs of primary culture of rat astrocytes, at 20 days of culture, plated in 100mm Petri dishes in a Waymouth medium supplemented with bovine serum albumin (0.5 mg/ml) and Insulin (5μg/ml) in absence (a) or presence (b) of bFGF (5ng/ml). Bar=14.3μm.

Northern blot analysis provides a unique specific poly(ADPR)P mRNA band of about 3.6Kb. The levels of poly(ADPR)P mRNA in long term rat astrocytes cultures from day 8 to day 39 are shown in Fig.2. The poly(ADPR)P mRNA pattern presents two increases periods. In absence of bFGF, the first one is located at day 15, before the increase in DNA amount (data not shown). In presence of bFGF, the level of this early mRNA peak is 1.5 fold higher and the maximum appears at day 11. Thereafter, the level of poly(ADPR)P mRNA decreases to a basal level. A second increase in poly(ADPR)P mRNA is reached at day 29 when the DNA increase is achieved. It appears also that, in presence of bFGF, the beginning of this peak occurs earlier and the mRNA total amount is increased 1.5 fold when compared to the control. In both cases at 32 days of culture, the poly(ADPR)P mRNA amount returns to a basal level.

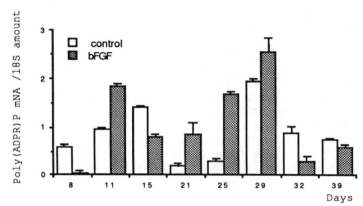

Figure 2: poly(ADPR)P mRNA levels in long term rat astrocytes cultures, in absence or presence of bFGF (5ng/ml). Poly(ADPR)P mRNA was estimated by Northern blot analysis. Briefly, 25µg of total RNA were separated by 1% Agarose/Formaldehyde gel electrophoresis and transfered on hybond C-extra membrane. The hybridization with the rat poly(ADPR)P cDNA (2.0)Kb was performed in 50% Formamide at 42°C overnight and the final wash in 0.5SSC, 0.1% SDS, 30mn at 50°C.

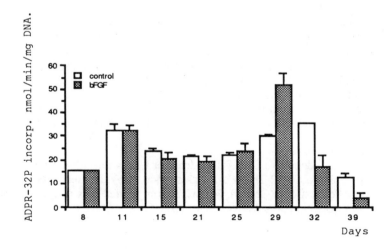

Figure 3: Nuclear poly(ADPR)P activity levels in rat astrocytes cultures in absence or presence of bFGF (5ng/ml). The enzymatic activity is measured by the incorporation of ^{32}P ADP-Ribose in permeability cells.

The enzymatic activity of the poly(ADPR)P has been recorded during the same period of time (Fig.3) and seems to correlate with the poly(ADPR)P mRNA levels. A first increase of enzymatic activity is observed at day 11 of culture; however in contrary to the mRNA values no difference of the activity levels between the control and the cells treated with bFGF is observed. The second peak of poly(ADPR)P activity is reached at day 32 in the absence of bFGF and at day 29 for the bFGF treated cells. At this delay, not only the poly(ADPR)P activity increase appears earlier, but it is also 1.5 fold higher in presence of bFGF, when compared to the control.

These results suggest either a direct effect of bFGF on poly(ADPR)P activity or a stimulation of poly(ADPR)P mRNA expression. It seems that the increases of the mRNA levels and enzymatic activities are not closely time related mainly during the proliferation phase. It seems likely that mRNA translation is not necessarily associated with gene transcription and that the steady-state enzyme level is sufficient in normal growth conditions. The cellular activators or inhibitors may play a major role in the control of nuclear poly(ADPR)P enzymatic activity. Let us remind that transient changes in poly(ADPR)P enzymatic activity associated both to early events preceding the onset of DNA synthesis and to a later phase, when the DNA synthesis is achieved, were reported in regenerating rat liver (Cesarone et al.,1990). Moreover Ittel & al ,1983, have demonstrated that the inhibition of ADP-ribosylation activity in human lymphocytes cultures, produces an inhibition of Phytohaemagglutinine induced DNA synthesis and cell proliferation. These results support the hypothesis that ADP-ribosylation reaction is a preliminary step in the initiation of DNA replication.

Like in rat astrocytes during the cell differentiation, an increase of poly(ADPR)P activity has been observed during the differentiation of Xenopus laevis embryos (Farzaneh et al.,1979) and in mouse L cells entering in a plateau phase (Berger et al.,1978). In contrast to our results, a depression in poly(ADPR)P activity and gene expression, during the differentiation, was reported in murine tumoral cells (Ohashi et al.,1984, Tanigushi et al.,1988[a]) and in Pheocromocytoma PC12 cells (Tanigushi et al.,1988[b]). It seems likely that the transient increase in poly(ADPR)P mRNA and enzymatic activity, during the differentiation, may be restricted to non tumoral cells.

Taking all together, it seems likely that a number of events occuring at the onset of cell growth or cell differentiation, is modulated by poly(ADPR)P already present and or by the potentiation of poly(ADPR)P transcription.

Acknowledgements

We are indebted to Dr. J. Thibodeau for kindly providing us the rat cDNA probe for poly(ADPR)P This work was supported by the Centre National de la Recherche Scientifique (CNRS). M.G.C. was supported by a doctoral fellowship from the Association pour la Recherche sur le Cancer (ARC).

References

Berger, N. A., G. Weber, A. S. Kaichi, and S. J. Petzold. 1978. Relation of poly(Adenosine diphosphoribose) synthesis to DNA synthesis and cell growth. B.B.A. 519:105-117.

Cesarone, C. F., L. Scarabelli, A. I. Scovassi, R. Izzo, M. Menegazzi, A. Carcereri De Prati, M. Orunesu, and U. Bertazzoni. 1990. Changes in activity and mRNA levels of Poly(ADP-ribose)polymerase during rat liver regeneration. B.B.A. 1087:241-246.

Farzaneh, F., and C. K. Pearson. 1979. The activity and properties of poly(Adenosine diphosphate ribose) polymerase *in vitro* during the embryonic development of South African clawed toad Xenopus laevis. Dev. Biol. 72:254-265.

Ittel, M.E., J. Jongstra-Bilen, C. Rochette-Egly, and P. Mandel. 1983. Involvement of poly ADP-ribose polymerase in the initiation of phytohemagglutinin induced human lymphocyte proliferation. B.B.R.C. 116:428-434.

Ohashi, Y., K. Ueda, O. Hayaishi, K. Ikai, and O. Niwa. 1984. Induction of murine teratocarcinoma cell differentiation by suppression of poly(ADP-ribose) synthesis. Proc. Natl. Acad. Sci. USA, 81:7132-7136.

Tanigushi, T., K. Yamauchi, T. Yamamoto, K. Toyoshima, N. Harada, H. Tanaka, S. Takahashi, H. Yamamoto, and S. Fujimoto. 1988a. Depression in gene expression for poly(ADP-ribose) synthetase during the interferon-γ-induced activation process of murine macrophage tumor cells. Eur. J. Biochem. 171:571-575.

Tanigushi, T., K. Morisawa, M. Ogawa, H. Yamamoto, and S. Fujimoto. 1988b. Decrease in the level of poly(ADP-ribose) synthetase during Nerve Growth Factor-promoted neurite outgrowth in rat pheochromocytoma PC12 cells.B.B.R.C. 154:1034-1040.

Thibodeau, J., G. Gradwhol, C. Dumas, S. Clairoux-Moreau, G. Burnet, C. Penning, G. Poirier, and P. Moreau. 1989. Cloning rodent cDNA encoding the poly(ADP-ribose)polymerase catalytic domain and analysis of mRNA levels during the cell cycle. Biochem. Cell Biol. 67:653-660.

Molecular Cloning of the Rat Poly (ADP-Ribose) Polymerase Gene and Preliminary Characterization of its Promoter and 5'-Flanking Regions

Frederic Potvin, Jacques Thibodeau, Guy G. Poirier and Sylvain L. Guérin

Laboratoire du Métabolisme du poly(ADP) ribose, Endocrinologie Moléculaire, Centre de Recherche du Centre Hospitalier de l'Université Laval (CHUL), 2705 boul. Laurier, Ste-Foy (Québec) G1V 4G2, Canada.

Introduction

Poly (ADP-ribose) polymerase (PARP) is a eukaryotic nuclear enzyme which has the ability to catalyzes post-translational modification of a variety of distinct proteins, including PARP itself. This is accomplished via the progressive addition of ADP-ribose moities from NAD to these specific proteins (1). Although the exact function of the enzyme has yet to be elucidated, it is clear, however, that it is involved in a number of physiologically critical functions such as DNA repair, cell proliferation and differentiation, gene transcription and carcinogenesis (2). In order to further complete the characterization of the rat PARP (rPARP) enzyme and to better understand the way by which its expression is regulated at the molecular level, we have cloned and sequenced the promoter and 5'-flanking sequence of the rat gene encoding this protein. Detailed analysis of the rPARP gene upstream sequence revealed the presence of potential cis-acting sequences for a number of known transcription factors. Most importantly, a unique conserved region (US1) located 185bp upstream from the potential RNA transcription start site and common to a number of genes including the human PARP (hPARP) gene and a number of other housekeeping genes was identified. Gel shift analysis clearly indicated that a specific nuclear protein binds to this element, suggesting that a common trans-acting factor might regulates the expression of a number of housekeeping genes, in which are included both the human and rat PARP genes.

Results and Discussion

Until very recently, no information was available regarding the structural organization of the PARP gene promoter and 5'-flanking sequences. With the recent works of Shizuta's (3) and Esumi's (4) groups, a limited amount of data were provided for the upstream region of the hPARP gene. Their works provided the evidences that the hPARP gene belongs to the category of genes lacking TATA and CCAAT-box motifs. In addition, 100bp of upstream sequence from the hPARP gene promoter was shown to be sufficient to confer responsiveness of the CAT reporter gene to both cAMP and TPA (3). However, despite these efforts, no such informations have been provided for the rPARP gene. The data presented in this study outline the first attempts to characterize cis-acting regulatory elements and transcription factors which altogether determine the level to which the rPARP gene is to be transcribed in various tissues.

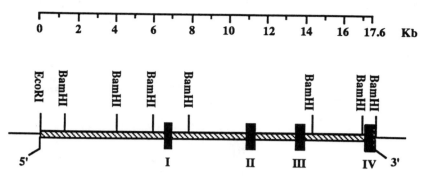

Fig.1. Genomic organization of the rat poly(ADP-ribose) polymerase gene deduced from the partial clone isolated. The organization of the rPARP gene-containing 17.6Kbp genomic fragment isolated is illustrated along with its partial restriction map. Positioning of the introns and exons has been deduced from sequence homology between the rPARP and hPARP genes.

Isolation of the genomic clone corresponding to the rat PARP gene. A [^{32}P]-labeled 125bp KpnI DNA fragment from the hPARP cDNA (5) was previously used as a probe to screen-out a rat λEMBL4 genomic library (6). Only one clone containing the promoter region and a substantial amount of DNA sequence from the 5'-flanking region of the rPARP gene turned out to be positive. The 17.6Kbp genomic insert from this clone contains only exon 1, 2 and 3 and part of exon 4, along with 6.6Kbp of upstream sequences (Fig.1). Sequence comparison with the hPARP cDNA revealed an overall 98% and 88% homology at both the amino acid and nucleotide level respectively (7), which comfirmed the identity of the rat gene. A RsaI fragment spanning all the first exon and 800bp of 5'-flanking sequence was completely sequenced using the dideoxy procedure. Nuclease S1 mapping experiments located the rPARP gene transcription start site at 103bp upstream from the first ATG (results not shown). This position is substantially different from that identified for the hPARP gene, which has been recently located 159bp upstream from the ATG (3).

Upstream sequence analysis and potential organization of the rPARP promoter. Although the coding region of both rPARP and hPARP genes are obviously highly conserved (with an overall homology of 88% at the DNA level), no long stretches of homology can be detected in the DNA sequence extending upstream from the first ATG of the rPARP gene. However, a substantial level of structural similarity is apparent in the overall organization of the promoter region of the rPARP, the hPARP and the human ß-polymerase genes. Indeed, potential binding sites for known transcription factors have been identified in the highly GC-rich promoter region of these three genes (Fig.2). Among the motifs identified are multiple GC-rich sequences likely to be bound by the well-known transcription factor Sp1 (5'-GGGGCGGGG-3') or by Sp1-like proteins (8,9) and a regulatory element which is only one base pair different from the cAMP responsive element (CRE: 5'-

NTGACGTCAN-3')(10). This potential CRE is located between two Sp1 binding

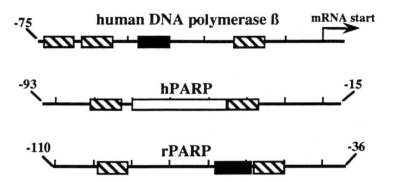

Fig.2. Promoter organization of the potential binding sites for the transcription factors Sp1 and CREB. The position of potential binding sites for the well known transcription factors Sp1 (hatched boxes) and CREB (black boxes) have been deduced from the comparison of their consensus binding site with the promoter sequence of the rPARP, hPARP and human ß-polymerase genes. Also indicated is a large 25bp almost perfect inverted repeat which does not present homologies with any known binding site (empty box). Numbers correspond to the position of the DNA sequences relative to the mRNA start site.

sites in both the human ß-polymerase and the rPARP genes. The functional significance of these closely associated sites is yet unknown. However, such complex promoter organization are not uncommon and have been described for a number of other genes. Whether these potential cis-acting regulatory sequences are truly binding nuclear proteins remains to be proven despite their very good homology with known binding sites for transcription factors. Experiments are presently underway to test the functionality of these DNA regions.

When one takes a closer look at the rPARP gene potential promoter, no TATA but two potential almost perfect CCAAT motifs (GCAAT: position -143; CCAAC: position -104) can be detected. The fact that no classical TATA motif has been detected in the promoter region of this gene raises the possibility that it belongs to the class of GC-rich promoters which are depleted of TATA-box but still are able to function normally without this element and which primarily includes housekeeping genes (11). Because of their high GC content, these promoters are likely to possess multiple binding sites for Sp1, which would presumably be sufficient to stimulate transcription of these genes by the RNA polymerase II complex. Ogura et al. (4) suggested that perfectly conserved TATA and CCAAT box motifs located 847 and 873bp upstream of the hPARP mRNA start site respectively, might regulate the transcriptional activity of this gene, as has been also suggested for the human and mouse DNA polymerase ß genes (11,12). However, deletion of this region in a CAT construct containing only 99bp of 5'-flanking sequence from the hPARP directed significant levels of

CAT activity (3) suggesting that these potential TATA and CCAAT-box motifs exert no significant effects on the transcription of this gene in transient transfection assays.

```
-197    AAAGGGGTGGCGCCTGT    -181    rPARP    US1
-200    AAAGGGGTGGCGCC-G-    -184    hPARP
+411    AAA-GGGTGG---CTGT    +427    hPolß
-84     -AAGGGGTGG-G-C---    -68     h17ßHSDII

-61     -A--GGGTGG-GCCTGT    -45     p12.A
```

Fig.3. Sequence homology between the rPARP US1 element and other similar regions from various genes. The rPARP US1 element (position -181 to -197) is aligned with sequences from other genes (hPARP (10), human polymerase ß (hPolß)(7) and human 17ß hydroxysteroid dehydrogenase (13)) that show significant homology with this element. Also shown is the cis-acting sequence (p12.A) from the mouse p12 gene which is bound by an Sp1-like positively acting transcription factor (6). Underlined Gs indicate the position of the guanines totally protected from methylation in the p12.A sequence following DMS methylation interference footprinting.

A unique motif (US1) is conserved within the 5'-flanking region of the rPARP gene. Detailed analysis of the first 700bp from the 5'-flanking region of the rPARP gene revealed none but a small 14bp region of homology with the hPARP flanking sequence. We designated this region as US1 for "Upstream Sequence 1". A more extensive comparison study revealed that regions highly similar to US1 can be identified in a number of other housekeeping genes, including the gene which encodes the human polymerase ß (Fig.3). To our surprise, the DNA binding site (p12.A) for a transcription factor known to exert a positive effect on the transcription of the mouse p12 secretory protease inhibitor gene and suggested to be an Sp1-like protein is found almost perfectly conserved within the rPARP US1 element (9) (Fig.3). Mutations introduced in this binding site reduced the level of CAT activity directed by the mouse p12 promoter up to 12 fold in a number of established tissue culture cell-lines. Furthermore, this protein, which generates a DNA/protein complex of low electrophoretic mobility in gel shift assay, has been detected in almost every tested nuclear extract prepared from tissue-culture cells (8 out of 9)(9), suggesting that it is likely to be a ubiquitous transcription factor. The pattern of protected Gs for this protein within the p12.A region, as determined by DMS methylation interference footprinting, is perfectly conserved within the rPARP US1 element strongly suggesting that a common transcription factor likely to be ubiquitous, regulates the expression of both genes and maybe of others as well. This observation is in agreement with the widely distributed expression pattern observed for both the mouse and rat PARP genes, which have been shown to be transcribed, although at various levels, in every tested tissue (4).

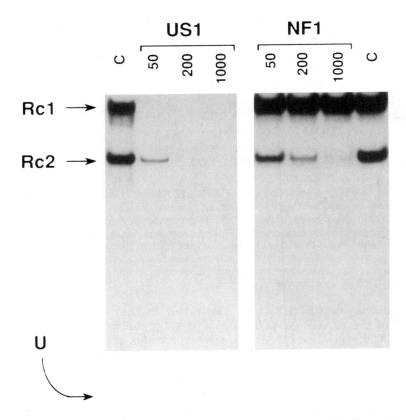

Fig.4. Gel shift analysis of the US1 element. A 20bp double-stranded oligonucleotide containing the rPARP US1 element was 5'-end-labeled and incubated with 5µg of a crude nuclear extract prepared from rat pituitary GH4C1 cells in the presence of increasing concentrations (50, 200 and 1000-fold molar excess) of unlabeled oligonucleotides containing either the US1 sequence (US1) or the unrelated binding site for NF1 (NF1). DNA/protein complexes were analyzed on a 4% non-denaturing polyacrylamide gel as described previously (14). The two distinct shifted DNA/protein complexes detected (Rc1 and Rc2) are indicated. The theoritical position of the free probe (U), which ran out of the gel because long run electrophoresis are needed to resolve the slow-migrating DNA-protein complexes, is also indicated. C: control reaction with no competitor added.

A nuclear protein binds to the rPARP US1 motif. We have used the gel retardation assay to assess whether or not the US1 motif can bind trans-acting proteins from the nucleus. For that purpose, we derived a double-stranded 20bp-long oligonucleotide (5'-ACAAAAGGGGTGGCGCCTGT-3') spanning the whole US1 element. The oligonucleotide was 5' end-labeled with T4 polynucleotide kinase and incubated with crude nuclear proteins prepared from a number of different cell lines from either human, rat or mouse origin (14). The DNA/protein complexes formed were then separated on a 4% polyacrylamide gel and analyzed by autoradiography as described (14). Two distinct DNA/protein complexes (see Fig.4, lanes C) have been detected in almost every tested nuclear extracts prepared from tissue-culture cell-lines (result not shown). The slower migrating complex (Rc1), which is in fact composed of two distinct complexes which co-migrate at almost the same position on the gel, appears to be abundant in all tested nuclear extracts. The faster migrating DNA/protein complex (Rc2) is far less abundant than the former and its formation appears to be inconsistent from one extract to another, as are other fast-migrating minor shifted bands often detected on the gel. The specificity of the DNA/protein complex formation corresponding to the Rc1 shifted band was comfirmed by its disappearance when increasing concentrations (50, 200 and 1000-fold molar excess) of an unlabeled US1-containing oligonucleotide but not of an unrelated DNA fragment (5'-TTATTTTGGATTGAAGCCAATATGAG-3') containing the binding site for the well known transcription factor NF1 (NF1) was added to the reaction mixture prior to its separation on polyacrylamide gel (Fig.4). The fact that any of the unlabeled oligonucleotides used competed for the formation of the Rc2 complex suggest that its formation results from the interaction of non-specific DNA binding protein with the labeled US1 element. This was further comfirmed by the disappearance the Rc2 complex when the final concentration of the unspecific competitor polydI:dC is progressively raised to 4µg.

A number of scientists believe that poly(ADP-ribose) polymerase might auto-regulate its own transcription through interactions with critical cis-acting regulatory elements likely to be located in the 5'-flanking region of its gene. However, gel shift analysis of highly purified preparations of rPARP incubated with a [^{32}P]-labeled US1 oligonucleotide probe have demonstrated that if rPARP regulates its own transcription by interacting with some unknown DNA sequences from the upstream region of its gene, this action is clearly not exerted through the US1 element (result not shown). Further experiments are presently underway to precisely map the nucleotides critical for the binding of the nuclear protein shown to interact with US1. These will help to define whether the protein interacting with the PARP gene US1 element is Sp1 or a related protein such as the one binding to the p12.A element from the mouse p12 gene (9). It will also be interesting to find out what the regulatory function of that protein is and whether this function vary from one tissue to another. The fact that PARP is an enzyme functionally critical for the survival of the cell stress out the need to complete the molecular characterization of the DNA sequences controlling its expression in the organism. These studies should also be useful in clarifying the roles of this enzyme in cell differentiation, DNA repair and cell transformation.

Acknowledgements

We would like to thank Pierre Gosselin (research assistant) who's help was very appreciated with the gel shift experiments and Dr. Girish Shah for critical reading of this manuscript. This work was supported by grants from the "Medical Research Council" (MRC) of Canada and the "Fonds pour la Formation de Chercheurs et l'Aide à la Recherche" (FCAR).

References

1. Ueda, K., and Hayaishi, O. (1985) Annu Rev Biochem 54:73-100
2. Althaus, F.R., and Richter, C.R. (1987) Mol Biol Biochem Biophys 37:1-125
3. Yokoyama, Y., Kawamoto, T., Mitsuuchi, Y., Kurosaki, T., Toda, K., Ushiro, H., Terashima, M., Sumimoto, H., Kuribayashi, I., Yamamoto, Y., Maeda, T., Ikeda, H., Sagara, Y., Shizuta, Y. (1990) Eur J Biochem 194:521-526
4. Ogura, T., Takenouchi, N., Yamaguchi, M., Matsukage, A., Sugimura, T., Esumi, H. (1990) Biochem Biophys Res Com 172:377-384
5. Gradwhol, G., Ménissier de Murcia, J., Molinete, M., Simonin F., Koken M., Hoeijmakers, J.H.J., De Murcia, G. (1990) Proc Natl Acad Sci USA 87: 2990-2994
6. Thibodeau, J., Potvin ,F., Kirkland, J.B., Dandeneault, B., Poirier, G. G. (1991) FEBS Letters (submitted)
7. Thibodeau, J., Gradwhol, G., Dumas, C., Clairoux-Moreau, S., Brunet, G., Penning, C., Poirier, G.G., Moreau, P. (1989) Biochem Cell Biol 67:653-660
8. Letovski, J., Dynan, W.S. (1989) Nucleic Acids Res 17: 2639-2653
9. Guérin, S.L., Pothier, F., Robidoux, S., Gosselin, P., Parker, M.G. (1990) J Biol Chem 265: 22035-22043
10. Widen, S.G., Kedar, P., Wilson, S.H. (1988) J Biol Chem 263:16992-16998
11. Seghal, A., Patil, N., Chao, M. (1988) Mol Cel Biol 8:3160-3167
12. Yamaguchi, M., Hirose, F., Hayashi, Y., Nishimoyo, Y., Matsukage, A. (1987) Mol Cel Biol 7:2012-2018
13. Luu-The, V., Labrie, C., Simard, J., Lachance, Y., Zhao, H.F., Couet, J., Leblanc, G., Labrie, F. (1990) Mol Endocrinol 4:268-275
14. Roy, R., Gosselin, P., Guérin, S.L. (1991) Biotechniques (In press)

Structure and Organization of the Mouse pADPRT Gene

H.Berghammer, M.Schweiger, B.Auer

Institut für Biochemie (Nat.Fak.), Universität Innsbruck, Peter-Mayr-Straße 1 a, A-6020 Innsbruck, Tel. 0043-512-507-3200, Telefax 0043-512-507-2544

Introduction

PolyADP-ribosylation is a very important process utilized by most living organisms to modify the structure of their proteins and ultimately alter their functions. The profound effect of this posttranslational modification on several biological important processes, such as DNA replication, recovery of cells from DNA damage, cellular transformation and differentiation was shown by inhibition of this reaction using NAD^+ analogues (for a review see 1). The apparent central role of ADP-ribosylation in the physiology of the cell led to the purification and cloning of nuclear NAD^+: protein ADP-ribosyltransferase (polymerizing) [pADPRT; poly(ADP-ribose)polymerase; EC 2.4.2.30] from human (2-5), murine (6) and rat tissues (7). Localization of the gene for human PADPRT on chromosome 1 bands q41-q42 (8) and determination of the organization of the gene (9) provided the prerequisites for the characterization of the human PADPRT gene and its regulation.

Nevertheless, how polyADP-ribosylation participates in these important biological mechanisms remains to be elucidated. Inhibitor studies suffer from the principal drawback that unspecific influences of the NAD^+ analogue on NAD^+ metabolism cannot be distinguished from specific action on ADP-ribosylation. Mutants which exhibit no or diminished pADPRT activity could give us new insights in the mechanism of pADPRT action and thereby lead to further understanding of the role of ADP-ribosylation. Detailed knowledge of this role could be the basis for understanding cellular differentiation and transformation.

Results and Discussion

Recent developments in molecular biology provided the tools to inactivate or mutate specific genes in the mouse germ line creating mutants in the respective gene (for a review see 10). The method is based on homologous recombination in embryonic stem cells, which, on fusion to recipient blastocysts, give rise to chimaeric mice that can transmit the mutant gene to their offspring. Inbreeding can then yield mice carrying the mutation in both alleles allowing the phenotypic analysis of recessive mutations. In addition to mice lacking a particular gene function, cell lines carrying defective alleles of normally expressed genes can be useful in establishing the function of the gene. These cell lines can either be obtained from homozygous animals or, should the mutation be lethal early in embryonic development, be generated by consecutive inactivation of both alleles by homologous recombination in cultured cells (11).

This method depends on the knowledge of the organization of the murine pADPRT gene, especially the promotor structure and the arrangement of exons and introns. We therefore screened a murine genomic library prepared in λ EMBL 4 using a DNA fragment which contains pADPRT cDNA (nucleotides 1-3066). From several positive clones, four overlapping ones were chosen and further characterized by restriction analysis. The overlapping region between λ mAG 17 and λ mAG 20 spans exactly 200 bp of identical sequence. Bam HI, Hind III, Eco RI restriction maps of the λ clones λ AG 20, λ mAG 3 and λ mAG 9 correspond exactly to each other in the respective overlapping region (Fig. 1).

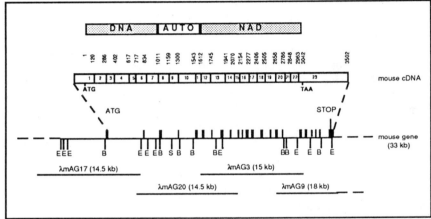

Fig.1:   **Restriction map and organization of the mouse nuclear pADPRT gene.** The 23 exons are shown as solid boxes (mouse gene), the corresponding position of the exons with respect to the cDNA is shown above with the numbers referring to nucleotides.

Each of the fragments detected in genomic southern blots hybridized with full length pADPRT cDNA could be associated with a fragment of the restriction map derived from the genomic clones (data not shown), indicating that the chosen clones span the entire and active mouse pADPRT gene. Southern blotting with pADPRT cDNA fragments as well as with specific oligonucleotides, subcloning of suitable restriction fragments containing exons and sequencing with standard primers as well as with internal pADPRT specific primers revealed the entire exon-intron organization of the mouse pADPRT gene. The mouse pADPRT open reading frame is split into 23 exons exactly at the same sites as the human pADPRT open reading frame (9) with the difference that they are spread over 33 kb instead of 43 kb. All the exon-intron borders correspond to a canonical splice consensus sequence (12).

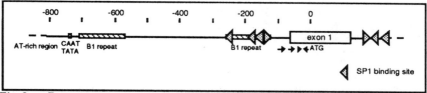

Fig.2.   **Promoter structure of the mouse pADPRT gene.** Exon 1 and
mouse repetitive sequences (B 1 alu like repeats) are depicted as boxes.
Short repeats are indicated by arrows. The position of the SP 1 binding
sites are given as triangles indicating the direction of the sequences.

Clone λ mAG 17 contained exon 1 and at least 10 kb 5′ flanking region.
About 1 kb of this sequence was analyzed (Fig. 2). Sequences identical to the
human promotor could be found very rarely, whereas the rat promotor showed
80 % sequence identity with some insertions and deletions of a few bases up to
500 bp upstream (Poirier, personal communication). A larger insertion in the
rat sequence appears there, but afterwards both sequences share again at least
80 % identity. Common to all three promotor sequences is a cluster of SP 1
binding sequences adjacent to the transcription start site. Interestingly, this
SP 1 cluster is located in a B 1 alu like repeat. Such a mouse repetitive
sequence is also found about 600 bases upward the translation start site. Both
mouse and human promotor contain a cluster of SP 1 binding sites also in the
first intron near the 3′end of the first exon. A CAAT-TATA box is found
780 bp 5′ to the beginning of the open reading frame. Several short direct and
inverted repeats are located in the analyzed region.
    Thus the primary structure of the promotor region resembles that of a
promotor which is not regulated by the common known transcription factors.
The role of the different structural elements remains to be elucidated by deletion
analysis of the promotor region. The gene for pADPRT is well expressed in
mouse NIH 3T3 fibroblasts and embryonic stem cells as shown by analyses of
mRNAs by Northern blots.
    This knowledge of the organization of the pADPRT gene and the
presumptive function of the respective promotor in mouse provides the
prerequisite for inactivation by disruption of the gene. Analysis of the thereby
created mutants on the level of cell cultures as well as living organisms will
hopfully provide new insights in the mechanism by which ADP ribosylation is
involved in DNA repair, cellular differentiation and cell proliferation.

References:
1. Althaus, F. R., Richter, C. (1987). ADP-ribosylation of proteins. Molecular
    Biology Biochemistry and biophysics 37. 1-122
2. Schneider, R., Auer, B., Kühne, C., Herzog, H., Klocker, H., Burtscher, H.J.,
    Hirsch-Kauffmann, M., Wintersberger, U., Schweiger, M. (1987). Isolation of
    a cDNA clone for human NAD⁺:protein ADP-ribosyltransferase. Eur. J. Cell
    Biol. 44. 302-307
3. Kurosaki, T., Ushiro, H., Mitsuuchi, Y., Suzuki, S., Matsuda, M., Katanuma,
    N., Kangawa, K., Matsuo, H., Hirose, T., Inayama, S., Shizuta H. (1987).

Primary structure of human poly(ADP-ribose)synthetaseas deduced from cDNA sequence.. J.Biol.Chem. 262. 15990-15997

4. Uchida, K., Morita, T., Sato, T., Ogura, T., Yamashita, R., Noguchi, S., Suzuki, H., Nyunoya, H., Miwa, M., Sugimura, T. (1987). Nucleotide sequence of a full-length cDNA for human fibroblast poly(ADP-ribose)-polymerase.. Biochem. Biophys. Res. Commun. 148. 617-622

5. Alkahatib, H. M., Chen, D., Cherney, B., Bhatia, K., Notario, V., Giri, C., Stein, G., Slattery, E., Roeder, R. G., Smulson, (1987). Cloning and expression of cDNA for human poly(ADP-ribose) polymerase. Proc. Natl. Acad. Sci. USA 84. 1224-1228

6. Huppi, K., Bhatia, K., Siwarski, D., Klinman, D., Cherney, B., Smulson, M., (1989). Sequence and organization of the mouse poly(ADP-ribose) polymerase gene. Nucleic Acids Res. 17. 3387-3401

7. Thibodeau, J., Gradwohl, G., Dumas, C., Clairoux-Moreau, S., Brunet, G., Penning, C., Poirier, G.G., Moreau, P. (1989). Cloning or rodent cDNA encoding the poly(ADP-ribose) polymerase catalytic domain and analysis of mRNA levels during the cell cycle. Biochem. Cell Biol. 67. 653-660

8. Herzog,H.,Zabel,B.U.,Schneider,R.,Auer,B.,Hirsch-Kauffmann,M.,Schweiger,M. (1989). Human nuclear NAD^+ ADP-ribosyltransferase: Localization of the gene on chromosome 1q41-q42 and expression of an active human enzyme in Eschrichia coli. Proc. Natl. Acad.Sci. USA 86. 3514-3518

9. Auer, B., Nagl, U., Herzog, H., Schneider, R., Schweiger, M. (1989). Human nuclear NAD^+ ADP-ribosyltransferase(polymerizing): Organization of the gene. DNA 8. 575-580

10. Capecchi, M.R. (1989). Altering the genome by homologous recombination. Science 244. 1288-1292

11. te Riele, H., Maandag, E.R., Clarke, A., Hooper, M., Berns, A. (1990). Consecutive inactivation of both alleles of the pim-1 proto-oncogene by homologous recombination in embryonic stem cells. Nature 348. 649-651

12. Brethnach, R., Chambon, P. (1981). Organization and expression of eukariotic split genes coding for proteins. Annu. Rev. Biochem. 50, 349-383

Strategies for Expressing Analogs of PADPRP in Eukaryotic Cells
Veronica Kang, Saikh J. Haque, Barry Cherney, and Mark Smulson
Department of Biochemistry & Molecular Biology
Georgetown University School of Medicine
Washington, D.C. 20007 USA

Introduction

Towards our goal of obtaining cell lines expressing altered levels of poly(ADP-ribose) polymerase (PADPRP), we have constructed expressing plasmids containing various mutagenized forms of PADPRP and have stably introduced these into NIH3T3 cells. We reasoned that it might be feasible to compete out or enhance the activity of the endogenous PADPRP in intact cells with a homologous or partial PADPRP protein. This might molecularly perturb the various functions of PADPRP *in vivo*. One recombinant plasmid was constructed with the heavy-metal inducible mouse metallothionein (MT) promoter upstream of the partial cDNA coding the PADPRP DNA-binding and part of the automodification domains. Another mutagenized PADPRP with a deletion of 132bp in the NAD-binding domain and shown in *in vitro* assay to be inactive, was subcloned downstream of SV40 early promoter. Each plasmid was stably co-transfected with pNeomycin vector into NIH3T3 cells.

Results and Discussion

Stable Transfectants Expressing the DNA-Binding Domain in NIH3T3. Induction of neomycin-resistant colonies with Zn^{2+} and Cd^{2+} for 18h and analysis of 2.6kb human PADPRP transcript by Northern analysis and of 60 kDa truncated PADPRP protein by immunoprecipitation indicated that the introduced DNA-binding domain was expressed in the NIH3T3 cells. However, the MT promoter was somewhat leaky; in the absence of exogenous metal addition, some truncated PADPRP mRNA and protein were expressed. Perhaps due to integration sites, certain clones (e.g. cell line E3) were under tighter promoter control. In subsequent experiments, we utilized those clones with the tightest promoter control. Kameshita *et. al.* (1) had shown that addition of the DNA binding domains of PADPRP *in vitro* to a purified preparation of the enzyme increased its *in vitro* automodification activity. This activation was suggested to be due to the ability of the truncated DNA binding domain to increase the affinity of the enzyme for DNA by protein-protein interaction. Accordingly, it was of interest to determine the status of poly ADP-ribosylation in cells *in vivo* under conditions of induction of this same region of PADPRP. The analysis of the catalytic activity of PADPRP in either cell extracts, isolated nuclei, chromatin, or intact cells is complicated by the complexity of the reaction itself.

For this reason we employed several assays to assess the influence of metal-induction of the truncated DNA binding domain of PADPRP in the transfected cells. In the experiment shown in Table 1, strain E3 which contained integrated 5' human PADPRP and appropriate control NIH3T3 cells were grown in conditioned medium, and induced for 18hr with Cd and Zn. The cells were then divided and assayed by either a sonicated cell-free extract assay or alternatively, the cells were made permeable to nucleotides (2) and ADP-ribose incorporation was determined in the cells. The sonication assay is a 40 second linear rate measurement and has been established to be independent of DNA strand breaks which activate PADPRP and is essentially proportional to the amount of PADPRP molecules in a given extract (3). No differences were detected in the specific activity by this method, either in the presence or absence of expression of the human 5' DNA binding domain. In contrast, significant differences in overall nuclear ADP-ribosylation patterns were detected by the permeabilized cell assay. The specific activity (in the absence of DNase) as assessed by this method was approximately 2-fold higher in the transfected cells as compared to controls, either in the presence or absence of induction conditions. Also, the induction of the human DNA PADPRP binding domain by metals increased the overall nuclear ADP-ribosylation activity. As observed above, the nuclear ADP-ribosylation reaction is complex *in vivo*. To eliminate the possibility that differences in activity *in vivo* may be due to differences in DNA strand break status of cells, the permeabilized assays were performed in the presence of DNase I, shown to stimulate activity under such conditions. The specific activities of ADP-ribosylation under the four conditions were essentially the same and suggested that the absolute amount of PADPRP was essentially equivalent and that the differences in activity were not due to increases or decreases in enzyme content *per se*.

Table 1. Specific activities of PADPRP in NIH3T3 cells expressing exogenous DNA-binding domain (methods in ref. 2 and 3)

Cells	Induction	pmol ADP-Rib/min/mg protein		Sonicated cell extracts
		Permeabilized Cells		
		-DNase	+DNase	
Control	–	181±22	3906±839	931±72
	+	173±15	3744±617	923±55
Experimental	–	258±21	2781±556	890±64
	+	319±29	3375±540	886±36

The growth rate of the transfected cell lines was compared with cells not expressing exogenous human PADPRP fragment. Studies with several strains of transfected cells indicated that the rate of growth of the transfected cells was slightly less than that of control cells. To determine whether the difference in growth properties was due to a function of reduced rates of DNA synthesis or alteration in S phase timing, an analysis of S phase was accomplished by cell synchronization and thymidine labelling (Fig.1A). The data suggested that there was no significant difference in the length of S phase between the transfected and nontransfected cells. We next performed fluorescent activated cell sorter analysis (Fig. 2B). The data was quite striking; an extremely high number of E3 clonal cells, expressing the analog of PADPRP, was arrested in G_2 phase as compared to NIH3T3 control cells. It was of interest that this G_2 accumulation was only evident when cells were compared during early exponential growth; by stationary growth stage, the distribution of the cells within the cell cycle was essentially equivalent in the two cell populations.

However, it should be noted that in several experiments, a *decrease* in PADPRP specific activity has also been observed. The reasons for these differing effects by expression of the PADPRP DNA binding domain is under investigation.

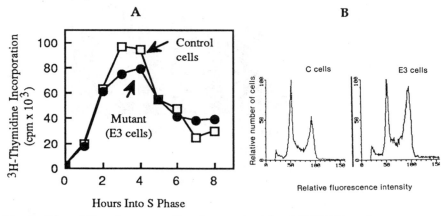

Fig. 1. Thymidine incorporation in synchronized NIH3T3 and exogenous DNA-binding domain expressing clone (E3) (A). Fluorescent activated cell sorter analysis of control and E3 cells (B).

Stable Transfectant Expressing Human PADPRP Mutated in the NAD-Binding Domain. Another PADPRP analog that was expressed in NIH3T3 cells was truncated by 132 bp (obtained by removal of the 2 *Cla*I sites) in the NAD-binding domain near a site of amino acid identity with the enzyme ricin A. A fusion protein of the analog conjugated to ubiquitin and expressed in *E. coli* resulted in a totally inactive PADPRP enzyme (data not shown). The mutagenized PADPRP cDNA was subcloned into a eucaryotic expression vector under SV40 early promoter and stably transfected into NIH3T3 cells. Expression of the mutant was confirmed by various analysis including Southern, Northern and immunoprecipitation. PADPRP activity was measured by sonication cell-free extract assay for several clonal cell lines (Fig. 2). The data indicated that the PADPRP activity in all clones tested was enhanced as compared to control cell lines transfected with pNeomycin vector only. We postulated that the enhanced activity may be due to the availability of the exogenous PADPRP as acceptor protein for poly ADP-ribosylation since the automodification domain is still intact in this PADPRP analog. In fact, preliminary *in vitro* studies with partially purified full-length PADPRP enzyme indicated that the addition of the mutant protein resulted in increase in PADPRP incorporation. Further studies are underway to determine whether the mutant is being poly ADP-ribosylated or whether a protein-protein interaction may be occurring to account for the stimulation in endogenous PADPRP activity.

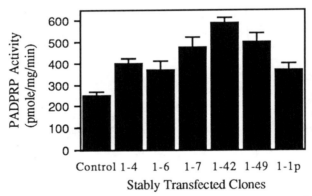

Fig. 2. PADPRP incorporation in NIH3T3 and clones expressing NAD-binding mutant. Assay was done as described in ref. 3.

Acknowledgements

This work was supported by a grant (AFOSR-89-0053) from the United States Air Force Office of Scientific Research and a contract (DAMD17-90-C-0053) from the U.S. Army Medical Research and Development Command.

References

1. Kameshita, I., Matsuda, M., Nishikimi, M., Ushiro, H. and Shizuta, Y. (1986) J. Biol. Chem. 261, 3863-3868.
2. Berger, N.A., Adams, J.W., Sikorski, G.W., Petzold, S.J. and Shearer, W.T. (1978) J. Clin. Invest. 62, 111-118.
3. Cherney, B.W., Midura, R.J., Caplan, A.I. (1985) Dev. Biol. 112, 115-125.

Expression and Characterization of PADPRP and a Novel
Glucocorticoid-PADPRP "Finger Swapped" Protein from *Escherichia coli*
Barry Cherney, Kishor Bhatia, Bushra Chaudhry, Tauseef Butt,
Mark Danielsen and Mark Smulson
Department of Biochemistry & Molecular Biology
Georgetown University School of Medicine
Washington, D.C. 20007 USA

Introduction

PADPRP is a DNA binding protein which becomes active
enzymatically when bound to single and double stranded breaks in
DNA. This recognition and activation by DNA strand breaks is
independent of DNA sequence (1). The enzyme is thought to ADP-
ribosylate proteins in the vicinity of the damaged DNA. We reasoned
that redirection of PADPRP to sequence specific areas of chromatin
might likewise ADP-ribosylate proteins localized to this specific area
of chromatin. Thus, this procedure could be used to label and identify
proteins associated with specific areas of DNA, as well as, test the
effects of PADPRP activity on DNA repair of damaged DNA at the
site of PADPRP binding and at sites distal to the polymerase.

To achieve this aim of redirecting PADPRP to sequence specific sites
chromatin we have exchanged the normal PADPRP zinc finger region
with the zinc finger, DNA binding domain of the glucocorticoid
receptor (GR). This domain of the GR binds specifically to
glucocorticoid responsive elements (GRE).

Results and Discussion

*Expression of PADPRP as a ubiquitin fusion protein from Escherichia
coli.* To examine the biochemical properties of the finger swapped
protein it was necessary to obtain a system which would efficiently
express PADPRP. With this in mind, the cDNA of human poly(ADP-
ribose) polymerase, encoding the entire protein was subcloned into the
Escherichia coli expression plasmid pYUb. In this expression system,
the carboxyl terminus of ubiquitin is fused to the amino terminus of a
target protein, in this case PADPRP, stabilizing the accumulation of
the cloned gene product. Following heat induction of the transformed
cells, the sonicated extract contained a unique protein corresponding to
the predicted mobility of the fusion protein in SDS-PAGE (Fig. 1A,
lanes 13, 14), which was immunoreactive with both PADPRP (Fig. 1B,
10, 11, 12) and ubiquitin (Fig. 1C, lanes 11, 12, 13) antibodies. Expression
of PADPRP, in the absence of fusion to ubiquitin, yielded polymerase

fragments as the major product with relatively small amounts of full-length PADPRP produced (Fig. 1B, lanes 6-9).

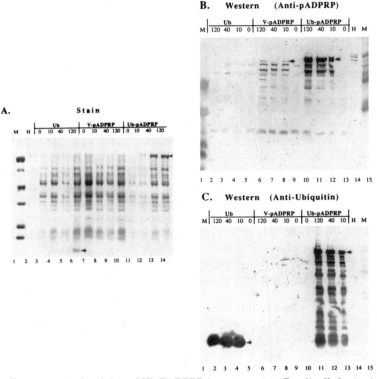

B. Western (Anti-pADPRP)

A. Stain

C. Western (Anti-Ubiquitin)

Fig. 1. Expression and stability of Ub-PADPRP fusion protein. *E. coli* cells harboring the indicated plasmids were heat induced at 42°C, aliquots removed at times indicated and subjected to SDS/PAGE on 7.5% gels. Coomassie stain (A) and immunoblot analysis with antibodies against PADPRP (B) and ubiquitin (C). Vectors: Ub-PADPRP-construct containing full-length PADPRP fused to ubiquitin; Ub-construct containing ubiquitin, V-PADPRP construct containing full-length PADPRP.

Catalytic activity of bacterially expressed eukaryotic PADPRP. Polymerase activity in the bacterial cell extracts was quantitated using a standard PADPRP assay (2). Using this assay, a high level of PADPRP activity was revealed in induced samples containing the Ub-PADPRP fusion protein (Table I). The observed incorporation of ^{32}P[ADP-ribose] into *E. coli* extracts was inhibited by the addition of 3-aminobenzamide, an established PADPRP inhibitor (Table I). No PADPRP activity was detected in uninduced cells containing the Ub-PADPRP fusion construct or in induced cells lacking a PADPRP vector. In V-PADPRP-induced samples, where ubiquitin was not fused to the enzyme, PADPRP activity was reduced by 90% when compared to the activity measured in sonicated extracts containing the Ub-PADPRP fusion protein (Table I). When optimized for maximum PADPRP expression the activity in sonicated extracts reached 45 nmol/min/mg protein. A biochemical characterization of the resulting recombinant fusion protein indicated the enzyme had catalytic properties which were nearly identical to native PADPRP obtained from mammalian tissues. These properties included specific activity, K_m for NAD, response to DNA strand breaks, response to Mg^{++}, inhibition by 3-aminobenzamide and activity in activity gel analysis.

Table I: Measurement of PADPRP Activity in *E. coli* Extracts Containing PADPRP Vectors[a]

cells	(pmol/min/mg protein)
uninduced Ub-PADPRP	4.0
induced Ub	8.0
induced Ub-PADPRP	3200
induced Ub-PADPRP + 3AB	228
induced V-PADPRP	330

[a] *E. coli* AR58 cells harboring the indicated plasmids were grown to an OD_{260} of 0.5, aliquots were removed and the remaining cells induced at 42°C for 20 mins. After sonication cell extracts were assayed for PADPRP activity.

Expression and catalytic activity of GR-PADPRP. Construction of the GR-PADPRP fusion protein involved deletion of the DNA encoding the first 232 amino acids of PADPRP from the Ub-PADPRP vector described above and replacing it with DNA encoding 106 amino acids of the GR previously shown to contain the protein domain responsible for specific binding to GRE (3). Starting from the amino-terminal end the expressed protein should contain 76 amino acids of ubiquitin, 106 amino acids of the GR DNA binding domain and 882 residues from the carboxyl end of PADPRP. Expression of this multi chimeric protein produced an intense band at the expected molecular weight (106 kDa) in SDS-PAGE, which was immunoreactive with both Ub and PADPRP antibodies and thus presumably contains the GR DNA binding domain. Activity measurement of this bacterial expressed GR-PADPRP fusion protein demonstrates a low level of polymerase activity (2600 pmol/min/mg protein) which is not stimulated by the addition of fragmented DNA (Fig. 2) or of an oligonucleotide containing a GRE.

To determine if the GR-PADPRP protein can recognize a GRE, a gel shift assay was performed using labelled oligomer containing a GRE, incubated with either bacterial cell extracts containing the GR or UbGR-PADPRP fusion construct. The results showed that the bacterially expressed GR protein (from a chemical induction) binds specifically to GRE, however, the fusion protein did not.

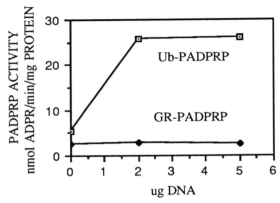

Fig. 2. *E. coli* cells harboring Ub-PADPRP or Gb-PADPRP plasmids were induced ad 42°C for 1 hr and the sonicated cell extract assayed for PADPRP activity with various amounts of sonicated salmon sperm DNA.

To explain the lack of binding two possibilities appear reasonable. First, the position of the GR portion between Ub and PADPRP may sterically hinder the recognition of the GRE. Alternatively heat induction of the fusion protein may prevent proper folding of the GR domain, as has recently been described (4). A salient question is whether the GR-PADPRP protein can transfer ADP-ribose units to other proteins. Currently we are addressing these problems to determine if this is a feasible approach toward creating a novel ADP-ribosylating enzyme.

Acknowledgements

This work was supported by grant CA13195 from the National Cancer Institute and by a grant (AFOSR-89-0053) from the United States Air Force Office of Scientific Research.

References

1. Menissier-de Murcia, J., Molinete, M., Gradwohl, G., Simonin, F., and de Murcia, G. (1989) *J. Mol. Biol. 210*, 229-233.
2. Cherney, B.W., Midura, R.J., and Caplan, A.I. (1985) Poly(ADP-ribose) synthetase and chick limb mesenchymal cell differentiation *Dev. Biol. 112*, 115-25.
3. Danielsen, M. (1990) Structure and Function of the Glucocorticoid Receptor, pp. 39-64 in Nuclear Hormone Receptors, edited by Malcolm G. Parker, Academic Press Harcourt Brace Jovanovich Publishers.
4. Wittliff, J. L., Wenz, L. L., Dong, J., Nawaz, Z., and Butt, T.R. (1990) *J Biol. Chem. 265*, 22016-22022.

Tumor Promoters, But Not EGF, Increase Nuclear Poly(ADP-Ribose) Polymerase Gene Expression in Rat Hepatocytes Initiated *in Utero* with DMN and Cultured After Birth in Low-Calcium Synthetic Medium

U. Armato[1], Lucia Testolin[1], Marta Menegazzi[2], Lia Menapace[1], Maria Ribecco[1], Alessandra Carcereri DePrati[2], and H. Suzuki[2]

Institutes of [1]Anatomy & Histology and [2]Biochemistry, School of Medicine, The University of Verona, I-37134, Verona, Italy

Introduction

Mixed, *in vivo-in vitro*, models of liver chemical carcinogenesis have been used only rarely [1,2]. Recently, we devised to utilize a new one of such mixed systems to clarify the role(s) played in the unfolding of liver malignancies by nuclear poly(ADP-ribosyl)ating reactions. Such metabolic events post-translationally modify manifold nuclear protein species, including the enzyme itself, i.e. poly(ADP-ribose) polymerase (pADPRP), by which they are catalyzed [3]. Nuclear poly(ADP-ribosyl)ations are involved not only in DNA repair, but even in the modulation of gene expression related to normal and abnormal cell proliferation, as the occurrence of malignancy is prevented by established inhibitors of pADPRP [4,5]. An increased activity of nuclear pADPRP is indispensable for the multiplication of fully transformed hepatocytes [6,7]. Likewise, both partial hepatectomy and the administration of lead nitrate increase the activity and genetic expression of nuclear pADPRP in rat hepatocytes *in vivo* [8]. Previously, we showed that the activation of nuclear pADPRP played a pivotal role in the enactment of the mitogenic effects elicited by several tumor promoters administered to bona fide normal, primary neonatal rat hepatocytes [9,10]. In this communication we relate that the *in utero* initiation with a transplacental genotoxic carcinogen like dimethylnitrosamine (DMN) [11,12] and the *in vitro* postnatal promotion of the offspring's initiated hepatocytes with divers xenobiotics [9,10] do increase the expression of the nuclear pADPRP gene.

Results

According to our protocol, an initiating single dose (15 mg/Kg) of DMN was administered i.p. to adult female rats on the 18th day of pregnancy; the livers of the corresponding progenies were collected and cultured *in vitro* 4 days after birth [9,10] . Unlike adult liver, neonatal rat liver is endowed with a discrete subpopulation of actively cycling hepatocytes. Such a proliferating hepatocellular fraction was detectable also in primary cultures of neonatal liver tissue [9,10], and its size was not significantly changed by a previous initiation with DMN *in utero*(data not shown). In keeping with this, Northern blot hybridization analysis of RNA species derived from DMN-initiated primary hepatocytes revealed that the amount of histone H3 transcripts did not differ from that found in bona fide normal, cultured hepatocytes; yet, the same analytical procedure unveiled that both c-myc and nuclear pADPRP mRNA copies were more abundant in DMN-initiated primary liver cells than in bona fide normal ones (data not shown). Further, the administration of a low dose (10^{-10} mol/L) of each of various tumor promoters (phenobarbital [PB], 12-O-tetradecanoyl 13-phorbolacetate [TPA], etc. [9, 10]) on day 4 in vitro elicited a specific mitogenic response in in utero DMN-initiated, after birth cultured rat hepatocytes kept in low-calcium (0.01 mmol/L), wholly synthetic HiWo5Ba2000 medium;

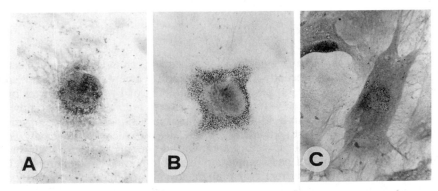

Fig. 1.   The expression of the gene encoding nuclear pADPRP detected by the in situ hybridization technique [13] with a specific cDNA probe [14] in prenatally DMN-initiated hepatocytes cultured after birth in low-calcium (0.01 mmol/L) synthetic medium. Specimens were treated with PB (10^{-10} mol/L) for: [A], 2 hours; [B], 16 hours; and [C], 20 hours prior to fixation with paraformaldehyde (4% w/v) for 4 min. at 4 C. Hybridization was carried out at conditions of the highest stringency with a [^{35}S]dCTP-labelled cDNA probe for rat nuclear pADPRP using Multiprime DNA Labelling Systems (0.5-1.0×10^9 c.p.m./ug cDNA). The specimens were next processed for autoradiography [9,10, 19]. Magnification, x1400.

conversely, the addition of EGF (10^{-10} mol/L) to this cation-deprived serumless medium could not induce de novo mitogenically activated hepatocytes to enter S phase (data not shown) [9]. Northern blot hybridization analysis of RNA species extracted from in utero DMN-initiated, in vitro PB-promoted neonatal hepatocytes grown in low-calcium synthetic medium disclosed that this treatment elicited both an earlier (2 hours) and persistent increase, peaking at the 4th hour, in the amount of nuclear pADPRP transcripts, and a later (4-8 hours) expansion of histone H3 mRNA copies; conversely, the DMN-elicited, preexisting amplification of c-myc transcripts was not further changed by the exposure to PB in these specimens (data not shown). Since neonatal rat liver cultures contained both stromal cells (10-40% of the total cell population) and hepatocytes, further cytological studies were warranted by in situ hybridization technique [13]. Using a specific nuclear pADPRP cDNA probe [14] we found that pADPRP transcripts were first detectable in the perinucleolar chromatin of primary hepatocytes (Fig. 1A), and thereafter diffused to the whole nucleoplasm. Only somewhat later (4-hours), did pADPRP mRNA copies start accumulating in the cytoplasm (Fig. 1B). A few pADPRP mRNA copies also accompanied the separating sets of chromosomes in mitotically dividing hepatocytes (Fig. 1C) Quantitative autoradiographic studies carried out on liver cultures hybridized *in situ* with the same specific cDNA probe showed that an 8-hour treatment with one of four xenobiotics (i.e., PB, TPA, nafenopin or DDT) added at 10^{-10} mol/L to the low-calcium (0.01 mmol/L) $HiWo_5Ba2000$ medium significantly increased the nuclear and cytoplasmic numbers of pADPRP mRNA copies in DMN-initiated hepatocytes (Fig. 2). On its own part, EGF (10^{-10} mol/L) only increased nuclear pADPRP gene expression when the level of calcium in the medium was normal (1.8 mmol/L).

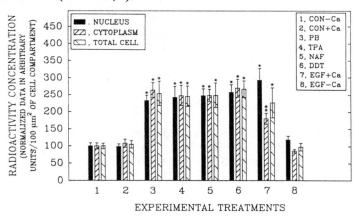

Fig. 2. The increase in specific pADPRP transcripts elicited by an 8-hour exposure to 4 tumor promoters (each at 10^{-10} mol/L) in prenatally DMN-initiated hepatocytes cultured after birth in low-calcium (0.01 mmol/L) synthetic medium; the incapacity of EGF (10^{-10} mol/L) to change the hepatocytic level of pADPRP mRNA in the same cation-devoid environment (EGF-Ca); the enhanced expression of such transcripts in EGF-treated hepatic cells kept in normal-calcium (1.8 mmol/L) medium (EGF+Ca); and the inability of the extracellular calcium deficiency by itself to change the numbers of pADPRP mRNA copies in DMN-initiated but otherwise untreated hepatocytes. The data, obtained by combining *in situ* hybridization with quantitative autoradiography [9,13], are expressed as radioactivity concentration (in arbitrary units), taking as 100 the respective values of nucleus, cytoplasm and total cell of untreated hepatocytes kept in low-calcium medium (CON-Ca). Bar heights are means + S.E.M. of the values from 3 distinct experiments. The level of significance between the values in the treated and control cultures is designated as: *, p<0.001; **, p<0.02; CON+Ca, untreated control specimens exposed to normal-calcium medium; NAF, nafenopin; DDT, dichlorodiphenyl trichloroethane.

Discussion

DMN is a potent hepatocarcinogen that crosses the placental barrier [11], and damages hepatocytes' DNA [12]. Such DMN-initiated cells are amenable to neoplastic promotion by PB [15]. Interestingly, c-myc proto-oncogene is reckoned as a cell growth regulator [16], its transcription being constitutively low in normal neonatal liver and, by contrast, quite high in fetal and regenerating livers and in hepatomas [16,17]. Hence, our data suggest that c-myc is significantly involved in the intrauterine initiation of hepatocytes by DMN. Studies are under course to establish whether the parallel increase in the basal level of expression of nuclear pADPRP gene also observed is only related to the repair of DMN-damaged DNA. Our findings also show for the first time that four distinct tumor promoters increase still further the enhanced expression of nuclear pADPRP gene in DMN-initiated primary hepatocytes. This genetic activation, occurring in low-calcium synthetic medium, i.e. under mitogenic conditions that are exclusive for xenobiotics, may integrally belong to a general mechanism of action of tumor-promoting agents [18]. It seems conceivable that this stimulation of both the activity [9] and the genetic expression of nuclear pADPRP elicited by xenobiotics is linked to an earlier induction of the *de novo* plasmalemmal generation of oxygen free radicals by the same agents [19], presumably via an increased new synthesis of polyamines.

Acknowledgments. This work was aided by grants from Italian Assoc. for Cancer Research, MURST (40%, 60%), and Natl. Res. Council (CNR; 88.03417.CT04, 88.00423.CT04, 89.02865.CT04, and 90.01416.CT04) to U.A., and from CNR's Target Project on Biotechnol. and Biostrument. and CISMI (MI-PA-VR) to H.S.

References

1. Kitagawa,T., Watanabe,R., Kayano,T. and Sugano,H. (1980), Gann, 71, 747-754.
2. Kauffmann,W.K., Ririe,D.G. and Kaufman,D.G., Carcinogenesis, 9, 779-782.
3. Ueda,K. and Hayaishi,O. (1985), Annu. Rev. Biochem., 54, 73-100.
4. Tsujiuchi,T., Tsutsumi,M., Denda,A., Kondoh,S., Nakae,D., Maruyama,H. and Konishi,Y. (1990), Carcinogenesis, 11, 1783-1787.
5. Borek,C., Morgan,W.F., Ong,A. and Cleaver, J.E. (1984), Proc. Natl. Acad. Sci. USA, 81, 243-247.
6. Gaal,J.C. and Pearson,C.K. (1986), TIBS, 11, 171-175.
7. Lea,M.A., Barra,R., Randolph,V. and Kuhr, W.G.(1984), Cancer Biochem. Biophys., 7, 195-202.
8. Menegazzi,M., Carcereri De Prati,A., Ledda-Columbano,G.M., Columbano,A., Uchida,K., Miwa,M., and Suzuki,H. (1990), Arch. Biochem. Biophys., 279, 232-236.
9. Romano,F., Menapace,L. and Armato,U. (1988), Carcinogenesis, 9, 2147-2154.
10. Armato,U., Andreis,P.G. and Romano,F. (1985), Carcinogenesis, 6, 811-821.

11. Williams,G.M. (1977), Cancer Res., 37, 1845-1851.
12. Martelli,A., Robbiano,L., Giuliano,L., Pino,A., Angelini, G. and Brambilla,G. (1985), Mutat. Res., 144, 209-211.
13. Davis,L.G., Dibner,M.D., and Battey,J.F. (1986), Basic Methods in Molecular Biology, Elsevier, New York, pp. 355-359.
14. Nyunoya,H., Ogura,T., Shiratori,Y., Makino,R., Hayashi,K., Sugimura,T., and Miwa,M. (1988), Proc. Jpn. Cancer Assoc., 47, 246.
15. Herren,S.L., Pereira,M.A., Britt,A., and Khouri,M.K. (1982), Toxicol. Lett., 12, 143-150.
16. Sobczack,J., Mechti,N., Tournier,M.-F., Blanchard,J.-M. and Duguet,M. (1989), Oncogene, 4, 1503-1508.
17. Zhang,X-K., Wang,Z., Lee,A., Huang,D-P. and Chiu,J-C. (1988), Cancer Lett., 41, 147-155.
18. Cohen,S.M. and Ellwein,L.B. (1990), Science, 249, 1007-1011.
19. Armato,U., Andreis,P.G. and Romano,F. (1984), Carcinogenesis, 5, 1547-1555.

Expression of the gene for poly(ADP-ribose) polymerase and DNA polymerase beta in rat tissues and in proliferating cells

Marta Menegazzi, Alessandra Carcereri De Prati, Tsutomu Ogura, Amedeo Columbano, Giovanna Maria Ledda-Columbano, Palmiro Poltronieri, Masanao Miwa , Massimo Libonati and Hisanori Suzuki

Introduction

On the basis of the structural characteristics of the 5'-flanking regions of genes encoding the two DNA repair-related enzymes, poly(ADP-ribose) polymerase and DNA polymerase beta, one may suggest that they can be active housekeeping genes (1, 2). The two enzymes are similarly distributed in a number of mouse tissues (3). This indicates that in some circumstances poly(ADP-ribose) polymerase and DNA polymerase beta can possibly be coexpressed. We have shown recently that poly(ADP-ribose) polymerase is induced in proliferating cells during G1/S phase (4, 5). An increase of DNA polymerase beta activity was shown to occur in mitogen-treated peripheral mononuclear cells (PBMC) after DNA synthesis (6). Data on the expression of DNA polymerase beta gene in proliferating cells are so far not available.

In order to obtain more informations on the regulation of the expression of the gene for poly(ADP-ribose) polymerase and DNA polymerase beta, the activity of poly(ADP-ribose) polymerase and DNA polymerase beta as well as the amount of their mRNAs in rat tissues have been measured with the activity gel method and Northern blot analyses, respectively. The change in activity of poly(ADP-ribose) polymerase and DNA polymerase beta, and the level of the mRNA for the two enzymes was also examined during the cell cycle progression using various normal cell proliferation systems.

Results and Discussion

Figure 1 shows a strikingly variable values of the activity of poly(ADP-

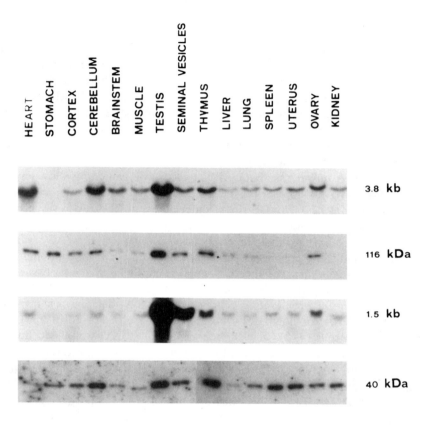

Fig. 1. Poly(ADP-ribose) polymerase and DNA polymerase beta activities and levels of their mRNAs in various rat tissues. Crude extracts containing 10 and 50 ug of proteins for poly(ADP-ribose) polymerase (116 kDa) and DNA polymerase beta (40 kDa) activities, respectively, were separated electrophoretically on polyacrylamide gel containing SDS and activated salmon sperm DNA. Enzymatic activities were detected in the gel after renaturation of proteins as described in (10). The gels were washed to remove unreacted substrates, dried and autoradiographed by exposing to X-ray film for 12 h. Total RNAs (50 ug) from each tissue was separated electrophoretically on a 1 % agarose gel containing 6 % formaldehyde, and then blotted to a membrane. Subsequently, mRNAs for poly(ADP-ribose) polymerase (3.8 kb) and DNA polymerase beta (1.5 kb) were hybridyzed with [^{32}P]dCTP labeled rat poly(ADP-ribose) polymerase cDNA (11) and rat DNA polymerase beta cDNA probes (12), and were visualized by X-ray film. Size of mRNAs are indicated on the right.

ribose) polymerase and DNA polymerase beta and different levels of the mRNA for the two enzymes in various rat tissues. The highest values of enzyme activity and of the quantity of mRNA were found in the testis, while the liver was poor in them. Also in other tissues such as thymus, cerebellum and ovary a high activity of poly(ADP-ribose) polymerase and DNA polymerase beta as well as correspondingly high levels of their mRNAs were shown to be present. In other tissues, like the brainstem, muscle, liver, and lung, low enzyme activities and correspondingly low levels of enzymes' mRNA were detected (Fig. 1). The generally good correlation between these variables was only contradicted in the seminal vesicles, where a very large amount of DNA polymerase beta mRNA was accompanied by a low level of mRNA for poly(ADP-ribose) polymerase, and in the heart tissue, in which a very high amount of poly(ADP-ribose) polymerase mRNA was evidenced in contrast with a quite modest level of mRNA for DNA polymerase beta. These data not only indicate that probably the amount of poly(ADP-ribose) polymerase and DNA polymerase beta are regulated at a transcriptional level but also the two enzymes could be expressed in parallel in many rat tissues.

We studied further the possible involvement of DNA polymerase beta in events occurring in proliferating cells, since previous data have shown the amount of mRNA for poly(ADP-ribose) polymerase increased in the G1/S phase of these cells (4, 5). In phytohemagglutinin (PHA)-stimulated human PBMC, a marked increase of the amount of the poly(ADP-ribose) polymerase mRNA (data not presented), which was strictly correlated with an increase of enzyme activity (Fig. 2), was observed before and during the synthesis of DNA. The higher stability of mRNA for poly(ADP-ribose) polymerase could contribute to increase its transient level: in fact, mRNA half-life increased from about 2 h in quiescent PBMC to more than 4 h in the same cells 1 day after PHA-treatment (data not presented). The possibility that the increase of the amount of mRNA may be also due to an enhancement of the transcriptional rate can not be escluded, since it appears very improbable that the two-fold stability increase of mRNA may account for the 10-15-fold increase of the transient mRNA level observed in T lymphocytes 1 day after PHA-stimulation (4). The increase of enzyme activity accompanied by that of the amount of the transcript was also observed in proliferating hepatocytes of partially hepatectomized liver and in lead nitrate-treated liver (5, Fig. 2).

Thus far no data are available on the involvement of DNA polymerase beta in DNA synthesis during cell proliferation (6-9). The results obtained in the present work show that the activity of DNA polymerase beta as well as the level of its mRNA increase with a pattern which is very similar to that observed for poly(ADP-ribose) polymerase in PHA-treated PBMC and in the two hepatocyte proliferation systems (Fig. 2). The increase of the amount of mRNA for DNA polymerase beta in PBMC seems to be due to an increased

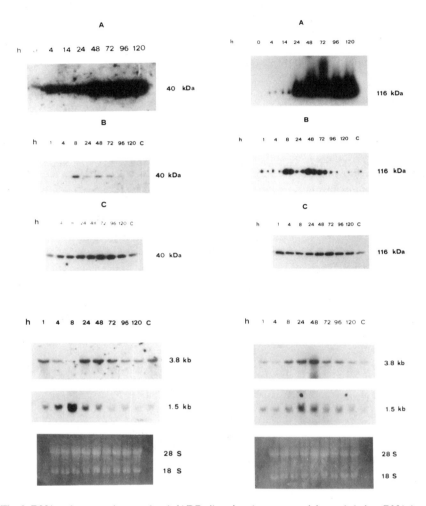

Fig. 2. DNA polymerase beta and poly(ADP-ribose) polymerase activity and their mRNA levels in phytohemagglutinin-stimulated human PBMC, and regenerating and mitogen lead nitrate-treated rat liver. (Upper) Enzymatic activity in crude extracts was detected in the gel as described in the legend to Fig. 1. (A) Phytohemagglutinin-treated PBMC. (B) Lead nitrate-treated rat liver. (C) Regenerating rat liver. The molecular weight of poly(ADP-ribose) polymerase (116 kDa) and DNA polymerase beta (40 kDa) is indicated on the right. (Lower) Level of mRNA for poly(ADP-ribose) polymerase (3.8 kb) and DNA polymerase beta (1.5 kb).

mRNA was determined as described in the legend to Fig. 1. (Left) Lead nitrate-treated rat liver. (Right) Regenerating rat liver. The amount of ribosomal RNAs in the agarose gel quantified by ethidium bromide staining is shown in the bottom part of the pannel. Positions of ribosomal RNAs (28 S and 18 S) are indicated on the right.

transcriptional rate, since the mRNA half-life was shown to be about 1.5 h both in quiescent and stimulated PBMC (data not shown).

The data presented suggest: 1) that poly(ADP-ribose) polymerase and DNA polymerase beta may be induced almost in parallel before DNA synthesis during the cell cycle progression; and 2) that the increase of the activities of the two enzymes might be triggered by the transient increase of the amount of their mRNAs.

In conclusion, we have produced possibly novel information concerning an involvement of DNA polymerase beta in the process of DNA synthesis in proliferating cells. Moreover, the parallel expression of poly(ADP-ribose) polymerase and DNA polymerase beta observed in many rat tissues and in various proliferating cells, is compatible with the structure of the 5'-flanking regions of the genes encoding these enzyme proteins, and suggests that the concerted action of these DNA repair-related enzymes may be significant also in physiological events in which DNA repair should not be a prevalent phenomenon.

Acknowledgements

This work was partly supported by grants of C.N.R's Target Project on Biotechnology and Bioinstrumentation and of the Centro Interuniversitario per lo Studio delle Macromolecole Informazionali (C.I.S.M., Milano, Pavia, Verona).

References

1. Ogura, T., Nyunoya, H., Takahashi-Masutani, M., Miwa, M.,Sugimura, T., Esumi, H. (1990) Biochem Biophys Res Commun 167: 701-710
2. Yamaguchi, M., Nishida, Y., Moriuchi, T., Hirose, F., Hui, C. -C, Suzuki, Y., Matsukage, A. (1990) Mol Cell Biol 10: 872-879
3. Ogura, T., Takeuchi, N., Yamaguchi, M., Matsukage, A., Sugimura, T., Esumi, H. (1990) Biochem Biophys Res Commun 172: 377-384
4. Menegazzi, M., Gerosa, F., Tommasi, M., Uchida, K., Miwa, M., Sugimura, T., Suzuki, H. (1988) Biochem Biophys Res Commun 156: 995-999

5. Menegazzi, M., Carcereri De Prati, A., Ledda-Columbano, G. M. Columbano, A., Uchida, T., Miwa, M., Suzuki, H. (1990) Arch Biochem Biophys 279: 232-236

6. Bertazzoni, U., Stefanini, M., Noy, G. P., Ginlotte, E., Nuzzo, F., Falaschi, A., Spadari, S. (1976) Proc Natl Acad Sci USA 73: 785-789

7. Zmudzka, B. Z., Fornace, A., Collins, J., Wilson, S. H. (1988) Nuc. Acid Res. 16: 9587-9596.

8. Spadari, S., Weissbach, A. (1974) J Mol Biol 86: 11-20

9. Chiu, R. W., Baril, E. F. (1975) J Biol Chem 250: 7951-7957

10. Scovassi, A. I., Izzo, R., Franchi, E., Bertazzoni, U. (1986) Eur J Biochem 159: 77-84

11. Nyunoya, H., Ogura, T., Shiratori, Y., Makino, R., Hayashi, K., Sugimura, T., Miwa, M. (1988) Proc Jpn Cancer Assoc 47: 246

12. Zmudzka, B. Z., Sen Gupta, D., Matsukage, A., Cobianchi, F., Kumar, P., Wilson, S. H. (1986) Proc Natl Acad Sci 83: 5106-5110

Expression of poly (ADP-ribose) polymerase in differentiating HL-60 cells.

Pierre Moreau, Guy Poirier and Scott Kaufmann

Poly (ADP-ribose) polymerase (PARP) is a zinc finger protein that is activated by binding to DNA single- or double- strand breaks (reviewed in Althaus et al., 1987). Studies with inhibitors have suggested that PARP plays a role in homologous and heterologous recombination as well as DNA repair. These processes are closely associated to cell differentiation and proliferation. To better understand the regulation of PARP activity during cellular differentiation, we have determined the PARP mRNA and protein levels during the granulocytic differentiation of HL-60 cells induced by DMSO.

Results

Differentiation of HL-60 toward granulocytes results in a marked decrease in PARP specific mRNA. Within 24 h of adding DMSO, PARP mRNA levels decreased to 70% of control. By 3 days and 5 days after addition of DMSO, PARP mRNA levels were ≈40% of control. Interestingly, two specific transcripts of 3.3 kb and 3.9 kb were detected 4 and 5 days after induction of differentiation, whereas only the 3.3 kb transcript was detected in control cells. This observation raises the possibility of differentiation-associated changes in PARP mRNA in these cells.

To directly address the question of how PARP polypeptide levels change during granulocytic differentiation of HL-60 cells, semi-quantitative western blotting was performed. Levels of PARP polypeptide diminished about 4-fold during the first 4 days of DMSO treatment and about 8-fold after 7 days of DMSO treatment.

Changes in cell cycle distribution were also analyzed in the DMSO treated HL-60 cells. Results of this analysis are shown in Table 1.

Discussion

We have observed a marked reduction in PARP polypeptide in HL-60 cells during DMSO-induced granulocytic maturation. This is consistent with previous observations that PARP is absent from mature granulocytes (Ikai et al., 1980, Ikai et al., 1982). Other nuclear polypeptides including topoisomerase II and the

Table 1 Cell cycle distribution of
DMSO-treated HL-60 cells.

	G1	S	G2/M
Day 0	43	38	19
Day 1	48	32	20
Day 2	65	22	13
Day 3	78	14	8
Day 4	85	8	7
Day 5	86	8	6

nucleolar protein B23 have been observed to decrease in amount during the
course of DMSO-induced granulocytic maturation in HL-60 cells, (Kaufmann et
al., 1991). On the other hand, other nuclear polypeptides have been observed to
remain at control levels (lamin B) or actually increase (lamins A and C) during
the same maturation process in vitro (op. ct.).

Our experiments have elucidated several possible explanations for the
maturation-associated decrease in PARP polypeptide levels (Table 2). First of
all, changes in cell cycle distribution might account for up to half of the decrease
in PARP levels. Previous studies (Leduc et al., 1988) indicated that G2 cells
contain twice as much PARP polypeptide as G1 cells. S phase cells contained
intermediate amounts. Flow cytometry of DMSO-treated HL-60 cells (Table 1)
revealed a marked decrease in the S and G2/M population within 24 h of addition
of DMSO. This decrease in S and G2/M cells would be expected to diminish
PARP levels. Second, PARP mRNA begins to diminish within 24 h after
addition of DMSO. This might reflect diminished transcription and/or increased
mRNA turnover.

Finally, since the cells (and nuclei) become markedly smaller during the
maturation process, the possibility of increased polypeptide degradation must be
mentioned. We did not, however, observe any immunoreactive degradation
products of PARP on western blots.

Table 2. Possible explanations of DMSO-induced
decrease in PARP polypeptide

1.   Changes in cell cycle distribution
2.   Changes in mRNA level
3.   Changes in protein turnover.

Four days after the addition of DMSO, we have observed the appearance of a novel 3.9 kb PARP transcript in addition to the control 3.3 kb transcript. Since there was no obvious alteration in the migration of the PARP band on western blots, it is unlikely that the 3.9 kb transcript contains a large insertion in the coding sequence for PARP. More likely, the second transcript represents an alteration in the 5' or 3' untranslated sequences. If this alteration occurs in the 5' sequences, it might alter the translation frequency of the transcript. If the alteration involves 3' sequences, it might affect the stability of the message. In this context, it is interesting to note that Kurosaki et al., (1987) and Cherney et al., (1987) have reported the isolation of human PARP cDNA with different 3' noncoding sequences.

References

Althaus, F.R.; Richter, C. ADP-Ribosylation of proteins: Enzymology and Biological Significance. Berlin: Springer Verlag. 1987

Cherney, B.; McBride, O.W.; Chen, D.; Alkhatib, H.; Bhatia, K.; Hensley, P.; Smulson, M.E. Proc. Natl. Acad. Sci. U.S.A. 84: 8370-8374; 1987.

Ikai, K.; Ueda, K.; Fukushima, M.; Hayaishi, O. Poly (ADP-ribose) synthesis, a marker of granulocyte differentiation. Proc. Nat. Acad. Sci. U.S.A. 77:3682-3685; 1980

Ikai, K.; Ueda, K.; Hayaishi, O.; Immunohistochemistry of poly(ADP-ribose). In: Hayaishi, O.; Ueda, K., eds. ADP-ribosylation reactions: Biology and Medicine. New York: Academic Press; 1982:p.339-360.

Kaufmann, S.H.; McLaughlin, S.M.; Kastan, M.B.; Liu, L.F.; Karp, J.E.; Burke, P.J. Topoisomerase II levels during granulocytic maturation *in vitro* and *in vivo*. Cancer Res. 51:3534-3543; 1991.

Kurosaki, T.; Ushiro, H.; Mitsuuchi, Y.; Suzuki, S.; Matsuda, M.; Matsuda, Y.; Katanuma, N.; Kangawa, K.; Matsuo, H.; Hirose, T.; Inayama, S.; Shizuta, Y. J. of Biol. Chem. 262:15990-15997; 1987.

Leduc, Y.; Lawrence, J.J.; DeMurcia, G.; Poirier, G.G. Cell cycle regulation of poly(ADP-ribose) synthetase in FR3T3 cells. Biochim. Biophys. Acta 968:275-282; 1988.

Part 2
Cancer, DNA Repair, and Metabolism

A Novel Model of Enzymatic Repair of UV-Induced DNA Damage in Human Cells

Malcolm C. Paterson, Michel Liuzzi, Anne M. Galloway, John R.A. Chan, Michael V. Middlestadt, Michael Weinfeld and Kevin D. Dietrich

Abstract

We have obtained evidence for the existence of a pre-incision step in the excision-repair pathway operating on UV-induced cyclobutyl pyrimidine dimers in cultured human cells. Demonstration of the occurrence of this reaction includes the following five observations. 1) In the majority (85%) of the dimer-containing excision fragments isolated from post-UV incubated normal fibroblasts, the intradimer phosphodiester linkage is cleaved, but the cleavage products remain joined together by the cyclobutane ring of the dimer. 2) Various human specimens (e.g., lymphocytes and liver tissue) contain an activity, termed intracyclobutyl pyrimidine dimer-DNA phosphodiesterase (IDP), capable of mediating this intradimer backbone-nicking reaction; the IDP activity is optimally expressed at pH 5.5 and is associated with a Mg^{++}-stimulated, 52 kDa protein, which has now been purified 3500-fold from human liver. 3) In the genomic DNA extracted from post-UV incubated xeroderma pigmentosum (XP) complementation group D cells, which are totally defective in dimer excision, a fraction (~15%) of the dimer sites have a hydrolysed internal phosphodiester bond. 4) These backbone-nicked dimers tend to accumulate in transcriptionally active (i.e., c-myc) genomic sequences in UV-damaged XP D cells. 5) Such modified dimer sites also appear transiently in normal fibroblasts at early times after UV treatment. Together, these findings suggest that the intradimer backbone cleavage reaction, perhaps by reducing local conformational stress, may promote transcription on a UV-damaged template and thereby defer the actual excision-repair event to a later time.

A novel nuclease digestion/HPLC assay has also been developed to characterize further human excision-repair pathways(s) active on two major classes of UV lesions, cyclobutyl dimers and (6-4) photoproducts. By analyzing isolated UV lesions which accumulate in excision fragments during incubation of UV-treated cells, the excision kinetics for these two classes of lesions have been shown to differ greatly. (6-4) photoproducts are removed in their entirety by 6 hr, whereas cyclobutyl dimers are processed much more slowly, with a retention half-life of ~16 hr. Unlike the situation for dimers noted above, (6-4) photoproducts are released with the internal phosphodiester linkage intact. Finally, fibroblast strains from XP groups A, C, D, and E exhibit a deficiency in the repair of cyclobutyl dimers and/or (6-4) photoproducts which is unique for each group, signifying that mutations at different loci underlie the different malfunctional repair patterns.

Introduction

The primary aim of the studies outlined in this chapter has been to clarify the mechanisms by which cultured human cells process (i.e., repair, bypass or otherwise circumvent) cyclobutyl pyrimidine dimers and $6' - 4'$ pyrimidin-$2'$-one pyrimidines [(6-4) photoproducts], two major classes of DNA lesions induced by solar UV (1). To realize this goal, considerable effort has been devoted to elucidation of the primary molecular defects in XP, a rare autosomal recessive affliction in which marked propensity to sunlight-related skin neoplasms and *in vitro* cellular hypersensitivity to the cytotoxic, mutagenic and carcinogenic actions of UV rays are causally linked to deficiencies in enzymatic systems that process UV photoproducts in cellular DNA (2). As demonstrated below, this line of experimentation on XP not only promises to clarify certain ill-defined biochemical anomalies in different genetic forms of this disorder (3) but may also lead to a clearer definition of early events in dimer repair in normal cells.

Discovery of Modified Dimer Sites in Post-UV Incubated Cells

Our studies were initiated in an effort to reconcile the longstanding inconsistency in the excision-repair properties of XP group D strains. These strains, although severely, if not totally, defective in recognizing dimer-containing sites (detected as UV-induced sites sensitive to the strand-incising activity of *Micrococcus luteus* UV endonuclease), nevertheless perform an appreciable amount of DNA repair replication (measure of repair capability) (4). To account for these peculiar repair properties, it was reasoned that, following UV exposure, group D cells may act aberrantly on a fraction of the dimer-containing sites, inserting "repair patches" while failing to excise the photoproducts themselves. To test whether a portion of the dimer-containing sites are in fact modified but not removed in XP D strains, we measured the photoreactivability (a well-established diagnostic probe of dimer authenticity) of such sites in DNA from normal and group D cells as a function of incubation time after each UV (254-nm) exposure (4). Following post-UV incubation, the cultures were lysed; their isolated DNA was subjected to exhaustive photoenzymatic reactivation (*Streptomyces griseus* photolyase plus visible light) to monomerize dimers, and was then analyzed for residual dimer-containing sites with an *M. luteus* extract containing UV endonuclease activity. This protocol revealed the appearance of single-strand breaks in the DNA from UV-treated and incubated XP fibroblasts that was not seen in similarly treated normal fibroblasts. The incidence of these unique sites, which were detected as strand nicks, peaked after 48 hr of post-UV incubation in the XP D strains; their maximal yield equalled 15% of the dimers initially introduced and roughly corresponded to the residual capacity (15-20% of normal) of the same strains to perform UV-stimulated DNA repair replication. A similar accumulation of such novel sites was observed in XP groups A, B, F, and G cells, whereas groups C, E, and variant cells behaved like normal controls (ref. 4; our unpublished results). The same incidence of sites was observed in XP A and D strains when the DNA extracted from post-UV incubated cultures was subjected to photoenzymatic reactivation treatment alone, that is, without subsequent UV endonuclease exposure. This unexpected observation led us to propose that during incubation of UV-treated strains from XP groups A and D, the phosphodiester bond between the two dimer-forming pyrimidines may be ruptured, and that at such altered sites, individual DNA chains are then held together solely by the cyclobutane bridge joining the two pyrimidine bases. The excision repair system presumably aborts at this stage in group A cells, it was surmised, since these cells carry out negligible repair synthesis. In contrast, the intradimer backbone cleavage in group D cells seems to be accompanied by the subsequent, presumably proximal, abortive insertion of a repair patch of normal size.

These XP data raised the intriguing possibility that this heretofore undetected reaction may also take place in normal cells. To test this proposal, we photochemically reversed the dimer-containing excision fragments isolated from post-UV incubated normal cells and observed the release of free thymidine (dThd) and thymidine monophosphate (TMP), but not thymine (5). The sum of the dT and TMP molecules released was equal to ~80% of the number of dimers photoreversed, suggesting that in most excision fragments, the dimer must both contain an internal backbone scission and be located at one end of the excised oligonucleotide.

Based on the above findings, we put forward a new model for the repair of pyrimidine dimers in human cells (Fig. 1). In this scheme, which is unlike that reported for any other biological system studied thus far, hydrolysis of the internal phosphodiester bond of dimerized pyrimidines was hypothesized to constitute the first step in the nucleotide excision-repair process that operates on these UV photoproducts. This reaction, it was reasoned, may then be followed by conventional strand-incision/lesion-excision/repair-synthesis/strand-ligation reactions. We speculated that the function of the putative intra-cyclobutyl

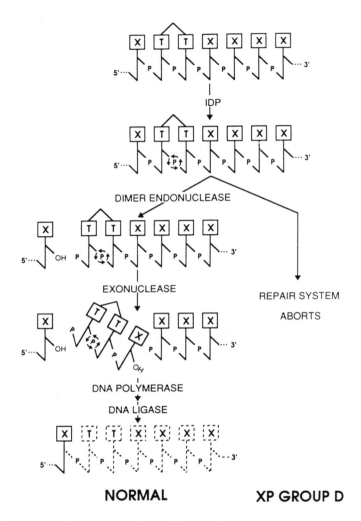

NORMAL   **XP GROUP D**

FIGURE 1.   Proposed model of the excision-repair system operating on pyrimidine dimers in human [normal (left) and XP D (right)] cells. The four arrows encircling the "P" indicate that the site at which the IDP hydrolyses the intradimer phosphodiester bond has yet to be determined. In XP group D cells repair apparently aborts following the action of IDP. See text for details.

pyrimidine dimer-DNA phosphodiesterase (IDP) may be to produce a localized structural change at the dimer-containing site such that the site is then recognized by a generalized "bulky lesion-repair complex" that possibly resembles the *Escherichia coli* UvrABC nuclease complex that acts on each member of an array of chemically disparate defects in DNA (6).

Intradimer Phosphodiesterase: Purification and Characterization

The fact that modified dimer sites accumulate in XP groups A, B, D, F, and G cells in the absence of dimer excision suggested that the activities of the putative IDP and "bulky lesion-repair complex" function separately and was consistent with the notion that the phosphodiesterase activity may reside in a low-molecular-weight protein, as is the case for various damage-recognizing DNA glycosylases (6). We thus set out to isolate and characterize the IDP enzyme. To design an assay for detection of the enzyme, a detailed study was undertaken to ascertain the reactivity of the phosphodiester linkage internal to a cyclobutane dimer or to a (6-4) photoproduct (7). For this purpose, synthetic oligonucleotides, containing in turn each of the two photoproducts, were treated with various nucleases. These experiments revealed that nuclease P1 (NP1) hydrolysed the intra-cyclobutane dimer phosphodiester linkage, but not the bond internal to a (6-4) photoproduct. All other nucleases tested including snake venom phosphodiesterase (SVP), calf spleen phosphodiesterase, mung bean nuclease and nuclease S1 proved incapable of attacking either of these internal bonds. Assay conditions for modified cyclobutane dimers were subsequently optimized in order to detect the presence of IDP. The enzymatic assay entailed overnight digestion of UV-irradiated, [^{3}H]dThd-labelled DNA with SVP and calf alkaline phosphatase (CAP) to generate trinucleotides with a modified or an intact dimer at the 3'-end, and quantitation of the reaction products by reverse phase HPLC. This methodology has led to the discovery, in the cytosol of human fibroblasts, of a protein with IDP activity (unpublished results). The IDP has been purified 3500-fold from human liver in a four-stage procedure employing affinity (Con-A Sepharose), ion exchange (Accell CM and QMA), and gel filtration (Sephadex G-100) chromatographies. The activity, which is optimally expressed at pH 5.5, is associated with a 52-kDa protein, is stimulated by Mg^{2+} or 2-mercaptoethanol, and can be inactivated by trypsin or heat. Preliminary experiments have shown that strains representing XP groups A-G and the variant contain normal, and in some cases perhaps elevated, levels of IDP activity, signifying that the UV hypersensitivity of these different strains does not result from defective IDP.

Genomic Distribution of Enzymatically Modified Dimer Sites

Recent reports demonstrating that dimers are preferentially repaired in transcriptionally active genes of mammalian cells (8) prompted us to determine the intragenomic distribution of modified dimer sites accumulating in post-UV incubated XP D cells. To measure backbone-nicked dimers in genomic DNA, we modified the protocol of Hanawalt and Bohr (8) for analysis of repair of pyrimidine dimers in defined genes, using *E. coli* DNA photolyase to quantitate, by conversion to single-strand breaks, dimers with cleaved interpyrimidine phosphodiester linkages. In short, following post-UV (15 J/m^2) incubation (4-25 hr) of XP D cells, total genomic DNA was extracted, digested with BamH1 restriction enzyme, and subjected to photoenzymatic treatment. DNA samples were then (i) electrophoresed in alkaline sucrose gels to separate full-length (modified dimer-free) restriction fragments from photolyase digestion products; (ii) transferred to membranes; and (iii) hybridized with either c-*myc* or c-*mos* ^{32}P-labelled probes. The membranes were autoradiographed and the films were assessed for amounts of full-length restriction bands by densitometry. Finally, the Poisson expression was applied to calculate the modified dimer incidence in the c-*myc* and c-*mos* sequences, the former transcriptionally active and the latter silent, as a function of post-UV cell incubation time. This analysis revealed that backbone-nicked dimers accumulated much more rapidly in the c-*myc* sequence than in the c-*mos* sequence. For example, by 12 hr post-UV ~40% of the dimers initially produced were converted to the backbone-nicked form in the c-*myc* fragment whereas only 16% underwent this conversion in the c-*mos* fragment. Moreover, preliminary observations on normal cells have disclosed the appearance, at early post-UV times, of these modified dimers in the c-*myc* sequence. Hence, enzymatic modification of pyrimidine dimers may arise in normal cells during the course of processing UV-induced DNA damage.

Enzymatic Digestion/HPLC Analysis of Excision Fragments

The enzymatic digestion/HPLC assay has been modified to study the excision-repair system(s) operative on cyclobutane dimers and (6-4) photoproducts (unpublished results). The modified technique involves digestion of the radioactively labelled excision fragments with SVP and CAP, followed by NP1 treatment. A set of dimerized dinucleotides (in addition to mononucleosides) is thereby created, which is subsequently separated by reverse phase chromatography. This results in the ability to monitor accurately the repair of not only cyclobutyl pyrimidine dimers but also the less common (6-4) photoproducts. In addition, this procedure provides a means to determine the relative abundance of the two distinct forms of cyclobutane dimers found in excision fragments, namely, those with and those without, an intact intradimer phosphodiester bond. Examination of excision fragments which accumulate during incubation of UV (40 J/m^2)-treated normal cells has disclosed that (6-4) photoproduct removal is complete by 6 hr, whereas cyclobutane dimers are eliminated with a half-life of ~16 hr. These results have been confirmed more recently using an independent assay in which [^{3}H]dThd-labelled genomic DNA is digested with SVP and CAP, followed by (i) dialysis to remove excess dThd, (ii) further digestion with NP1 and CAP, and (iii) quantitation by HPLC. An additional finding from this study is that whereas ~80% of cyclobutyl dimers recovered from excision fragments are modified (thus confirming our earlier finding), (6-4) photoproducts found in these oligonucleotides always contain an intact phosphodiester bond. Analysis of excision fragments from repair-deficient strains representing XP groups A, C, D, E and the variant has yielded repair kinetics for both dimers and (6-4) photoproducts which are unique for each genetic form of the disease studied. XP A cells are totally repair-deficient. XP C cells, which are partially defective in the removal of both photolesions, display ~50% of normal (6-4) photoproduct repair by 24 hr. In contrast, XP D cells, while unable to excise cyclobutane dimers (Fig. 1), retain ~40% of normal proficiency at repairing (6-4) photoproducts by 6 hr. XP E cells exhibit normal kinetics for dimer excision, but act sluggishly on (6-4) photoproducts, resulting in a protracted half-life of ~ 12 hr. XP variant cells show normal repair kinetics for both photoproducts.

Conclusion and Prospectus

Together, the results outlined here contribute substantially to clarification of the excision-repair mechanisms(s) operative on UV photoproducts in human cells. First, our data clearly indicate that cyclobutane dimers and (6-4) photoproducts are both repaired by the so-called nucleotide mode of excision repair (4). Second, the disclosures that (i) most of the excised cyclobutyl dimers contain a cleaved internucleotide bond while (6-4) photoproducts are released intact and that (ii) the two classes of photolesions appear in excision fragments with markedly different kinetics strongly imply that human fibroblasts possess at the very least two different damage recognition/incision steps, one operating on intact (6-4) photoproducts and the other on enzymatically modified cyclobutane dimers. A priori, intact dimers (i.e., those excised without undergoing intradimer-backbone cleavage) may be processed by either mechanism or possibly by a third pathway. Third, the observation that (6-4) photoproducts are recovered in their entirety from the cytosol by 6 hr post-UV, whereas few cyclobutane dimers are excised within this period signifies that many studies conducted over the last two decades have assessed effects due to the metabolic fate of (6-4) photoproducts rather than cyclobutane dimers. It will therefore be important that the results of these earlier investigations be reinterpreted in view of our finding. Fourth, the longstanding paradox regarding the excision-repair properties of XP D cells is resolved by the discovery that the residual level of UV-induced DNA repair replication displayed by these mutant cells derives completely from removal of (6-4) photoproducts, rather than, as stated in our original working hypothesis, faulty action on a fraction of the cyclobutane dimers wherein "repair patches" may be inserted without dimer removal. Fifth, the demonstration that enzymatically modified dimers not only tend to accumulate in transcriptionally active (i.e. c-*myc*) genomic sequences in UV-damaged XP-D cells but may also be formed transiently in these same active sequences during routine processing of UV damage in normal cells suggests that the intradimer sugar-phosphate hydrolysis reaction, possibly by reducing local conformational stress, may serve to promote transcription on a UV-damaged template and thereby permit relegation of the actual excision event to a later time. This current model further predicts that excision fragments (10-20%) containing intact dimers are released before the transcription (or perhaps DNA replication) apparatus passes the dimer-containing sites in genomic DNA, whereas the same machinery presumably circumvents the damaged sites before dimer excision in the case of those fragments containing a modified dimer. And last, but not least, the detection and purification of a human IDP activity capable of hydrolysing the intradimer phosphodiester bond provides the starting point for an in-depth investigation of dimer excision-repair in human cells. Inquiry into how the introduction of a modified dimer in a defined DNA sequence affects the activity of various DNA-processing enzymes (e.g., RNA and DNA polymerases) should lead to the elucidation of the biological function(s) of the IDP.

Acknowledgements

This work was supported by operating grants from the Medical Research Council and National Cancer Institute of Canada. M.C.P. is a Medical Scientist of the Alberta Heritage Foundation for Medical Research.

References

1.   Mitchell, D.L. The relative cytotoxicity of (6-4) photoproducts and cyclobutane dimers in mammalian cells. Photochem. Photobiol. 48:51-57; 1988.

2. Cleaver, J.E.; Kraemer, K.H. Xeroderma pigmentosum. Scriver, C.R.; Beaudet, A.L.; Sly, W.S.; Valle, D., eds. The Metabolic Basis of Inherited Disease, Vol. II, 6th edit. New York: McGraw-Hill; 1989:2949-2971.

3. Paterson, M.C. Accumulation of non-photoreactivable sites in DNA during incubation of UV-damaged xeroderma pigmentosum group A and group D cells. Prog. Mutation Res. 4:183-192; 1982.

4. Paterson, M.C.; Middlestadt, M.V.; MacFarlane, S.J.; Gentner, N.E.; Weinfeld, M. Molecular evidence for cleavage of intradimer phosphodiester linkage as a novel step in excision repair of cyclobutyl pyrimidine photodimers in cultured human cells. J. Cell Sci. Suppl. 6:161-176; 1987.

5. Weinfeld, M.; Gentner, N.E.; Johnson, L.D.; Paterson, M.C. Photoreversal-dependent release of thymidine and thymidine monophosphate from pyrimidine dimer-containing DNA excision fragments isolated from ultraviolet-damaged human fibroblasts. Biochemistry 25: 2656-2664; 1986.

6. Van Houten, B. Nucleotide excision repair in *Escherichia coli*. Microbiol. Rev. 54:18-51; 1990.

7. Liuzzi, M.; Weinfeld, M.; Paterson, M.C. Enzymatic analysis of isomeric trithymidylates containing ultraviolet light-induced cyclobutane pyrimidine dimers. I. Nuclease P1-mediated hydrolysis of the intradimer phosphodiester linkage. J. Biol. Chem. 264:6355-6363; 1989.

8. Bohr, V.A.; Phillips, D.H.; Hanawalt, P.C. Heterogeneous DNA damage and repair in the mammalian genome. Cancer Res. 47: 6426-6436; 1987.

POLY(ADP-RIBOSE)POLYMERASE
IN Xenopus laevis

B. SAULIER, F. SIMONIN, G. GRADWOHL,
G. de MURCIA and M. PHILIPPE

INTRODUCTION

Among the different biological systems in which poly(ADP-ribosyl)ation has been studied, cell lines have been widely used, especialy to investigate the role of poly(ADP-ribose)polymerase (PARP) in DNA repair(1). Other systems such as regenerating rat liver (2) or mitogen stimulated lymphocytes (3) permitted several authors to suggest the implication of the PARP in DNA replication and cell proliferation. Even so, one has to admit that the precise biological function of poly(ADP-ribosyl)ation of proteins acceptors is still unclear.

Xenopus laevis oogenesis and early embryonic development appear to be a very interesting and suitable system to investigate *in vivo*, the role of poly(ADP-ribose)polymerase. There are two main reasons for this:
1-Translation is always active during oogenesis and early development, but this is not the case for RNA synthesis and DNA replication. During oogenesis, transcription is very active, but there is no DNA replication. In full-grown oocytes (Stage VI) (4), which are blocked in first meiotic prophase, there is absolutely no DNA synthesis . After stimulation by progesterone, Stage VI oocytes undergo the first meiotic division. This process, known as oocyte maturation, occurs in the absence of gene transcription and DNA replication. The mature oocyte (or egg) is laid, and after fertilization an embryo is formed. During the first twelve synchronous divisions of the embryo, there is no RNA synthesis while DNA replication is very active(5).

X. laevis appears to offer a unique biological system to search for any possible correlation between the presence or level of PARP activity and a situation where only RNA synthesis (Stage VI oocytes) or DNA synthesis (early embryos) is active or neither DNA nor RNA synthesis occur (unfertilized eggs).

2-Fertilization of the eggs and development of the embryos occur outside of the animal and oocyte maturation can be obtained *in vitro* by incubation with progesterone. Moreover, stage VI oocytes, eggs and early embryos are about 1.0-1.5 mm in diameter. This fact alone makes it reasonably easy to microinject DNA (plasmid DNA or oligonucleotides), RNA and even protein into these cells(6).

Poly(ADP-ribosyl)ation in *Xenopus laevis* has been previously studied by several authors. In 1976, Burzio & Koide first detected the enzyme in *Xenopus* oocytes, and showed later that poly(ADP-ribose)polymerase activity is stimulated in the germinal vesicle (nucleus) of stage VI oocytes previously incubated *in vitro* with progesterone (7).
Some of the biochemical properties of the *Xenopus* enzyme were studied by Farzaneh & Pearson in 1979 (8). The authors also measured *in vitro* changes in poly(ADP-ribose)polymerase activity during embryogenesis and suggested a correlation between an increase in PARP activity and differentiation events.
Microinjection of radioactive NAD in developping embryos was used by Williams *et al.* to labell poly(ADP-ribose). (9).
However, almost nothing is known about the possible function of poly(ADP-ribose)polymerase in oocyte maturation and embryogenesis.
In order to use all of the potentialities of the *Xenopus* system, to better understand the biological function of the PARP, it appeared necessary to clone the cDNA for the *Xenopus* enzyme. The experiments reported here were performed to define the best cloning strategy.

METHODOLOGY.

X. laevis oocytes, unfertilized eggs and embryos.
Oocytes were obtained by incubating fragments of ovaries in F1 medium(10) containing 0.15% collagenase.

Free oocytes ranging from previtellogenic oocytes to full-grown oocytes, described respectively as stage I to VI by Dumont (4) were sorted into three groups, stages I+II, stages III+IV, stages V+VI.

Unfertilized eggs(UFE) were squeezed out of females previously injected with 1000 UI of human chorionic gonadotropin. Embryos were obtained by *in vitro* fertilization, and staged according to Nieuwkoop and Faber (11).

Preparation of crude extracts.

Oocytes, dejellied UFE and embryos were homogenized in 10 volumes of extraction buffer (100 mM Tris-HCl pH 7.5, 500 mM KCl, 1 mM DTT, 0.1 mM EDTA, 0.1% NP40) containing 1 µg/ml leupeptin, 1 µg/ml pepstatin and 1 mM PMSF. Homogenates were sonicated (3 sec) and centrifuged at 12 000 rpm, 15 min at 4°C. Supernatants were collected and the lipids removed by 1,1,2 trichlorotrifluoroethane extraction (12).

Activity blot.

Activity blots and analysis of the reaction product were performed as described by Simonin *et al.* (13).

PARP was assayed in oocytes of different stages, UFE and embryos by "activity slot-blot". Crude extracts were directly blotted in a native form, on nitrocellulose (previously soaked in extraction buffer) with a slot-blot apparatus. The slot-blot was then processed as a typical activity blot. Appropriately exposed autoradiograms of the blots were scanned by densitometry for relative quantification.

Western blot.

Western blots were performed according to Mazen *et al.* (14). Proteins were revealed by means of anti-rabbit globulins conjugated to alkaline phosphatase. *X. laevis* PARP was detected with three polyclonal antibodies: anti PARP directed against calf thymus PARP (13); anti F2 directed against the second zinc finger (residues 122 to 165) of human PARP (13), and anti 40K developped against the 40 KDa C-terminal fragment of human PARP (unpublished).

RESULTS AND DISCUSSION.

The activity blot is a convenient method to detect

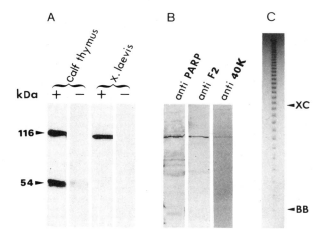

Fig.1 A. Activity blot analysis of PARP in a *X. laevis* UFE crude extract (100 µg proteins) compared to the activity blot of 300 ng of a trypsic digest of purified calf thymus PARP. The activity blot was performed with (+) or whithout (-) DNase I activated DNA (see Methodology).
B. Immunological cross reaction between *X. laevis* PARP and calf thymus PARP antibodies (anti PARP, 1/500), or human PARP antibodies (anti F2, 1/5000, anti 40 K, 1/2000) (see Methodology for description of antibodies).
C. Analysis of the reaction product of the *X. laevis* PARP. Poly(ADP-ribose) was extracted from the activity blot (13) and loaded on a 13% sequencing gel. XC is for xylene cyanol, BB for bromophenol blue.

intact PARP enzyme and its active proteolytic fragments.

Activity blot of a limited trypsic digest of purified calf thymus PARP (Fig. 1.A) revealed two bands at 116 and 54 KDa corresponding respectively to the whole enzyme and a C-terminal fragment containing the catalytic domain (13). In the absence of activated DNA, PARP activity is strongly reduced (13).

The activity blot of a *X. laevis* UFE crude extract (Fig; 1-A) revealed the presence of a labelled protein whith an apparent molecular weigh of 100 KDa. In the absence of activated DNA, this activity is not detected This strongly suggests that this protein is poly(ADP-ribose)polymerase.

In order to confirm our presumption, we analysed the nature of the radioactive product. The labelled band was excised from the nitrocellulose, the bound polymer extracted by mild alcaline treatment and processed for sequencing gel electrophoresis. The result displayed in Fig. 1-C demonstrates that the radioactive material is resolved in a typical ladder characteristic of authentic poly(ADP-ribose) (13).

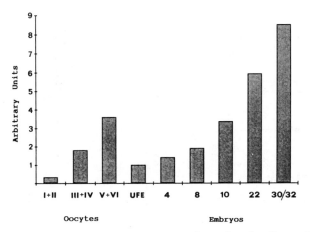

Fig.2 PARP assay throughout oogenesis and early development. PARP was assayed in crude extracts of Stage I+II; III+IV; V+VI oocytes, UFE and embryos of stage 4; 8; 10; 22 and 30/32 by activity slot-blot.(see Methodology). Absolute values varied in two experiments, but relative enzyme activities were constant as shown. Ordinate: PARP activity per oocyte, UFE or embryo in arbitrary units.

Moreover, this result shows that polymers as long as 30 residues can be synthesized by the *Xenopus* PARP immobilized onto nitrocellulose.

In oocytes from stage I to VI, and in embryos, PARP is also detected as a single band with a constant apparent Mr of 100 KDa (data not shown).

Detection of poly(ADP-ribose)polymerase by activity blot was confirmed by immunodetection. After drying and autoradiography, the activity blot of a UFE crude extract was immunostained with a polyclonal antibody directed against calf thymus PARP (Fig. 1-B; anti PARP). This antibody revealed a band of the same Mr as the one detected by activity blot. The two signals obtained in activity blot and Western blot are exactly superimposable. An antibody directed against the second zinc finger of human PARP also revealed a strong band at 100 KDa in a UFE crude extract (Fig. 1-B, anti F2). In an oocytes (St III+IV) crude extract, the same protein was also recognized by an antibody directed against the 40 KDa C- terminal fragment of human PARP.

These results taken together strongly support the idea that the *Xenopus* PARP might not be very different from the bovine enzyme, even if it seems to be slightly smaller. Good immunological cross reaction of the *Xenopus* enzyme with antibodies raised against mammalian PARP plays in favor of a high degree of homology between the amino-acid sequence of mammalian and amphibian PARP. However, this high degree of homology does not seem to be conserved at the nucleotide sequence level. Cross hybridization of *Xenopus* mRNA with the human cDNA probe was totaly unsuccessfull, even in low stringency conditions.

Extracts from oocytes and embryos at different stages were prepared, and the poly(ADP-ribose)polymerase was assayed *in vitro*, by activity slot-blot (see Methodology). The results displayed in Fig.2 show a larger amount of PARP in stage V+VI oocytes (x 9) than in stage I+II, suggesting a *de novo* synthesis, or an activating modification of the PARP during oogenesis.
In UFE, this amount drops by about 3.5 fold. From UFE to stage 30/32 embryos, PARP amount increases about 9 fold, which is in agreement with the results reported by Farzaneh & Pearson (8).

Concluding remarks.

Taking into account all the results reported here, we can plan two cloning strategies; 1- antibody screening of a cDNA expression library and 2- PCR cloning of the cDNA using degenerated oligonucleotides. Both approaches are bared on the good immunocrossreactivity found between *Xenopus* and mammalian enzymes.
Because PARP is present in large amount in stage V+VI oocytes and seems to be synthesized throughout oogenesis, an oocyte (stage I to stage VI) cDNA library should permit us to clone a cDNA for the *Xenopus* poly(ADP-ribose)polymerase.

REFERENCES

1 FARZANEH, F. MELDRUM, R. SHALL, S.(1987) Nucleic Acids Res. 15, 3493-3502.
2 CESARONE, X.F., SCARABELLI, L., SCOVASSI, A.I., IZZO, R., MENEGAZZI, M., CARCERERI, A. DE PRATI, A., ORUNESU, M. and BERTAZZONI, U. (1990) Biochim. Biophys. Acta 1087, 241-246.
3 MENEGAZZI, M., GEROSA, F., TOMMASI, M., UCHIDA, K., MIWA, M., SUGIMURA, T. and SUZUKI, H. (1988) Biochem. Biophys. Res. Commun. 156, 995-999.
4 DUMONT, J.N. (1972) J. Morphol. 136, 155-179.

5 KIRSCHNER, M., NEWPORT, J. and GERHART, J. (1985) Trends Genet. 1, 41-47.
6 HEIKKILA, J.J.(1989) Biotech. Adv. 7, 47-59.
7 BURZIO, L.O. and KOIDE, S.S. (1977) Ann. N.Y. Acad. Sci. 286, 398-407.
8 FARZANEH, F. and PEARSON, C.K. (1979) Dev. Biol. 72, 254-265.
9 WILLIAMS, G.T., SHALL, S. and FORD, C. (1983) Bioscience Reports 3,
  461-467.
10 WEBB, D. and CHARBONNEAU, M. (1987) Cell. Differ. 20, 33-44.
11 NIEUWKOOP, P. and FABER, J. (1967) In "Normal Table of *Xenopus laevis*
  (Daudin)" North-Holland Publ., Amsterdam.
12 GURDON, J.B. and WICKENS, M.P. (1983) Methods Enzymol. 101, 370-386.
13 SIMONIN, F., BRIAND, J.P., MULLER, S. and de MURCIA, G. (1991) Anal.
  Biochem. (in press).
14 MAZEN, A., MENISSIER-de MURCIA, J., MOLINETE, M., GRADWOHL, G., POIRIER,
  G. and de MURCIA, G. (1989) Nucleic Acids Res. 17, 4689-4698.

ADP-RIBOSYLATION IS INVOLVED IN THE INTEGRATION OF EXOGENOUS DNA INTO THE MAMMALIAN CELL GENOME, BUT IS NOT REQUIRED FOR THE EPISOMAL REPLICATION OR EXPRESSION OF AUTONOMOUSLY REPLICATING PLASMIDS

Bhushan D Hardas, Marcus J Braz and Farzin Farzaneh
Molecular Medicine Unit, King's College School of Medicine and Dentistry, Denmark Hill, London SE5 8RX, UK.

INTRODUCTION

The stable maintenance and expression of exogenous DNA molecules introduced into mammalian cells is dependent on either their integration into the host cell chromosomes or their episomal maintenance. Episomal maintenance is dependent on the presence of cis-acting autonomously replicating sequences (ARS), and the trans-acting factors which interact with this sequence.

We have previously demonstrated that inhibition of poly(ADP-ribose) polymerase by its competitive inhibitors blocks the integration into the mammalian cell chromosomes of exogenous DNA molecules introduced into these cells by calcium phosphate mediated DNA transfection (1). This inhibition was shown to be specific to the integration process and poly(ADP-ribose) polymerase activity did not appear to be required for the transfection process itself (plasmid entry into the cells), or the expression of the transfected genes either before or subsequent to integration into the host cell genome. In the present study we have investigated the possible involvement of poly(ADP-ribose) polymerase in the episomal maintenance and expression of autonomously replicating DNA sequences.

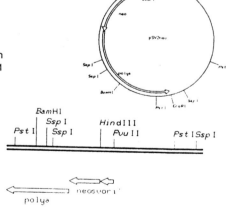

Fig. 1. The schematic representation of covalently closed and EcoR1 linearise pSV2.neo plasmid.

COS cells, derived from CV-1 cells by the introduction of a replication defective SV40 genome, constitutively express the large-T antigen required for the trans-activation of the SV40 origin of replication (ori). Covalently closed plasmids such as pSV2.neo (Fig. 1), carrying ori, can replicate episomally in COS cells. Therefore, in these cells the stable maintenance and expression of ori containing circular plasmids is not dependent on their integration into the cell genome. However, the stable maintenance of linearised plasmids (Fig. 1) in these cells requires either circularisation or integration into the genome. By contrast, in the parental CV-1 cells, which lack the trans-acting factors required for the episomal replication of ori containing plasmids, stable maintenance and expression of genes encoded by either linear or covalently closed plasmids is dependent on their integration into the host cell genome.

Studies presented here demonstrate that inhibitors of poly(ADP-ribose) polymerase block the stable maintenance and expression of both circular and linear plasmids containing SV40 ori sequences in the parental CV-1 cells. However, in COS7 cells the episomal maintenance and stable expression of circular plasmids containing the ori sequence is not blocked by the inhibitors. Treatment of COS7 cells with the inhibitors does however block the maintenance and stable expression of linear plasmid DNA. These observations confirm and extend our previous demonstration that inhibitors of poly(ADP-ribose) polymerase activity, block the integration of exogenous DNA molecules into the mammalian cell genome by specifically inhibiting their integration. In addition they show that poly(ADP-ribose) polymerase activity is not required for the episomal replication and expression of SV40 ori containing plasmids in COS7 cells.

RESULTS

The plasmid pSV2.neo contains the SV40 early promoter, including the ori sequence, and the neomycin phosphotransferase gene (neo). The stable maintenance of pSV2.neo and expression of the neo gene, confers to the mammalian cells resistance to the neomycin analogue, G418. When COS7 cells were transfected with covalently closed pVS2.neo approximately 770 neo^R colonies were isolated after two weeks of culture in the presence of 1mg/ml G418 (Fig. 2). In COS7 cell cultures which were transfected in the continuous presence of 2mM 3-methoxybenzamide, a competitive inhibitor of poly(ADP-ribose) polymerase, there was a small reduction of approximately 45% in the number of neo^R colonies. There was no significant inhibition of neo^R colony formation in cultures treated with the less inhibitory analogue

Fig. 2. Effect of poly(ADP-ribose) polymerase inhibition on the formation of neo^R colonies in COS7 cells transfected with pSV2.neo. 10^6 Cells were transfected either in the absence (control), or presence of 2mM 3-methoxybenzamide (3MB), or 2mM 3-methoxybenzoic acid. 48 hours after transfection 1mg/ml G418 was added to each culture and maintained either in the absence or continued presence of the indicated compounds for approximately 2 weeks, at which time the number of neo^R colonies were counted.

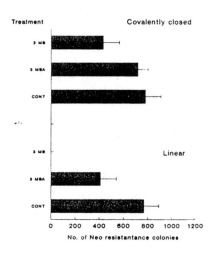

3-methoxybenzoic acid. By contrast in COS7 cells which were transfected with the linearised pSV2.neo plasmid, 2mM 3-methoxybenzamide completely abolished the formation of neo^R colonies. In these cultures, the presence of 2mM 3-methoxybenzoic acid reduced the number of neo^R by approximately 50% (Fig. 2). Therefore, the stable expression of the linearised, but not the covalently closed, pSV2.neo required the activity of poly(ADP-ribose) polymerase.

In CV-1 cells, which lack SV40 trans-acting factors required for the episomal replication of SV40 ori containing plasmids, the presence of 2mM 3-methoxybenzamide reduced the number of neo^R colonies by approximately 99%. This high level of inhibition was observed with both covalently closed and linearised pSV2.neo (Fig. 3). There was no significant reduction in the number of neo^R colonies in CV-1 cell cultures treated with 2mM 3-methoxybenzoic acid. Therefore, the stable expression of neo in cells transfected with either the covalently closed or linearised pSV2.neo required poly(ADP-ribose) polymerase activity.

Plasmids isolated from dam⁻ E. coli are methylated in the N^6 position of the adenine residues in the sequence GATC. However, the replication of such plasmids in eukaryotic cells results in the loss of the methyl groups, due to the absence of the dam methylase. Therefore, methylation sensitive restriction enzymes which contain the sequence GATC in their recognition site could be used to detect eukaryotic cell replication of these plasmids. In order to directly examine the effect of poly(ADP-ribose) polymerase inhibition on the episomal replication of pSV2.neo, small molecular weight DNA was

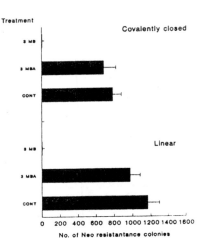

Fig. 3. Effect of poly(ADP-ribose) polymerase inhibition on the formation of neo^R colonies in CV-1 cells transfected with pSV2.neo. The transfection and selection conditions are as described in Fig. 2.

isolated, by the Hirt procedure, from the COS7 cells transfected with the pSV2.neo plasmid. The isolated plasmids were then digested with either DpnI which digests only the methylated DNA (indicative of eukaryotic cell replication), with MboI which digests only the un-methylated DNA (indicative of un-replicated DNA), or with Sau3A which digests the DNA independently of its methylation status. Analysis of the digested DNA (Fig. 4) demonstrates that the covalently closed pSV2.neo plasmid is digested with both Sau3A and MboI, but not with

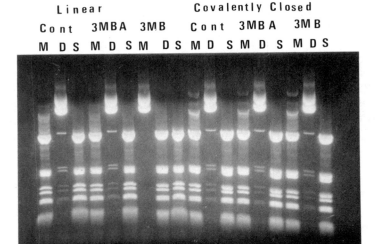

Fig. 4. Analysis of plasmid replication in COS7 cells transfected with pSV2.neo. Small molecular weight DNA was isolated from the neo^R transfected cells by the Hirt procedure and incubation with the restriction enzymes MboI (M), DpnI (D) or Sau3A (S). Appearance of the low molecular weight DNA bands is indicative of plasmid digestion.

*Dpn*I. This is the case in the control, 3-methoxybenzamide and 3-methoxybenzoic acid treated cells. This indicates the replication of the covalently closed pSV2.neo in the *COS7* cells, independently of the activity of poly(ADP-ribose) polymerase. By contrast the pSV2.neo plasmids isolated from *COS7* cells which were transfected and maintained in the continuous presence of 2mM 3-methoxybenzamide, were digested by *Sau*3A and *Dpn*I, but not by *Mbo*I. However, plasmids isolated from either the control or 3-methoxybenzoic acid treated cells were digested with *Sau*3A and *Mbo*I, but not with *Dpn*I. Therefore, the replication of the linearised pSV2.neo was inhibited by 3-methoxybenzamide treatment, but there was clear evidence of replication both in the control and 3-methoxybenzoic acid treated cells.

DISCUSSION

Inhibition of poly(ADP-ribose) polymerase activity, by the competitive inhibitor 3-methoxybenzamide, blocks the stable expression of the *neo* gene in *CV-1* cells transfected with either the covalently closed or the linearised pSV2.neo, and in *COS7* cells transfected with the linearised plasmid. These observations confirm and extend our previous demonstration of the involvement of this enzyme in the integration of exogenous DNA molecules into the mammalian cell genome (1). However, the stable expression of the *neo* gene in *COS7* cells transfected with pSV2.neo is not blocked by the inhibition of poly(ADP-ribose) polymerase. Therefore, the stable maintenance and expression of genes encoded by autonomously replicating plasmids does not require this enzyme activity. The suggestion that episomal replication is independent of poly(ADP-ribose) polymerase activity is confirmed by the replication dependent de-methylation of the GATC sites in the pSV2.neo plasmid. The fact that inhibition of poly(ADP-ribose) polymerase blocked not only the replication of the covalently closed pSV2.neo but also the linearised plasmids demonstrates that the replication of the linear plasmid requires its covalent ligation prior to replication.

In conclusion, the present data demonstrates the involvement of poly (ADP-ribose) polymerase in the integration of exogenous DNA molecules into the mammalian cell genome. However, this activity is not required for the episomal maintenance and replication of autonomously replicating plasmids in suitable hosts, or for gene expression either prior or subsequent to the integration of exogenous DNA molecules into the mammalian cell genome.

Reference:
1. Farzaneh F. *et al.*, (1988). Nucleic Acids Res. 16; 11319-11326.

INHIBITORS OF POLY(ADP-RIBOSE) POLYMERASE BLOCK THE INFECTION OF MAMMALIAN CELLS BY RETROVIRAL VECTORS.

Farzin Farzaneh[*][1], **Manoochehr Tavassoli**[*][2], **Mark Boyd**[3], **Mary K.L. Collins** [3], **David Darling**[1], **Joop Gäken**[1], **and Sydney Shall**[2]
[1]Molecular Medicine Unit, King's College School of Medicine and Dentistry, Denmark Hill, London, SE5 8RX; [2]Cell and Molecular Biology Laboratory, Biology Building, University of Sussex, Brighton, BN1 9QG, England; [3]Institute of Cancer Research, Fulham Road, London, SW3 6JB.

Introduction

Early events in retroviral infection entail interaction between the viral envelope and specific host-cell receptors, entry and uncoating of the enveloped virion, reverse transcription of the viral RNA genome by reverse transcriptase into the double-stranded proviral DNA, followed by the integration of a proportion of the proviral DNA molecules into the host cell chromosomal DNA (4, 10). Inhibitors of reverse transcription (6), and "soluble" receptor molecules which hinder uptake of virus (1, 2, 7) have been developed for intervention in the retroviral life cycle, and inhibition of viral replication. However, the molecular components of provirus integration into the host cell genome have so far evaded successful intervention (4, 9).

Provirus integration into the host cell genome involves a coordinated set of DNA strand breakage and ligation events (3). This process requires the function of a virus-encoded protein called IN, which has both endonuclease and DNA ligase activities (3). However, although IN is both necessary and sufficient for provirus integration in a model system (5), in the absence of, as yet unidentified cell-encoded factors, this process is relatively inefficient (5). Therefore, high efficiency viral infection requires both the viral IN protein and cellular components.

Because of the known involvement of the nuclear enzyme, poly(ADP-Ribose) polymerase in a range of cellular processes requiring DNA strand ligation (see Ref. 8 for a recent review), the possible involvement of this enzyme in the integration of provirus into the mammalian cell genome was investigated.

The data presented here demonstrate that competitive inhibitors of poly(ADP-ribose) polymerase can block the successful infection of rodent fibroblasts (NIH/3T3) and human epithelial (HeLa) or 'pro-myelocytic' leukaemia (HL-60) cells by two recombinant retroviral vectors (pWeTAC and MPSV.sup[28]). The inhibition appears to be specific to provirus integration and affects neither the viral entry into the host cells, reverse transcription and synthesis of the proviral

---

* Footnote: The first two authors have contributed equally to this paper. The remaining authors are listed alphabetically.

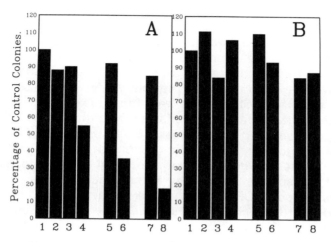

FIG. 1. Inhibition of infection by recombinant retroviral vectors due to the inhibition of poly(ADP-ribose) polymerase activity, as assessed by formation of G418 resistant colonies. (A) NIH/3T3 cells were infected with the pWeTAC in the presence of the indicated compounds. Three days later cells were washed and placed in fresh medium containing 1mg/ml G418 and the number of neo[R] colonies was determined 7 days later. Infection in the absence of inhibitor or analogue, control (1); or in the presence of either 3mM 4-aminobenzamide (2); 3mM 3-aminobenzoic acid (3); 3mM 3-aminobenzamide (4); 1mM 4-formylaminobenzamide (5); 1mM 3-formylaminobenzamide (6); 2mM 4-methoxybenzamide (7); or 2mM 3-methoxybenzamide (8). (B) The protocol was the same as in (A) except that previously infected and selected NIH/3T3 cells were mixed with uninfected NIH/3T3 cells, mock-infected, and then re-selected for G418 resistance.

genome, nor the post-integration expression of the viral genes and production of infective virions.

RESULTS

The recombinant retroviral vector pWeTAC contains the neomycin phosphotransferase gene (neo), the expression of which confers resistance to the neomycin analogue, G418. Successful infection of NIH-3T3 cells by this vector was assayed by selecting the G418 resistant cells. This long-term selection requires the stable integration and expression of the neo gene. The presence of the competitive inhibitors of poly(ADP-ribose) polymerase, 3-aminobenzamide, 3-formylaminobenzamide or 3-methoxybenzamide, significantly inhibited the formation of neo[R] colonies (Fig. 1A). By contrast, the non-inhibitory chemical analogues of the inhibitors, 4-aminobenzamide, 3-aminobenzoic acid, 4-formylaminobenzamide, 4-methoxybenzamide, had little or no inhibitory effect on the formation of neo[R] colonies (Fig. 1A). In these experiments the poly(ADP-ribose) polymerase inhibitors were applied for only the first three days of infection. After this period, the medium was changed and G418 was added to allow selection of neo[R] cells. Therefore, poly(ADP-ribose) polymerase inhibition and G418 selection were temporally separated to avoid the possible cytotoxicity due to long-term exposure of cells to combinations of G418 and enzyme inhibitors. The inhibition of colony formation by the inhibitors was due neither to non-specific cytotoxicity nor to inhibition of expression of *neo* resistance. This was demonstrated by the absence of an effect on the formation of *neo*[R] colonies by cells which had been previously infected and selected (in the absence of the inhibitors) for *neo*[R], then mixed with un-infected NIH/3T3 cells, mock-infected in the presence of the enzyme inhibitors and once again selected for *neo*[R] (Fig. 1B). It was therefore concluded that poly(ADP-ribose) polymerase inhibitors do not block the expression of the *neo* gene, do not selectively kill the infected cells, nor exhibit non-specific cytotoxicity at the concentrations used.

Inhibition of poly(ADP-ribose) polymerase activity also blocks the infection of NIH/3T3, HL-60 or HeLa cells by another recombinant vector, MPSV.sup[28]. This vector, which is derived from the murine myeloproliferative sarcoma virus, also carries the *neo* gene. Inhibition of viral infection was assayed both by counting of *neo*[R] colonies and by Southern hybridization analysis of high molecular weight DNA. The results show that the number of *neo*[R] colonies obtained after infection were markedly reduced by the addition of the inhibitors; but again not by the non-inhibitory analogues (Fig. 2). Analysis of the high molecular weight genomic DNA isolated from NIH/3T3 cells infected either in the presence or absence of 2mM 3-methoxybenzamide demonstrates that

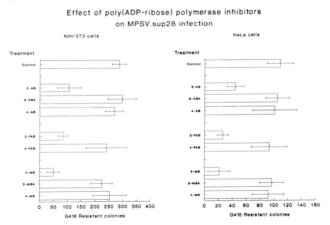

FIG. 2. Effect of poly(ADP-ribose) polymerase inhibitors on MPSV.sup[28] infection of NIH/3T3 and HeLa cells. Cells were infected with the MPSV.sup[28] vector either in the absence or presence of one of the poly(ADP-ribose) polymerase inhibitors or the corresponding non-inhibitory analogue. The infected cells were selected in 1mg/ml G418 and after two weeks the neo[R] colonies were counted. Control indicates cells infected in the absence of inhibitors; 3mM 3-AB (3-aminobenzamide), 3-ABA (3-aminobenzoic acid) or 4-AB (4-aminobenzamide); 1mM 3-FAB (3-formylaminobenzamide) or 4-FAB (4-formylaminobenzamide); or 1.5mM 3-MB (3-methoxybenzamide), 3-MBA (3-methoxybenzoic acid) or 4-MB (4-methoxybenzamide). Results show the mean and standard deviation of three experiments.

FIG. 3. Inhibition of MPSV.sup[28] provirus integration in NIH/3T3 or HL-60 cells. (A) Southern blot analysis of genomic DNA isolated from NIH/3T3 cells infected with MPSV.sup[28] vector, either in the absence (-), or in the presence (+) of 2mM 3-methoxybenzamide. At the indicated times (hours) total genomic DNA was digested with Sstl which cuts in the two LTR's releasing a 4.9kb proviral DNA fragment containing the neo gene. After size fractionation and Southern transferred the blots were hybridized to a 1.6kb Xbal/HindIII fragment of the vector containing 1.2kb of the neo cDNA and 400bp of flanking (gag gene) virus sequences. (B) The filter used in A was stripped and re-probed with a 1.1kb mouse ß-actin cDNA. (C) Dot-blot analysis of genomic DNA isolated from HL-60 cells infected with MPSV.sup[28], in the absence (-), or presence (+) of 1.5 mM 3-methoxybenzamide. The transferred DNA was hybridized first to a 762bp PvuII neo specific cDNA fragment (left), then stripped and reprobed with the 1.1kb mouse ß-actin cDNA (right).

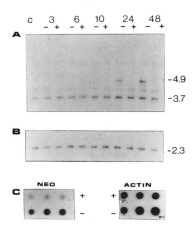

inhibition of poly(ADP-ribose) polymerase activity blocks provirus integration into the host cell genome. This is shown by the presence of a 4.9 kb proviral DNA fragment, which is released by Sst1 digestion, only in cells infected in the absence of 3-methoxybenzamide (Fig. 3A). In these hybridizations a number of other bands, generated by hybridization of the probe to the mouse endogenous retroviral sequences, are also visible and serve as an internal control for digestion, loading, transfer and hybridization (Fig. 3A). Re-probing of this blot with a neo specific probe identifies only the 4.9kb MPSV.sup[28] band (data not shown). The inhibition of proviral DNA integration is clearly visible at 24 and 48 hours after infection, although a clear effect is evident as early as 10 hours (Fig. 3A). Similarly, dot-blot analyses of high molecular weight DNA obtained 72 hours after infection of HL-60 cells shows clearly that 1.5mM 3-methoxybenzamide blocks the integration of proviral DNA into high molecular weight genomic DNA of HL-60 cells (Fig. 3C).

In contrast to the inhibition of provirus integration, poly(ADP-ribose) polymerase inhibitors had no detectable effect on virus/host cell receptor interaction, host cell entry, uncoating of the viral capsid or the reverse transcription and synthesis of the provirus DNA. This was demonstrated by the detection of similar levels of proviral DNA in the low molecular weight DNA fraction isolated by Hirt extraction from the NIH/3T3 cells infected either in the presence or absence of 2mM 3-methoxybenzamide (Fig. 4).

In order to study the possible involvement of poly(ADP-ribose) polymerase in the post-integration transcription of the proviral genome and the generation of infective virions, the effect of enzyme inhibitors on virus production by MPSV.sup[28] producer cells was examined. Poly(ADP-ribose) polymerase inhibitors, or their non-inhibitory

FIG. 4. Effect of 3-methoxybenz-amide on the synthesis of provirus DNA.

Analysis of low molecular weight DNA, isolated by Hirt extraction, from NIH/3T3 cells infected with MPSV.sup[28] vector either in the absence (-), or presence (+) of 2mM 3-methoxybenzamide for the indicated time (hours). The presence of the 5.5kb provirus DNA was assessed by hybridization to a 762bp PvuII fragment of the neo gene carried by the vector.

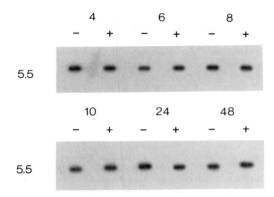

analogues, had little or no effect on the titre of the virus shed by the producer cells which were cultured in the presence of these compounds for three days (Table 1). Poly(ADP-ribose) polymerase inhibitors had no effect on the formation of neo[R] colonies by cells which had already been infected with retrovirus (see Fig. 1B). Therefore, poly(ADP-ribose) polymerase activity is required neither for the post-integration expression of virally encoded genes nor for the transcription, packaging and release of infective retrovirions. This is shown by the observation that inhibitors had no effect on the formation of neo[R] colonies by the already infected cells (see Fig. 1B), and by the absence of an effect on virus production (Table 1).

		Virus titre
Control		$1.6 \pm 0.9 \times 10^5$
3-aminobenzamide	(3mM)	$1.4 \pm 0.8 \times 10^5$
3-aminobenzoic acid	(")	$0.9 \pm 1.4 \times 10^5$
4-aminobenzamide	(")	$1.7 \pm 1.9 \times 10^5$
3-formylaminobenzamide	(1mM)	$0.9 \pm 1.7 \times 10^5$
4-formylaminobenzamide	(")	$1.1 \pm 1.2 \times 10^5$
3-methoxybenzamide	(2mM)	$1.2 \pm 1.2 \times 10^5$
4-methoxybenzamide	(")	$1.1 \pm 0.9 \times 10^5$

Table 1. Effect of poly(ADP-ribose) polymerase inhibitors on retrovirus production by MPSV.sup28 producer cells. MPSV.sup[28] producer cells, were plated at 5 x 10^5/90mm plate in DMEM containing 10% new born calf serum. Approximately 24 hours later the medium was removed and fresh medium containing the indicated concentration of each compound was added. Three days later the titre of the virus shed into the culture supernatant was determined by infection of NIH/3T3 cells and determination of the number of G418 resistant colonies.

DISCUSSION

The data presented here demonstrate that inhibitors of nuclear poly(ADP-Ribose) polymerase block recombinant retroviral infection. This inhibition appears to be mediated through a process, or processes, which intervene between the synthesis of the proviral DNA and its integration into the host cell chromosomes. The enzyme inhibitors do not seem to affect the necessary interactions between the virus envelope and host cell receptors, virus entry into the cell, uncoating of the viral RNA genome or its reverse transcription and synthesis of the proviral DNA. The inhibitors also appear not to affect the post-integration expression of the virally encoded genes, transcription of the viral genome, or the packaging and release of infective virions by the producer cells. The inhibition is therefore specific to processes required for the efficient integration of proviral DNA into the host cell genome.

Poly(ADP-ribose) polymerase enzyme assays in cell-free systems, and direct measurement of the enzyme product, poly(ADP-ribose), in intact cells, has demonstrated the dose dependent inhibition of this reaction by 3-aminobenzamide and a number of its related derivatives. However, the acid analogues, or 4-substituted derivatives of these compounds, are much less inhibitory (the Ki for the latter compounds is at least 100 fold higher than it is for the inhibitors). This dose-dependency and the rank order of inhibition is reflected in the inhibition of recombinant retroviral infection, which suggests that an ADP-ribosylation reaction is required at some stage of the viral infection. Recent studies with purified and recombinant IN protein have demonstrated the presence of both endonuclease and DNA ligase activities in this protein (3, 5). These studies have also shown that IN is both necessary and sufficient for the integration of retrovirus, or retrovirus-like DNA molecules, into target DNA (3, 5). These studies have shown that the efficiency of IN-mediated provirus integration is greatly enhanced by the presence of other cellular factors. On the basis of the present data the nuclear enzyme poly(ADP-ribose) polymerase, appears to be a strong candidate for such a cellular factor. This is consistent with the known characteristics of this enzyme, namely its dependence on DNA strand-breaks and its requirement in a number of eukaryotic cellular processes involving DNA ligation (8).

Whatever the mode of action of the 3-substituted benzamides, it is clear that they have the property of inhibiting recombinant retroviral infection. The compounds used in this study should be regarded as first generation "lead" compounds in this respect; it is probable that a systematic structure-activity study would produce much more effective compounds. Therefore, although the effect of these compounds on replication-competent retroviruses is yet to be examined, the 3-substituted benzamides and perhaps, inhibitors of nuclear poly(ADP-Ribose) polymerase activity in general, do represent a novel class of potential anti-viral compounds, at least for retrovirus infection. These inhibitors may, in practice, be best used in combination with other inhibitors of retroviral infections. Given that the inhibitors do not affect recombinant virus production by the producer cells, they are likely to be most effective in blocking retroviral infection, rather than elimination of an already established infection.

ACKNOWLEDGEMENTS

We are grateful to Dr Barbara Skene (NIMR, Mill Hill, London) for useful discussions, and to Dr Carol Stocking (University of Hamburg) for kindly providing the retroviral vector MPSV.sup28; also to Nigel Gorman and Sean Wells for technical assistance. This study was supported by the Cancer Research Campaign, the Leukaemia Research Fund, the Medical Research Council and the King's College Academic Strategy Fund.

REFERENCES:

1. Capon, D.J., S.M. Chamow, J. Mordenti, S.A. Marsters, T. Gregory, H. Mitsuya, R.A. Byrn, C. Lucas, F.M. Wurm, J.E. Groopman, S. Broder, and D.H. Smith. 1989. Designing CD4 immunoadhesins for AIDS therapy. Nature (London) 337: 525-531.

2. Clapham, P.R., J.N. Weber, D. Whitby, K. McIntosh, A.G. Dalgleish, P.J. Maddon, K.C. Deen, R.W. Sweet, and R.A. Weiss. 1989. Soluble CD4 blocks the infectivity of diverse strains of HIV and SIV for T Cells and monocytes but not for brain and muscle cells. Nature (London) 337: 368-370.

3. Craigie, R., T. Fujiwara, and F. Bushman. 1990. The IN protein of Moloney murine leukaemia virus processes the viral DNA ends and accomplishes their integration in vitro. Cell 62: 829-837.

4. Grandgenett, D.P., and S.R. Mumm. 1990. Unravelling retrovirus integration. Cell 60: 3-4.

5. Katz, R., G. Merkel, J. Kulkosky, J. Leis, and A.M. Skalka. 1990. The avian retroviral IN protein is both necessary and sufficient for integrative recombination in vitro. Cell 63: 87-95.

6. Mitsuya, H., and S. Broder. 1987. Strategies for antiviral therapy in AIDS. Nature (London). 325: 773-778.

7. Sattentau, Q.J., and R.A. Weiss. 1988. The CD4 antigen: physiological ligand and HIV receptor. Cell 52: 631-633.

8. Shall, S. 1989. ADP-ribosylation of proteins: a ubiquitous cellular control mechanism. Biochem. Soc. Trans. 17: 317-322.

9. Skalka, A.M. 1988. Integrative recombination of retroviral DNA. p.701-724. In R. Kucherlapati and G.R. Smith (eds), Genetic recombination. American Society for Microbiology, Washington, D.C.

10. Varmus, H., and P. Brown. 1989. Retroviruses. p53-108. In D.E. Berg and M.M. Howe (eds), Mobile DNA. American Society for Microbiology, Washington, D.C.

Detection and Analysis of NAD Binding Proteins Including Poly(ADP-Ribose) Polymerase Immobilized on The Membrane

Masanao Miwa, Masako Uchida, Tadakazu Akiyama, and Kazuhiko Uchida

Introduction

NAD is a coenzyme required for the oxidation-reduction reaction in respiratory chain as well as the substrate for cellular ADP-ribosylation reactions (1-3). Pyridine nucleotide metabolism plays an important role in regulating cell growth through ATP synthesis. Regulating mechanisms of cellular NAD levels have not been clarified. The depletion of cellular NAD levels and activation of poly(ADP-ribose) polymerase after the treatment of DNA damaging agents in vitro and in vivo is well known (4,5). Many data suggest that poly(ADP-ribosyl)ation of nuclear proteins play an important role in DNA repair processes. Mono ADP-ribosyl transferase in cytoplasm might be involved in regulation of activation of G regulatory proteins (6). Thus various proteins within nuclei and cytoplasm should interact with NAD and regulate a wide variety of cellular functions.

Recently, the catalytic site for poly(ADP-ribose) polymerase was determined by the "activity blot"(7). Poly(ADP-ribose) polymerase purified from tissue or expressed in E. coli was detected on the nitrocellulose membrane with its enzyme activity. As yet no blotting procedure for detecting various NAD binding proteins has been reported. And there is no available method for detection of poly(ADP-ribose) polymerase from crude cell extract by protein blotting method when specific antibody is not available.

In this article, we developed a protein blotting method to detect the NAD binding proteins on the membrane with defined molecular weight. This method could detect not only poly(ADP-ribose) polymerase but also other NAD binding proteins with molecular weights of 77 kDa and 62 kDa, which showed different characters from poly(ADP-ribose) polymerase.

Results and Discussion

1. Detection of NAD binding proteins with crude extract from Jurkat cell on the nitrocellulose membrane

The procedure is summarized in Fig. 1. Proteins in the extract from Jurkat cells, a human T-cell line, were separated on a SDS-polyacrylamide gel electrophoresis. The proteins on the gel were electrotransferred onto a nitrocellulose membrane. The membrane was incubated with [32P]NAD and subjected to autoradiography. As the result, a main protein band of 113 kDa (p113) and two minor bands of 77 kDa (p77) and 62 kDa (p62), were revealed by NAD binding assay. The band of p77 had so small amout of radioactivity that p113 and p62 were further analyzed.

Jurkat cells

↓

Suspended with sample buffer

↓

Sonication

↓

Centrifugation 13,000g, 5 min.

↓

Supernatant

↓

SDS/7.5% Polyacrylamide gel
electrophoresis

↓

Electroblotting onto
nitrocellulose membrane

↓

Incubation with [^{32}P]NAD

↓

Wash

↓

Autoradiography

Fig.1 The Method for detection of cellular NAD binding proteins from crude cell extract.

2. Characterization of p113 and p62

The p113 was suggested to be poly(ADP-ribose) polymerase from its molecular weight. To prove this, Jurkat cell extract was adsorbed with anti-human poly(ADP-ribose) polymerase antiserum or preimmune serum, and then the extract was subjected to NAD blot assay. The p113, was adsorbed with anti-human poly(ADP-ribose) polymerase antiserum but p62 and p77 were not. Preimmune serum had no effect. This result indicates that p113 is poly(ADP-ribose) polymerase bound to [32P]NAD and p62 and p77 are not degradation products of poly(ADP-ribose) polymerse.

Next, we have determined the subcellular localization of these NAD binding proteins. NAD blot was performed using the nuclear fraction and the cytoplasm-membrane fraction. The p113 was mainly detected in the nuclear fraction and very faintly in the cytoplasm-membrane fraction. The localization of p113 is consistent with that of poly(ADP-ribose) polymerase.

3. Effect of NAD Analogues on NAD Binding to p113 and p62.

To know the specificity of NAD binding to p113 and p62, NAD analogues were included in the reaction mixture and analysed with NAD blot. As shown in Fig.2, the band of p113 but not that of p62 disappeared, when 1 mM 3-aminobenzamide (3-AB) was included in the reaction mixture. 5'-ADP, ADP-ribose at 1 mM and sodium pyrophosphate at 50 mM did not change the intensity of p113, but reduced or diminished the intensity of p62.

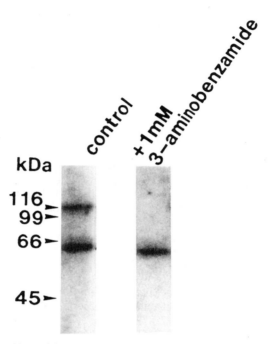

Fig.2   Effect of 3-aminobenzamide on NAD binding.
The molecular weight standards; b-galactosidase(116kDa),
phosphorylase b (99 kDa), bovine serum albumin (66 kDa), and
ovalbumin (45 kDa).

4. Enzyme activity of NAD binding proteins immobilized on the nitrocellulose membrane.

To characterize the nature of interaction of [32P]NAD on the membrane, the radioactive product was analysed. The membrane was cut and the radioactive product was eluted from the membrane by incubation in 0.1 N NaOH at 55∞C for 113 kDa band. Snake venom phosphodiesterase digestion and thin layer chromatography revealed that the product at 113 kDa was identified as poly(ADP-ribose). This data clearly showed that enzymatic reaction of p113 took place on the membrane.

Adsorption with anti-human poly(ADP-ribose) polymerase, inhibition of NAD binding by 3-AB, and product analysis of radioactive product demonstrated that p113 immobilized on the membrane was poly(ADP-ribose) polymerase, which retained its enzymatic activity. The nature of p62 is under further investigation.

The present work demonstrated that NAD binding protien could be identified and poly(ADP-ribose) polymerase was detected from crude cell extract without usage of specific antibody by this NAD blot assay. Furthermore, this method would be useful to identify thus far uncharacterized NAD binding proteins from crude cell extracts.

Acknowledgment: This work was supported in part by Grants-in-aid for Cancer Research and for the University of Tsukuba Project Research from the Ministry of Education, Science and Culuture, and for Cancer Research from the Sagawa Foundation, Japan.

References

1. Olivera, B. M.; Ferro, A. M. Pyridine nucleotide Metabolism and ADP-ribosylation. In: Hayaishi, O.; Ueda, K., eds. ADP-ribosylation Reactions Biology and Medicine. New York: Academic Press.; 1982: p19-38.
2. Miwa, M.; Sugimura, T. ADP-ribosylation and carcinogenesis. In: Moss, J.; Vaughan, M., eds. ADP-ribosylating toxins and G proteins: insights into signal transduction. Washington,DC: American Society for Microbiology.; 1990: p. 543-560.
3. Althaus, F.R. Poly-ADP-ribosylation reactions. In: Althaus, F. R.; Richter, C., eds. ADP-ribosylation of proteins; enzymology and biological significance. Mol. Biol. Biochem. Biophys. Berlin, Springer-Verlag K.G.; 1987: 1-126
4. Whish, W. J.; Davies, M. I.; Shall, S. Stimulation of poly(ADP-ribose) polymerase activity by the anti-tumour antibiotic, streptozotocin. Biochem. Biophys. Res. Commun. 65: 722-730; 1975.
5. Uchida, K.; Takahashi, S.; Fujiwara, K.; Ueda, K.; Nakae, D.; Emi, Y.; Tsutsumi, M.; Shiraiwa, K.; Ohnishi, T.; Konishi, Y. Preventive effect of 3-aminobenzamide on the reduction of NAD levels in rat liver following administration of diethylnitrosamine. Jpn. J. Cancer Res. 79: 1094-1100; 1988.
6. Tanuma, S.; Kawashima, K.; Endo, H. Eukaryotic mono(ADP-ribosyl)transferase that ADP-ribosylates GTP-binding regulatory Gi protein. J. Biol. Chem. 263: 5485-5489; 1988.
7. Simonin, F.; Menissier-de Murcia, J.; Poch, O.; Muller, S.; Gradwohl, G.; Molinete, M.; Penning, C.; Keith, G.; De Murcia. G. Expression and site-directed mutagenesis of the catalytic domain of human poly(ADP-ribose)polymerase in Escherichia coli; Lysine 893 is critical for activity. J. Biol. Chem. 265: 19249-19256; 1990.

AN *IN VITRO* REPLICATION SYSTEM FOR AUTONOMOUSLY REPLICATING MAMMALIAN ORIGIN-ENRICHED SEQUENCES

G. B. Price, C. E. Pearson, and M. Zannis-Hadjopoulos
McGill Cancer Centre, McGill University, 3655 Drummond Street, Montréal, Québec H3G 1Y6, Canada.

ABSTRACT

An *in vitro* system for mammalian origin-rich sequences(*ors*) has been developed. The system uses HeLa cell extracts and is dependent upon the presence of an *ors* template. *In vitro* replication has been shown to initiate in the *ors*, is semi-conservative, bidirectional, aphidicolin-sensitive, and independent of SV40 T-antigen. This *in vitro* system affords an opportunity to study the essential factors necessary for initiation of replication from mammalian ors. One structural feature which we have studied as potential contributor to the regulation of DNA replication is an extruded inverted repeat yielding a cruciform structure. Monoclonal antibodies specific for cruciform DNA, reacting with the "elbow" at the base of the stem-loop of the cruciform, have been prepared and tested in permeablized nuclei and in this *in vitro* system of mammalian DNA replication. Addition of anti-cruciform antibody to the *in vitro* reaction mixture results in 1 to 2 orders of magnitude enhancement of overall *in vitro* DNA synthesis.

In the following discussion we will describe the essential methodology for establishment of an *in vitro* replication system for mammalian ors and summarize some of our experience with the the characterization of the system.

MATERIALS AND METHODS

Preparation of Nuclear and Cytoplasmic Extracts

The essential components for *in vitro* replication are contained in nuclear and cytoplasmic extracts of HeLa cells(Pearson et al., 1991; Zannis-Hadjopoulos et al., 1991). We initially used log phase HeLa cell monolayers to prepare extracts, as described previously by Decker et al.(1986). Subsequently, we found that extracts prepared in a similar fashion from suspension cultures of HeLa S3 cells were superior in apparent concentration of the essential components necessary for *in vitro* replication activity. All steps in the preparation of the extracts were performed at 4°C with cells at mid-log phase (4-5 x 10^5 cells/ml). The difference in the preparation of suspension cell extracts from that of the monolayer cells was principally in the washing of the cells. The HeLa S3 cells were resuspended in the isotonic buffer described by Decker et al. (1986), and then washed and resuspended at 7x 10^7 cells/ml in hypotonic buffer (20 mM HEPES, pH 7.8, 5mM potassium acetate, 0.5 mM magnesium chloride, and 0.5 mM dithiothreitol). Nuclei were prepared by Dounce homogenization and centrifugation. The cytosol was spun at 100,000 x g for 1 hr., and the resultant supernatant (cytoplasmic extract) was aliquoted and stored at -70°C. Nuclear pellets were resuspended in 2.5 X volume of the pellet with hypotonic buffer plus 500 mM potassium acetate; the pellet was extracted for a period of 90 min. on ice with occasional vortex mixing. The nuclear extract was then centrifuged at 300,000 x g for 1 hour, and the supernatant (final nuclear extract) was aliquoted and stored at -70°C.

The *In Vitro* Reaction Mixture

Typically, the reaction volume is 50 μl. The mixture consists of 15 μl cytoplasmic extract, 8 μl nuclear extract and 250 ng of template DNA. Reactions also contained as a final concentration 45 mM HEPES, pH 7.8, 5 mM magnesium chloride, 0.4 mM dithiothreitol, 1 mM EGTA, 60 mM sucrose, 240 mM ethylene glycol, 5% polyethylene glycol(M_r12,000), 6 mM phosphoenolpyruvate, 0.3 U pyruvate kinase, 2mM ATP, 100 mM each CTP, GTP, UTP, dATP, dGTP, and dTTP, 10 μM dCTP and approximately 10 μCi of [α-^{32}P]dCTP(3000Ci/mM). Depending upon plasmid DNA concentrations, the amount available for addition of other test substances which might modify the *in vitro* DNA replication was up to 5 μl. In experiments to test the anti-cruciform antibody effect, 1-4 μl (15-60 ng) of 2D3 anti-cruciform antibody (Frappier et al., 1987; Frappier et al., 1989) was added to the mixture after the DNA and before the extracts. In the case of addition of a putative modifying agent, such as for example 2D3, the mixture is left on ice for 15-30 min. prior to incubation at 30°C for up to 60 min.   Reactions were terminated by the addition of one volume of 1% SDS, 30 mM EDTA.   The reaction mixture is subsequently digested with proteinase K, extracted with phenol and ether to remove protein, and precipitated with ethanol in the presence of ammonium acetate three times to remove free radioactivity. The final pellet is dried and then resupended in water and loaded on agarose-gel for electrophoretic separation of the replication products. The dried gel is then analyzed by autoradiography.

RESULTS

Some of the results have been previously described by Pearson et al., 1991, and by Zannis-Hadjopoulos et al., 1991. Firstly, the DNA provided as template must contain a mammalian sequence with an active origin of replication (*ors*), detected as controlling the initiation of replication of *ors* plasmid (Frappier and Zannis-Hadjopoulos, 1987; Landry and Zannis-Hadjopoulos, 1991); bacterial plasmids (vectors) alone and randomly cloned sequences of mammalian DNA in these vectors fail to undergo replication in this *in vitro* system (Pearson et al., 1991; Zannis-Hadjopoulos et al., 1991).

Briefly, the products of the *in vitro* reaction include forms I, II, and III as well as replicative intermediates (migrating slowly above form II). Topoisomeric molecules are most readily observed in those reaction mixtures which contain both cytoplasmic and nuclear extracts. Cytoplasmic extracts alone provide a certain minimal number of factors for a modest level of *in vitro* replication of mammalian *ors*, but do not generally contain topoisomerase activity.

We had previously shown that the anti-cruciform antibody could enhance DNA synthesis in permeablized cells by 2- to 11-fold (Zannis-Hadjopoulos et al., 1988). The addition of the anti-cruciform antibody to the *in vitro* replication system achieved similar results; the *in vitro* replication of a series of *ors*-containing bacterial plasmids was enhanced in the presence of 2D3 monoclonal antibody, relative to no antibody or control antibody P3, from 1 to 2 orders of magnitude (total of all replication products). The magnitude of the enhancing effect of the antibody was proportional to the amount of basal *in vitro* replication products that were obtained in the absence of anti-cruciform antibodies.

The SV40 *in vitro* replication system by comparison to the *in vitro* replication system described herein is estimated to produce from 2 to 3 orders of magnitude more DNA products than those observed when mammalian DNA templates containing origins of replication are used (Berberich et al., 1991; Zannis-Hadjopoulos et al., 1991). The SV40 system, however, requires the addition of purified T antigen, the viral initiator protein, for the optimal initiation and replication of SV40 DNA (Decker et al., 1986; Stillman and Gluzman, 1985). The inclusion of anti-cruciform antibody in our *in vitro* system augments the replication level of some *ors* to near the lower limits of production of the SV40 system. Thus, the anti-cruciform antibody would seem to both recognize a cruciform-like structure associated with active origins of DNA replication and to stabilize it, providing a sustained, essential replication signal that results in an overall increase in number of initiations per template and in resultant replication products (forms I, II, and III and intermediates).

Recently, we have used the anti-cruciform antibody to affinity-purify genomic sequences from mammalian DNA, and shown that at least 4 of 9 such sequences have autonomous replicating activity (Bell et al., 1991). These observations concur with our viewpoint that origins of DNA replication contain inverted repeats (Landry and Zannis-Hadjopoulos, 1991; Rao et al., 1990; Zannis-Hadjopoulos et al., 1984), which may be capable of transient extrusion into cruciform structures to serve as recognition signals for initiation of mammalian DNA replication.

ACKNOWLEDGEMENTS

This work was supported by grants from the Medical Research Council of Canada (MA-7965) and the Natural Sciences and Engineering Research Council of Canada.

REFERENCES

Bell D, Sabloff M, Zannis-Hadjopoulos M, Price G (1991) Anti-cruciform DNA affinity purification of active mammalian origins of replication. Biochim. Biophys. Acta in press

Berberich S, Daniel D, Trivedi A, Johnson E, Leffak M (1991) In: Regulatory Mechanisms of DNA replication, Human Frontier Science Program Workshop, Les Arcs, France, 1991. Abstract.

Decker RS, Yamaguchi M, Possenti R, DePamphilis ML (1986) Initiation of simian virus 40 DNA replication *in vitro* : aphidicolin causes accumulation of early-replicating intermediates and allows determination of the initial direction of DNA synthesis. Mol. Cell. Biol. 6:3815-3825

Frappier L, Price GB, Martin RG, Zannis-Hadjopoulos M (1987) Monoclonal antibodies to cruciform DNA structures. J. Mol. Biol. 193:751-758

Frappier L, Price GB, Martin RG, Zannis-Hadjopoulos M (1989) Characterization of the binding specificity of two anti-cruciform DNA monoclonal antibodies. J. Biol. Chem. 264:334-341

Frappier L, Zannis-Hadjopoulos M (1987) Autonomous replication of plasmids bearing monkey DNA origin-enriched sequences. Proc. Natl. Acad. Sci. USA 84:6668-6672

Landry S, Zannis-Hadjopoulos M (1991) Classes of autonomously replicating sequences are found among early-replicating monkey DNA. Biochim. Biophys. Acta 1088:234-244

Pearson CE, Frappier L, Zannis-Hadjopoulos M (1991) Plasmids bearing mammalian DNA-replication origin enriched (*ors*) fragments initiate semiconservative replication in a cell-free system. Biochim. Biophys. Acta in press

Rao BS, Zannis-Hadjopoulos M, Price GB, Reitman M, Martin RG (1990) Sequence similarities among monkey ori-enriched (*ors*) fragments. Gene 87:233-242

Stillman BW, Gluzman Y (1985) Replication and supercoiling of simian virus 40 DNA in cell extracts from human cells. Mol. Cell. Biol. 5:2051-2060

Zannis-Hadjopoulos M, Kaufman G, Martin RG (1984) Mammalian DNA enriched for replication origins is enriched for snap-back sequences. J. Mol. Biol. 179:577-586

Zannis-Hadjopoulos M, Frappier L, Khoury M, Price GB (1988) Effect of anti-cruciform DNA monoclonal antibodies on DNA replication. The EMBO J. 7:1837-1844.

Zannis-Hadjopoulos M, Pearson CE, Bell D, Mah D, McAlear M, Price GB (1991) Structural and functional characteristics of autonomously replicating mammalian origin-enriched sequences (*ors*). In: Control mechanisms of DNA synthesis, Human Frontier Science Program Workshop, Les Arcs, France, 1991. Springer-Verlag, New York, in press

Variation in Poly(ADP-ribose) Polymerase Activity and 2', 5'-Oligoadenylates Core Concentration in Estrogen-Stimulated Uterus and Liver of Immature and Adult Rats

H. Suzuki,D. Tornese Buonamassa,L. Cicatielloand A. Weisz

Introduction

Since the finding that some 2', 5'-oligoadenylates ($ppp(A2'p)_nA$, $1 \leq n \leq 6$, referred to as $ppp2',5'A_n$) and their core ($2',5'A_n$) inhibit poly(ADP-ribose) polymerase (pADPRP) activity in vitro (1), it has been postulated that these substances could act as endogenous inhibitors of pADPRP, modulating the activity of the enzyme in response to different stimuli (1-3). Several results suggest that pADPRP activity and the concentration of $ppp2',5'A_n$ and $2',5'A_n$ can be regulated by steroid hormones (4-10). Estrogen induced an enhancement of pADPRP activity (3-5), while at the same time a decrease of the concentration of $ppp2',5'A_n$ and/or $2',5'A_n$ occurred (3, 6, 8). On the other hand, glucocorticoids were found to induce the opposite effect in various cells and organs (7, 9, 10).

To examine further the relationship between poly(ADP-ribose) polymerase activity and the 2',5'-oligoadenylate system, we analyzed the effects of estrogen on of pADPRP activity and $2',5'A_n$ content in the uterus and liver of immature and adult, castrated rats.

Results and Discussion

I.M. injection of 0.15 ml sesame oil containing a long acting estrogen (estradiol valerate, 10 µg/100 g body weight) in immature female Sprague Dawley rats resulted in the gradual increase of pADPRP activity in the uterus and liver, which started 6 h after the injection and went on for 4 days (Fig. 1). Instead, the content of $2',5'A_n$ (n=2-4) decreased by about 50 % within 6 h, reaching 10-20 % of its starting value 4 days after the treatment (Fig. 1). In hormone-treated, adult castrated rats pADPRP activity also increased in the uterus and liver but this event was preceded by a rapid and transient decrease of enzyme activity occurring about 2 h after injection (Fig. 2). We postulated that glucocorticoid hormones, released during the stress reaction consequent to the manipulation of the animals, could be responsible for this sudden drop in pADPRP activity, since it has been shown that these hormones affect negatively the pADPRP activity (9, 10). We thus performed the same experiment using ovariectomized and adrenalectomized adult rats. In this case estrogen injection resulted in the enhancement of pADPRP activity in the uterus and liver without the initial drop of enzyme activity (data not shown). When the concentration of $2',5'A_n$ (n=2-4) was measured in the uterus and liver of estrogen-injected castrated, adult rats, a two-fold increase of the content of $2',5'A_n$ was detected in both organs in coincidence with the drop of pADPRP

Fig.1. Variation of pADPRP activity and of $2',5'A_n$ content in the uterus and liver of immature female rats after estrogen administration. Determination of pADPRP activity and $2',5'A_n$ content was performed as already described in (3) and (12), respectively. (A), pADPRP activity (△) and concentration of $2',5'A_n$ (▲) in the uterus. (B), pADPRP activity (△) and $2',5'A_n$ content (▲) in the liver. Enzyme activity is exressed as pmol [^{14}C]NAD incorporated in trichlo-roacetic acid insoluble fraction / min (units) / mg DNA. The concentration of $2',5'A_n$ is the sum of $2',5'A_2$, $2',5'A_3$ and $2',5'A_4$ and is ex- pressed as pmol / mg DNA. Values are the mean ± S.D. of the determination performed in tissue samples from three animals.

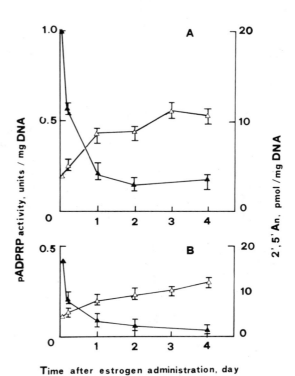

Time after estrogen administration, day

activity (Fig. 2).This initial increase of the concentration of $2',5'A_n$ was followed by a gradual decrease of its concentration, which was inversely correlated to the increase of pADPRP activity (Fig. 2). In ovariectomized and adrenalectomized adult rats a gradual decrease of $2',5'A_n$ content was observed in the uterus and liver without an initial increase of the concentration of $2',5'A_n$ (data not shown). Thus, the response of immature and adult rat uterine and liver cells to estrogen stimulation consisted in an increase of pADPRP activity accompanied by reduction of the cellular concentration of $2',5'A_n$. These data are compatible with previous reports on the decrease of $2',5'A_n$ synthetase activity in the liver of estrogen-treated ovariectomized rats (8) and on the increase of pADPRP activity in endometrial nuclei after estrogen treatment of ovariectomized rats (11). Furthermore, the present data seem to suggest that glucocorticoids counteract the effect of estrogen in uterus and liver of rats. These results are also in line with previous data on the decrease of pADPRP activity induced by glucocorticoid in chick embryo liver (10) and on the enhancement of $2',5'A_n$ synthetase activity in lymphoblastoid cells after glucocorticoids treatment (7).

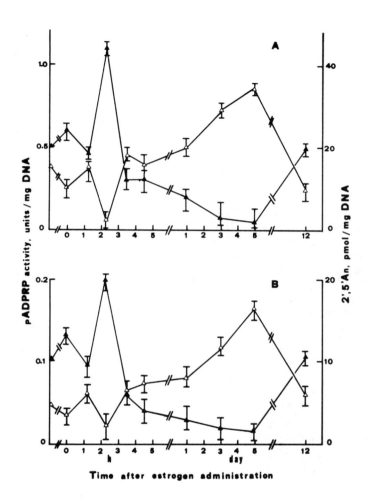

Fig. 2. Variation of pADPRP activity and of 2',5'A_n content in the uterus and liver of ovariectomized rats after estrogen administration. Hormonal treatment was performed ten days after ovariectomy of the animals. Measurement of pADPRP activity and 2',5'A_n content was performed as described in (3) and (12), respectively. (A), pADPRP activity (△) and concentration of 2',5'A_n (▲) in the uterus. (B), pADPRP activity (△) and 2',5'A_n content (▲) in the liver. Enzyme activity and the concentration of 2',5'A_n are expressed as in Fig. 1. Values are the mean ± S.D. of determinations performed in tissue samples from three animals.

In conclusion, the data presented here seem to support the possibility that endogenous pADPRP activity is regulated by cellular $2',5'A_n$, at least in some physiological circumstances, which are controlled by the concerted action of estrogen and glucocorticoids.

Acknowledgements
This work was partly supported by grants of C.N.R's Target Project on Biotechnology and Bioinstrumentation and of the Centro Interuniversitario per lo Studio delle Macromolecole Informazionali (C.I.S.M., Milano, Pavia, Verona).

References
1. Pivazian, A. D., Suzuki, H., Vartanian, A. A., Zhelkovsky, M., Farina, B.,Leone, E., Karpeisky, M. Ya. (1984) Biochem Int 9: 143-152
2. Tornese Buonamassa, D., Malanga, M., Coccia, E. M., Romeo, G.,Fabris, E., Farina, B., Suzuki, H. (1989) FEBS Lett 258: 163-165
3. Suzuki, H., Tornese Buonamassa, D., Weisz, A. (1990) Mol Cell Biochem 99: 33-39
4. Muller, W. E.G., Zahn, R. K. (1976) Mol Cell Biochem 12: 147-159
5. Ciarcia, G., Lancieri, M., Suzuki, H., Manzo, C., Vitale, L., Tornese Buonamassa, D., Botte, V. (1986) Mol Cell Endocrinol 47: 235-241
6. Stark, G. R.,Dower, W. J.,Schimke, R. T., Brown,R. E.,Kerr, I. M. (1979) Nature 278: 471-473
7. Krishnan, I., Baglioni, C. (1980) Proc Natl Acad Sci USA 7: 6506-6510
8. Smekens, M., Dumont, J. E., De Geyter, A., Galand, P. (1986) Biochem Biophys Acta 887: 341-344
9. Tanuma, S. , Johnson, L. D.,Johnson, G. S.(1983)J Biol Chem 258:15371-15375
10. Kitamura, A., Tanigawa, Y., Doi, S., Kawakami, K., Shimoyama, M. (1980) Arch Biochem Biophys 204: 455-463
11. Cummings, A. M. (1989) Endocrinology 124: 1408-1416
12. Suzuki, H., Tornese Buonamassa, D. (1991) Biomed. Chromat. in press.

Possible involvement of poly(ADP-ribose) polymerase in the brain function

Grassi Zucconi, G., Menegazzi, M., Cosi, C., Carcereri De Prati, A. , Miwa, M., Suzuki, H.

Introduction

Poly(ADP-ribose) polymerase (pADPRP) is involved in all those processes which include chromatin rearrangement such as replication and repair of DNA, and gene expression (1). We have previously reported studies on [^{3}H]methyl-thymidine incorporation into the adult rat brain (2) as a measure of DNA synthesis. In the past this DNA synthesis has been ascribed to neurogenesis (3), turnover (4) or repair (5). By protein blotting and immunoprecipitation techniques Mandel et al. (6) first demonstrated the presence of a high pADPRP activity in the brain. More recently high levels of pADPRP mRNA, as compared to several other organs, have been detected in the brain (7). To understand the physiological role of this enzyme in the CNS we have studied its regional and cellular distribution by immunocytochemistry and in situ hybridization.

Results

Hybridization signals are observed in cingulate and piriform cortex, in hippocampal regions CA1, CA3 and dentate gyrus, and in cerebellar cortex. High resolution autoradiography obtained by nuclear emulsion demonstrated that hybridization occurred in a high fraction of neuronal perikarya (Fig. 1). The most intense specific hybridization signal was observed in dentate granule cells and in piramidal cell layers CA1 and CA3. Although labeled neurons were detected diffusely in the cerebral cortex, in parietal and occipital cortices only few cells were labeled with a grain density above the background level. However, in the cingulate, temporal and peripheral cortices a larger number of cells appeared labeled. A lower fraction of cells with neuronal morphology were labeled in thalamus and striatum. Light but specific labeling appears along the ependyma of lateral ventricles and third ventricle. High level of pADPRP mRNA appeared in the large neurons of nuclei of brainstem. Intense labeling was found in Purkinje cells, while the granule cells of cerebellar cortex revealed a weak labeling in scattered cells.

The immunocytochemical localization of pADPRP was performed on sections adjacent to those used for in situ hybridization. The overall staining pattern did not differ qualitatively and quantitatively from that seen for the

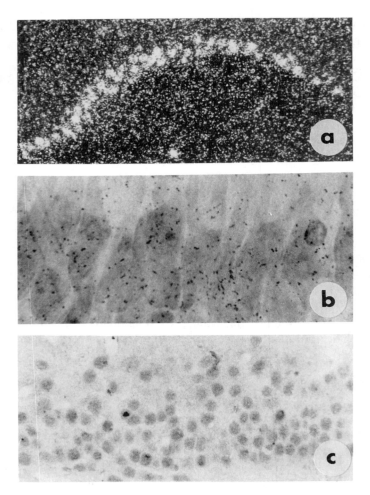

Fig. 1. The photomicrographs show neurons from the brain regions expressing the highest level of pADPRP mRNA. (a) Darkfield photomicrographs of Purkinje cells of cerebellar cortex. All the Purkinje cells of the cerebellum appear labeled, whereas very few granule cells show grains. (b) Brightfield photomicrographs of pyramidal cells of hippocampal CA1. In the hippocampal regions the pADPRP positive neurons were approximately 50 % of whole neuronal population. For $\underline{in}$ $\underline{situ}$ hybridization, ^{35}S cDNA probe ($1x10^{6}$ cpm/slide) in 100 ul standard buffer was applied to sections and incubated for 18-20 h at 50°C. Following washing at high stringency (52°C) the slides were dipped in K5 Ilford emulsion and exposed for 2-3 weeks. Counter staining with hematoxylin. (c) pADPRP stained sections of dentate granule cells of hippocampal formation. The overall staining pattern did not qualitatively differ from that seen for mRNA level.

Fig. 2. RNA blot analysis of mRNA prepared from three different regions. Fifty ug of total RNAs were separated electrophoretically on a 1 % agarose gel containing 6 % formaldehyde, and then blotted onto a membrane. Then total RNA was hybridized with ^{32}P-labeled cDNA probes for rat pADPRP (0.4 kb) and visualized by a X-ray film. The amount of ribosomal RNAs in the agarose gel quantified by ethidium bromide staining is shown in the bottom part of the panel. Positions of ribosomal RNAs (28 S and 18 S) are indicated on the right.

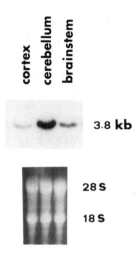

enzyme mRNA. The pADPRP positivity ranges from a maximum (40-50 % of whole neuronal population) in piriform cortex, CA1, CA3 and dentate gyrus of hippocampal formation, to a minimum (10-20 %) in parietal and occipital cortex, thalamus, striatum and granule cells of cerebellum. Intense staining was found on Purkinje cells of cerebellum.

Northern blot analysis using [32]P cDNA probe (Fig.2) showed the regional distribution of the pADPRP mRNA. The comparison of the different areas indicates that cerebellum has the highest expression, while a lower but significative hybridization signal is present in the cortex and brainstem .

Discussion

Our data indicate that pADPRP is expressed by discrete neuronal populations such as pyramidal cells of the hippocampus and piriform cortex, granule cells of the dentate gyrus and Purkinje cells of the cerebellum. The mRNA localization coincides with the distribution of the protein, suggesting that the level of the enzyme can be controlled by its mRNA, despite a uniformly low level of the transcript per cell.

These data demonstrate that a gene associated in all the processes involving chromatin reorganization is also transcribed in the CNS. Several observations suggest that neuronal chromatin is not, or at least not always, in a quiescent state. First, conformational changes involving a decrease of ADP-ribosylation are reported to occur in neuronal chromatin when it passes from proliferative to non proliferative state (8). Second, the brain has transcriptional activity two or three times higher than other organs (9); in addition, several steps of the transcription are associated with the presence of pADPRP. Finally, a recent study (10) reports that a gene involved in site-specific recombination in the immune system (RAG1) is expressed in the brain. These authors suggest the RAG1 might also function to protect the neuronal genome from the deleterious effect of aging. pADPRP is also involved in the recombination processes (11).

In conclusion, the results presented in this work confirm that the pADPRP gene coding for an enzyme involved in conformational changes of the DNA molecule is expressed in the brain, and that this protein is localized in preferential areas. This suggests that the presence of pADPRP could be used to map areas of the brain where functional changes of the DNA molecule are taking place.

We are presently investigating brain functional conditions in which the expression and activity of pADPRP change. This may help elucidate the role of this enzyme in the nervous tissue.

Acknowledgements.

This work was partly supported by grant Target Project on Biotechnology and Bioinstrumentation from Consiglio N.le Ricerche and of the Centro Interuniversitario per lo Studio delle Macromolecole Informazionali (C.I.S.M.I., Milano, Pavia, Verona).

References

1. Ueda, K., Hayaishi, O. (1985) Annu Rev Biochem 54: 73-100
2. Grassi Zucconi G., Menichini, E., Castigli, E., Belia, S., Giuditta, A. (1988) Brain Res 447: 253-261
3. Kaplan, M. S., Bell, D. H. (1983) Exp Brain Res 52: 1-5
4. Perrone Capano, C., D'Onofrio, G., Giuditta, A. (1982) J Neurochem 38: 52-56
5. Vilenchik, M. M., Tretjak, T. M. (1977) J Neurochem 29: 1159
6. Mandel, P., Niedergang, C., Ittel, M. E., Thomassin, H., Masmoudi, A. (1986) Role of RNA and DNA in Brain Function, A. Giuditta, B. B. Kaplan and C. Zomzely-Neurath (eds.), Martinus Nijhoff, Boston pp. 233-246
7. Ogura, T., Takenouchi, M., Yamaguchi, M., Matsukage, A., Sugimura,T., Esumi, H. (1990) Biochem Biophys Res Commun 172: 377-384
8. Das, B.R., Kanungo, M.S. (1986) Biochem Intern 12: 303-311
9. Sutcliffe, J. G. (1988) Annu Rev Neurosci 11: 157-198
10. Chun, J. M., Schatz, D. G., Oettinger, M. A., Jaenisch, R., Baltimore, D. (1991) Cell 64: 189-200
11. Waldman, B.C., Waldman, A.S. (1990) Nucleic Ac Res 18: 5981-8.

Inhibition of interferon-γ-dependent induction of major histocompatibility complex class II antigen by expressing exogenous poly(ADP-ribose) synthetase gene.

Taketoshi Taniguchi, Takashi Tomoda, Ichiro Nomura, Takanobu Kurashige, Hiroshi Yamamoto, and Shigeyoshi Fujimoto.

Introduction

We previously observed down-regulation of poly(ADP-ribose) synthetase gene during the interferon-γ-induced activation process of murine macrophage P388D1 tumor cells. In order to clarify whether or not the down-regulation of the synthetase is a necessary step of the interferon-γ-dependent induction of Ia antigen, we transfected an expression plasmid of the exogenous synthetase gene into the cells and examined the effect on the induction of Ia antigen. We demonstrated that the expression of the exogenous synthetase gene inhibited the interferon-γ-mediated induction of Ia antigen.

Results and Discussion

We isolated a full length poly(ADP-ribose) synthetase cDNA by screening the Okayama-Berg cDNA expression library. The cDNA(pcD-ARS), thus isolated, contained 20 nucleotides upstream from the initial ATG codon. The pcD-ASR plasmid DNA was co-transfected with pSV2-neo gene into P388D1 cells and clones resistant to G418 were selected. The synthetase activity of the the synthetase gene-transfected clones was twice as high as that of the control cells. TABLE I shows the changes in the enzyme activity of the synthetase gene-transfected clones and control clones by the treatment with interferon-γ. The activity decreased to approximately half of the initial value in control clones 2 days after treatment with interferon-γ, whereas the decrease of the activity in synthetase gene-transfected clones, especially I-B5, was less than that of the control clones. The results indicate that the increase in the synthetase activity is due to the expression of the exogenous synthetase gene. The synthetase gene-transfected clones and control cells were cultured in the absence or presence of 100 units/ml interferon-γ for 2 days and cytoplasmic RNAs were analyzed with the cDNA for the synthetase (Figure 1A), and the Ia antigen gene (Figure 1B). The expression of the synthetase was markedly higher in the transfected clones than in P388D1 cells when the cells

TABLE I. Changes in the enzyme activity by the treatment with interferon-γ. The control P388D1 cells, and clones (I-B5, I-D2, I-D3) transfected with simian virus 40 early promoter-regulated synthetase gene were treated with 100 unit/ml interferon-γ for 2 days, and the enzyme activity was examined in the presence of DNase I.

interferon-γ	Relative Activity (%)			
	control	I-B5	I-D2	I-D3
–	100 ± 7	199 ± 16	192 ± 7	185 ± 11
+	47 ± 8	182 ± 15	141 ± 6	141 ± 5

Mean ± SD (n=3) are shown.

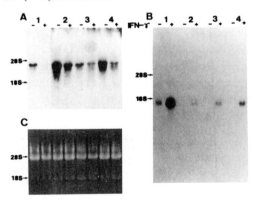

Figure 1. RNA blot analysis of control cells and cells transfected with simian virus 40 early promoter-derived synthetase gene. The RNA was hybridized with the ^{32}P-labelled DNA probe for the synthetase (panel A) and for the Ia antigen (panel B). Lanes 1, 2, 3 and 4 represent the RNA isolated from untransfected P388D1 cells, transfected clones I-B5, I-D2 and I-D3, respectively, cultured in the absence(-) or presence(+) of interferon-γ. The lower panel depicts the ethidium bromide-stained gel showing the relative amounts of 28S and 18S rRNA in the samples.

143

were cultured in the absence of interferon-γ. The synthetase gene continued to be expressed in the transfected clones 2 days after the cells were treated with interferon-γ. Figure 1B shows the mRNA level of the Ia antigen in P388D1 cells and the transfected clones when the cells were cultured for 2 days with or without interferon-γ. The mRNA of Ia antigen was induced clearly in P388D1 cells(Figure 1B, lane 1) and neo gene-transfected control clones(data not shown) by the treatment with interferon-γ, while it was induced much less in the transfected clones(Figure 1B, lane 2,3,4). The levels of mRNA for β-actin, as the control, did not change by treatment with interferon-γ (data not shown). The results indicate that the Ia antigen was induced in P388D1 cells and the neo gene-transfected control clones, but much less in the clones transfected with the synthetase gene.

The simian virus 40 early promoter-regulated synthetase gene, however, was affected by the treatment of interferon-γ, and the expression of the synthetase was not so high in murine macrophage P388D1 cells. Thus, we

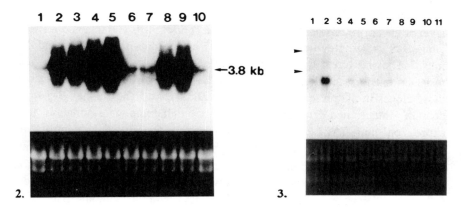

Figure 2. Expression of metallothionein promoter-regulated synthetase gene in transfected clones. Total RNA was isolated from P388D1 control cells (lane 1,10) and transfected clones A-2 (lane 2,3), B-2 (lane 4,5) B-5(lane 6,7) and D-1 (lane 8,9), cultured for 1 day in the absence (lane 1,2,4,6,8,10) or presence (lane 3,5,7,9) of 80 μM ZnSO4 and 2 μM CdCl2, and then hybridized with³²P-labelled cDNA of the synthetase.
Figure 3. Effects of the exogenous synthetase gene expression on interferon-γ-dependent induction of Ia antigen. Total RNA was isolated from P388D1 control cells (lane 1,2), exogenous synthetase gene-transfected clones A-2 (lane 3,4,5), B-2 (lane 6,7,8) and D-1 (lane 9,10,11). Cells were cultured in the absence (lane 1-11 except 5,8,11) or presence (lane 5,8,11) of 80 μM ZnSO4 and 2 μM CdCl2 for 15 h, then treated with (lane 2,4,5,7,8,10,11) or without (lane 1,3,6,9) interferon-γ for 2 days. The RNA was hybridized with ³²P-labelled Ia antigen gene.

TABLE II Activities of poly(ADP-ribose) synthetase in control cells and clones

metal inducer	control	Relative Activity (%)		
		A-2	B-2	D-1
−	100 ± 10	133 ± 23	201± 23	150 ± 27
+	69 ± 10	230 ±23	249 ± 46	217 ± 32

Mean ± SD (n=3) are shown.

constructed an expression plasmid which contained the metallothionein I
promoter-regulated synthetase gene and thymidine kinase promoter-derived
neo gene. The metallothionein promoter-regulated synthetase gene was
transfected into P388D1 cells and permanent transformants were selected by
culturing in G418 selection medium. The level of mRNA for the synthetase in
a transfected clone was analyzed by RNA blot analysis(Figure 2). The mRNA
level of the synthetase did not change in the control cells and in a neo-resistant
clone B-5, when they were cultured for 15 h with or without Cd^{2+} and Zn^{2+}
ions.   The mRNA level of the synthetase, however, increased in neo-
resistant transformant clones, A-2,B-2 and D-1. The mRNA level enhanced in
those clones in the presence of Cd^{2+} and Zn^{2+} ions. The clone B-5 may
contain intact neo-gene expression unit but incomplete synthetase gene
expression unit. TABLE II shows the synthetase activity of the control and
neo-resistant clones 24 h after culturing with or without Cd^{2+} and Zn^{2+} ions.
The activity of transfected clones increased 2 to 3-fold over that of control.
The synthetase gene-transfected clones were cultured for 15 h with or without
Cd^{2+} and Zn^{2+} ions and then cultured for 2 days with or without interferon-
γ. Total RNA of the cells was analyzed by hybridizing with the Ia antigen
gene. The results indicate that Ia antigen was induced clearly in control cells by
treating with interferon-γ but it was induced much less in the synthetase gene-
transfected clones. The induced Ia antigen on the surfaces of control cells and
the transfected clones was analyzed by flow-cytometry, and the Ia antigen
molecule was induced in much less amount on the surface of the transfected
clone B-2. These findings indicate that continuous high expression of the
synthetase gene strongly inhibits the interferon-γ-dependent induction of Ia
antigen. This may suggest that down-regulation of the synthetase gene may be
a step of the signal transduction from interferon-γ to Ia antigen induction.

ADP-ribose Polymers Bind Specifically and Non-covalently to Histones

Phyllis L. Panzeter, Claudio Realini, and Felix R. Althaus

University of Zürich-Tierspital, Institute of Pharmacology and Biochemistry, Winterthurerstrasse 260, CH-8057 Zürich, Switzerland

Poly ADP-ribosylation of nuclear proteins may induce alterations in chromatin structure which affect chromatin function (for reviews see 1, 2). Covalent, post-translational modification of histones by poly(ADP-ribose) has been particularly investigated as the most probable mode of influencing DNA structure (3-6). However, under DNA damage conditions, the enzyme responsible for polymer formation, poly(ADP-ribose) polymerase, is itself the most highly modified protein *in vivo* (7). An apparent paradox of post-translational protein modification arises: the modifying enzyme becomes the predominant, modified protein. If the main function of poly(ADP-ribose) polymerase is to modify itself, what then is the function of modified poly(ADP-ribose) polymerase? We propose that the polymers covalently bound to poly(ADP-ribose) polymerase themselves influence chromatin structure by non-covalently interacting with chromatin proteins, i.e. histones (8). This paper describes the non-covalent interactions between histones and poly(ADP-ribose) *in vitro*.

The Phenol Partitioning Assay for Protein Binding to Poly(ADP-ribose)
As all polynucleotides, ADP-ribose polymers remain in the aqueous phase upon extraction with phenol and/or chloroform. However, we discovered that non-covalent binding of certain proteins to poly(ADP-ribose) causes the polymers to enter the organic phase (Figure 1). This novel and simple method for partitioning protein-polymer complexes from free polymers, together with high-resolution polyacrylamide gels (9), allowed us to: a) determine which proteins bind to polymers; b) quantify polymer:protein binding stoichiometries; and c) show to which subset of a heterogeneous polymer population the protein prefers to bind.

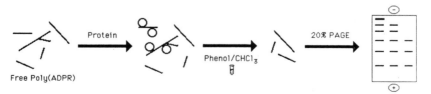

Free Poly(ADPR) — Protein — Phenol/CHCl₃ — 20% PAGE

Figure 1: The Phenol Partitioning Assay for Protein Binding to Poly(ADP-ribose). A given' protein is incubated with free poly(ADP-ribose) and subsequently extracted with phenol/chloroform. Polymers in the aqueous and organic phases can be recovered and analyzed by polyacrylamide gel electrophoresis.

Histones Bind to Poly(ADP-ribose)

The phenol partitioning assay was used to test for binding of various proteins to poly(ADP-ribose). Of 25 proteins tested, only histones and protamine bound to ADP-ribose polymers. Proteins not binding to poly(ADP-ribose) included: bovine serum albumin, cytochrome C, lysozyme, proteinase K, DNase I, nuclease S1, T4 DNA ligase, DNA gyrase, DNA polymerases α, δ, ε, Klenow fragment, RNase H, single-strand DNA binding protein, micrococcal nuclease, Rec A protein, serum amyloid P component, and unmodified poly(ADP-ribose) polymerase. Histone-polymer interactions were resistant to 1 M NaCl, 0.2 M HCl or H_2SO_4, 1 M acetic acid, 4 M urea, 0.25% Triton X-100, and sodium dodecyl sulfate concentrations below 2%, all conditions used to demonstrate putatively covalent modification of histones. DNA effectively released histones from poly(ADP-ribose) at a nucleotide:ADP-ribose ratio of 1.

Polymers of ADP-ribose bound histones specifically but not equally (Figure 2A). One molecule of H1, H2A, H2B, H3, or H4 bound an average of 7.5, 5, 4, 4, or 2.6 ADP-ribose residues, respectively. All histone-polymer interactions were completely saturable, as shown in a titration experiment for H1 (Figure 2B); however, more histones, relative to ADP-ribose residues, were necessary to bind shorter polymers than longer polymers and polymers at the gel origin.

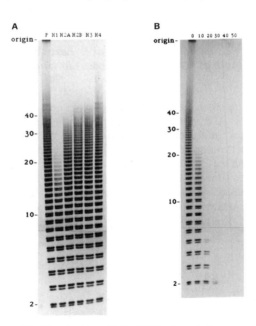

Figure 2: Histone Binding to Free Poly(ADP-ribose). Ten pmoles of the indicated histone, A, or indicated pmoles of H1, B, were added to 147 pmoles of ADP-ribose in the form of polymers. Protein binding to poly(ADP-ribose) was analyzed by the phenol partitioning assay. Lane P, in A, represents polymers to which no protein was added. The lengths of the polymers in terms of ADP-ribose residues are indicated on the left.

Histones Bind Specifically to Branched Poly(ADP-ribose)
We observed that each histone bound ADP-ribose polymers which remain at the gel origin before binding polymers which migrate into the gel (Figure 2). In order to determine whether poly(ADP-ribose) at the gel origin represents a population of polymers with unique structural characteristics or simply polymers of extreme length, HPLC analyses were performed (10) on the separated populations, and the results are summarized in Figure 3. Interestingly, the average size of polymers remaining at the gel origin was only 54.9 residues - a size which should easily migrate into the gel (9). However, this polymer fraction contained the branched polymers, whereas the polymers migrating into the gel were unbranched. We conclude that histones show a preference for binding to branched polymers of ADP-ribose.

Histones Protect Branched Polymers from Glycohydrolase Digestion
To further characterize the affinity of histones for branched polymers, poly(ADP-ribose) was incubated with poly(ADP-ribose) glycohydrolase (11) in the presence and absence of H2B (Figure 4). Free polymers were digested rapidly and completely by 60 minutes (Figure 4, Lanes 6-10). Upon addition of H2B, polymer digestion was severely retarded, and some polymers remained undigested even after 120 minutes (Figure 4, Lanes 1-5). While H2B did not appear to retard glycohydrolase digestion of linear polymers, degradation of branched polymers was markedly inhibited. These results, together with those above, show that histones specifically protect branched polymers from glycohydrolase digestion by binding to them.

		Average Polymer Size	Average Chain Length	Branches per Polymer
	A	20.4 ± 0.7	14.8 ± 0.4	0.36 ± 0.02
	B	12.8 ± 1.6	12.6 ± 0.8	0.00 ± 0.02
	C	54.9 ± 3.3	22.9 ± 1.1	1.33 ± 0.03

Figure 3: Characterization of Poly(ADP-ribose) at the Gel Origin. Enough H1 was incubated with poly(ADP-ribose) to result in the phenol extraction of only those polymers which would remain at the gel origin. The aqueous phase was collected. The polymers in the organic phase were recovered upon extraction with 0.1 mg/ml DNA. Total poly(ADP-ribose), A, poly(ADP-ribose) which enters the gel, B, and poly(ADP-ribose) from the gel origin, C, were subjected to polyacrylamide gel electrophoresis and HPLC analysis.

148

Figure 4: Histone Protection of Poly(ADP-ribose) from Glycohydrolase Digestion. 120 pmoles of free poly(ADP-ribose) were incubated with 0.9 ng of poly(ADP-ribose) glycohydrolase in the absence or presence of 6.5 pmoles of H2B. Incubation times (min) are indicated above each lane. Time 0 represents polymers incubated for 120 min in the presence of heat-inactivated glycohydrolase. Digestions were terminated by heating at 60°C for 5 min. After proteinase K digestion of protein followed by chloroform extraction, polymers were analyzed on a polyacrylamide gel.

In conclusion, it seems likely that not only covalent, but also non-covalent interactions between poly(ADP-ribose) and histones play an important role in altering chromatin structure.

Acknowledgements

This work was supported by Grant 31-31.20391 (to F.R.A.) from the Swiss National Foundation for Scientific Research, financial support from the Krebsliga des Kantons Zürich, and the Jubiläumsstiftung of the University of Zürich.

References

1. Althaus, F.R. and Richter, C. (1987) *ADP-Ribosylation of Proteins: Enzymology and Biological Significance.* Springer Verlag, Berlin.
2. Jacobson, M.K. and Jacobson, E.L. (1989) *ADP-ribose Transfer Reactions: Mechanisms and Biological Significance.* Springer Verlag, Berlin.
3. Jump, D.B., Butt, T.R., and Smulson, M. (1979) *Biochemistry 18,* 983-990.
4. de Murcia, G., Huletsky, A., Lamarre, D., Gaudreau, A., Pouyet, J., Daune, M., and Poirier, G.G. (1986) *J. Biol. Chem. 261,* 7011-7017.
5. Huletsky, A., de Murcia, G., Muller, S., Hengartner, M., Menard, L., Lamarre, D., and Poirier, G.G. (1989) *J. Biol. Chem. 264,* 8878-8886.
6. Boulikas, T. (1990) *J. Biol. Chem. 265,* 14638-14647.
7. Ogata, N., Kawaichi, M., Ueda, K., and Hayaishi, O. (1980) *Biochem. Int. 1,* 229-236.
8. Althaus, F.R., Bachmann, S., Braun, S.A., Collinge, M.A., Höfferer, L., Malanga, M., Panzeter, P.L., Realini, C., Richard, M.C., Waser, S., Zweifel, B. (1991) This volume.
9. Panzeter, P.L. and Althaus, F.R. (1990) *Nucleic Acids Res. 18,* 2194.
10. Alvarez-Gonzalez, R. and Jacobson, M.K. (1987) *Biochem. 26,* 3218-3224.
11. Thomassin, H., Jacobson, M.K., Guay, J., Verreault, A., Aboul-ela, N., Menard, L., and Poirier, G.G. (1990) *Nucleic Acids Res. 18,* 4691-4694.

Poly(3'-deoxyADP-ribosyl)ation of Proteins in Liver Chromatin Isolated from Rats Fed with Hepatocarcinogens

Rafael Alvarez-Gonzalez, Phyllis Panzeter, & David P. Ringer, & and Hilda Mendoza-Alvarez&

Department of Microbiology and Immunology, Texas College of Osteopathic Medicine, The University of North Texas, Fort Worth, Texas, 76107-2690, U.S.A.

We have recently found that 3'-deoxyNAD is a good substrate for poly(ADP-ribose)polymerase (PADPRP) (1,2). In fact, we observed that PADPRP makes small linear oligomers of 3'-deoxyADP-ribose with an average size of 4 ADP-ribose residues. The main advantage of this approach is that the highly branched and complex polymers of ADP-ribose synthesized with NAD (3) are not observed. Therefore, the electrophoretic identification of poly(ADP-ribosyl)ated-polypeptides following incubation of biological samples possessing PADPRP activity with [^{32}P] 3'-deoxyNAD is facilitated.

Stimulation of PADPRP activity by DNA-damaging agents that result in the formation of DNA-strand breaks is very well documented (4). However, identification of the critical protein targets for ADP-ribose polymer modification following exposure of mammalian cells to DNA-damaging agents remains to be accomplished. Here, we have utilized a number of chemicals that are known to induce DNA-damage to rat liver cells *in vivo* (hepatocarcinogens) and subsequently incubated chromatin fractions isolated from these animals with [^{32}P] 3'-deoxyNAD.

Results and Discussion

Exposure of Sprague-Dawley rats to diets containing potentially carcinogenic amounts of several xenobiotics. DNA-damaging agents that elicit DNA repair mechanisms are typically considered "initiators" of cancer or carcinogens. Several xenobiotics behave as hepatocarcinogens. Most chemical hepatocarcinogens are identified by their ability to cause the formation of hyperplastic nodules in the liver after a feeding regimen similar or identical to the Teebor-Becker model (5). Thus, we fed male Sprague-Dawley rats with basal diets containing: a) 0.3% ethionine; b) 0.0067% 3'-methyl cholantrene; c) 0.012% benzidine; d) 0.06% 4'-fluoro, 4'-methylaminoazobenzene; e) 0.06% p-aminoazobenzene; f) 0.06% fluorene; and g) 0.0012% aflatoxin.

& Biomedical Division, The Samuel Roberts Noble Foundation, Inc., Ardmore, Oklahoma, 73402.

150

The Teebor-Becker model typically calls for feeding cycles with a continuous administration of hepatocarcinogen in the diet for three weeks followed by a fourth week of regular diet (5). After 4 or 5 cycles of hepatocarcinogen feeding, surgical removal of the liver usually shows the appearance of multiple hyperplastic nodules which are in turn indicative of tumor formation. Administration of chemical carcinogens to mammalian organisms typically result in their conversion to ultimate carcinogens in liver cells by the cytochrome P450 pathway (6,7). Ultimate carcinogens subsequently induce DNA-damage and elicit DNA-excision repair mechanisms which require protein-poly(ADP-ribosyl)ation (2,8) reactions for higher efficiency. Therefore, this study was initiated in order to test and compare the patterns of protein-poly(ADP-ribosyl)ation in chromatin fractions isolated from rats fed with 5 hepatocarcinogens (ethionine, 3'-methyl-cholantrene, benzidine, 4'-fluoro, 4'-methyl aminoazobenzene and aflatoxin) and 2 non-carcinogenic analogs of hepatocarcinogens (fluorene and p-aminoazobenzene) (9).

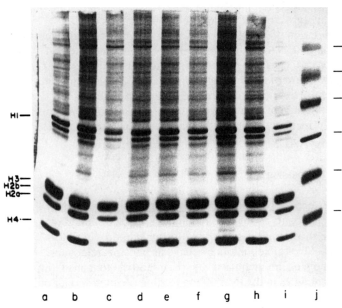

Fig. 1. Protein composition of chromatin fractions isolated from male Sprague-Dawley rats fed for three weeks with either normal diet or normal diet containing xenobiotics. Lanes a; b; c; d; e; f; g; h; and i correspond to chromatin proteins from rats fed with Chow; Basal Diet; Ethionine; 3'-Methyl cholantrene; Benzidine; 4'-Fluoro, 4'-methyl aminoazobenzene; p-Aminoazobenzene; Fluorene; and Aflatoxin, respectively. Lane j corresponds to the molecular weight markers phosphorylase b (97K), bsa (67K), albumin, carbonic anhydrase (31K), soy bean trypsin inhibitor (21K) and lysozyme (14K), respectively.

The protein composition of chromatin isolated from male rats fed the indicated xenobiotics following incubation with 100 μM [^{32}P] 3'-deoxyNAD for 5 min at 25 °C is shown in Fig. 1. The electrophoretic migration of histone proteins is indicated to the left and the molecular weight markers phosphorylase b, bovine serum albumin, ovalbumin, carbonic anhydrase, soybean trypsin inhibitor, and lysozyme are indicated to the right of the Coomassie blue stained gel. The results show no significant changes in the protein composition of chromatin between the control animals fed with either Chow or basal diet and those fed xenobiotics. Lane i showing chromatin isolated from animals fed with aflatoxin contained about half the amount of protein amount of the other samples and thus non-histone proteins are hardly visible.

The autoradiograph of the same gel is shown in Fig. 2. As mentioned above, the electrophoretic identification of the oligo(3'-deoxyADP-ribosyl)ated polypeptides is facilitated under these conditions of modification due to the inability of PADPRP to synthesize highly branched and complex polymers of ADP-ribose with this substrate analog.

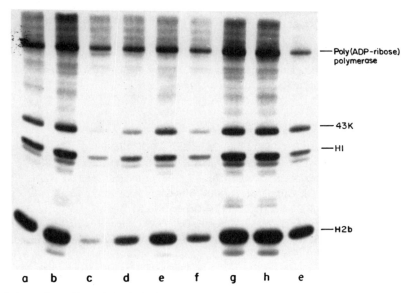

Fig. 2 Autoradiographic analysis of poly(3'-deoxyADP-ribosyl)ated-proteins in chromatin isolated from male Sprague-Dawley rats following incubation with 100 μM [^{32}P] 3'-deoxyNAD for 5 min at 25 °C. Lanes a; b; c; d; e; f; g; h; and i correspond to chromatin proteins from rats fed with Chow; Basal Diet; Ethionine; 3'-Methyl cholantrene; Benzidine; 4'-Fluoro, 4'-methyl aminoazobenzene; p-Aminoazobenzene; Fluorene; and Aflatoxin. The dash lines to the left of the autoradiograph show the electrophoretic migration of the molecular weight markers.

The results of these experiments show that the main acceptors for oligo(3'-deoxyADP-ribosyl)ation were the same in all samples analyzed. Prominent protein acceptors co-migrated with PADPRP, histones H1 and H2b, as well as a 42-43 kD protein. However, the overall levels of oligo(3'-deoxyADP-ribosyl)ation were significantly lower in chromatin isolated from animals fed hepatocarcinogens than those fed with regular diet and non-hepatocarcinogens, e.g., p-aminoazobenzene and fluorene. The lowering of oligo(3'-deoxyADP-ribosyl)ation was particularly more evident in the case of the 43 kD protein and other minor acceptors. Minor acceptors co-migrated with histone H2a and molecular weights between 14 and 17 kD. These may correspond to the non-histone high mobility group (HMG) proteins 14 and 17. Interestingly, it has previously been suggested that these HMG proteins are good acceptors for ADP-ribose polymers in intact cells (10). It has also been shown that their extent of poly(ADP-ribosyl)ation decreases in mouse mammary tumor virus infected cells following treatment with glucocorticoids (11). These polypeptides have previously been associated with actively transcribing genes (12). Thus, our observations are consistent with a higher poly(ADP-ribosyl)ation of minor protein acceptors in male rats fed with hepatocarcinogens since our *in vitro* incubation with [^{32}P] 3'-deoxyNAD appears to result in lower levels of HMG-poly(ADP-ribosyl)ation.

Acknowledgments. This project was supported by a grant from the Samuel Roberts Noble Foundation, Inc. to Rafael Alvarez-Gonzalez.

References

1. Alvarez-Gonzalez, R. (1988) J Biol Chem 263: 17690-17696
2   Pedraza-Reyes, M., & Alvarez-Gonzalez, R. (1990) FEBS Letters 277: 88-92
3. Alvarez-Gonzalez, R., & Jacobson, M., K. (1987) Biochem 26: 3218-3224
4. Benjamin, R. C., & Gill, D. M. (1980) J Biol Chem 255: 10502-10508
5. Teebor, G. W., & Becker, F. F. (1971) Cancer Res 31: 1-3
6. Miller, J. A., Cramer, J. W., & Miller, E. C. (1960) Cancer Res 20, 950-962
7. Thorgeirsson, S. S., Felton, J. S., & Nebert, D. W. (1975) Mol Pharmacol 11: 159-165
8. Durcakz, B. W., Omidiji, O., Gray, D. A., & Shall, S. (1982) Nature 283: 593-596
9. Kizer, D., Clouse, J. A., Ringer, D. P., Hanson-Painton, O., Vaz, A. D., Palakodety, R. B., & Griffin, M. J. (1985) Biochem Pharmacol 34: 1795-1800
10. Tanuma, S.-I., & Johnson, G. (1983) J Biol Chem 258: 4067-4070
11. Tanuma, S.-I., Johnson, L. D., & Johnson, G. S. (1983) J Biol Chem 258: 15371-15375
12. Weisbrod, S., & Weintraub, H. (1979) Proc Natl Acad. Sci (USA) 76: 631-635

ADP-ribose Polymer Metabolism: Implications for Human Nutrition

Elaine L. Jacobson, Viyada Nunbhakdi-Craig, Debra G. Smith, Hai-Ying Chen, Bryan L. Wasson, and Myron K. Jacobson

Research which has led to our current understanding of the relationship between niacin and human health can be segregated into three distinct periods (Figure 1). The first period was concerned with the study of the killer disease, pellagra, and culminated with the discovery by Elvehjem and coworkers of nicotinate and nicotinamide as anti-pellagra factors (1). The second period led to the understanding that nicotinate and nicotinamide were converted within cells to NAD and NADP and that these pyridine nucleotides play a fundamental role in the hydride transfer reactions central to cellular energy metabolism. The third period involves the study of the involvement of NAD as the donor in ADP-ribose transfer reactions. While multiple classes of ADP-ribose transfer reactions occur in cells (2-4), the ADP-ribose metabolism of primary focus in the context described here is the utilization of NAD for the synthesis of polymers of ADP-ribose. As will be described below, the function of ADP-ribose polymer metabolism as a protective response of cells to carcinogen-induced DNA damage leads us to postulate that optimal niacin nutriture may be a preventive factor in carcinogenesis.

A large body of epidemiological evidence has concluded that diet is a major risk factor in cancer (5). Studies on individual components of the diet have identified initiators and promoters that increase cancer risk. However, a variety of evidence has shown that the diet also contains anti-initiation and anti-promotion factors that confer protection from carcinogenesis. While the molecular mechanisms of carcinogenesis are still poorly understood, DNA damage and the manner in which cells respond to DNA damage are key factors (6).

One of the most rapid cellular responses to carcinogen induced DNA damage is the synthesis of ADP-ribose polymers (7). Studies with cultured cells have revealed many features of ADP-ribose polymer metabolism (4). In cells that have not been subjected to injury, ADP-ribose polymer content is low. However, DNA-damaging agents cause a rapid increase in polymer content. Increases of as much as 1000-fold in a few minutes are not uncommon. Indeed, polymer synthesis can exceed the maximal rate of NAD biosynthesis by at least a factor of 10, which can result in depletion of the cellular NAD pool. A compelling body of eidence has led to the conclusion that the common factor of DNA damage that stimulates ADP-ribose polymer synthesis is the occurrence of DNA

Phase 1 (1700's to 1930's):
Niacin and the pellagra puzzle

Phase 2 (1940's to 1960's):
Niacin, NAD, NADP, and energy metabolism

Phase 3 (1970's to present):
NAD and ADP-ribose transfer reactions

Figure 1. Phases of Niacin Research

strand breaks. Another interesting feature of this metabolism is that polymers made in response to DNA damage are present very transiently. Individual polymer residue half-lives range from a few seconds to a few minutes.

Recently, methods to allow characterization of the size distribution of endogenous polymers and their location within subfractions of chromatin have been developed in our laboratory (8). A wide range of polymer sizes, from a few to over 250 residues, has been detected (9). Considering that each polymer residue carries two negative charges, the possible function of a polyanion of this size raises many interesting questions. Chromatin fractionation studies have led to the finding that a major portion of endogenous polymers fractionated with the nuclear matrix (10). This finding has led us to propose a model in which ADP-ribose polymer metabolism effects recovery from DNA damage by mediation of the association of a damaged region of chromatin with the nuclear matrix during repair of DNA damage.

Several studies have shown that inhibitors of ADP-ribose polymer metabolism increase the cytotoxic effects of DNA alkylating agents (11-14) and increase the frequency of malignant transformation by these agents (14-16). Thus, our current knowledge of ADP-ribose polymer metabolism suggests that this system is an integral part of cellular defenses that limit cytotoxicity and malignant transformation.

The basic feature of ADP-ribose polymer metabolism that initially led us to consider the possible nutritional consequences of this system was that the K_m of poly(ADP-ribose) polymerase for NAD is in the same range as the intracellular concentration of NAD (2). Therefore, polymer synthesis should be highly sensitive to cellular NAD content and thus the NAD content of tissues may play an important role in determining the response of target cells to DNA damage. In that vein, it is interesting that individuals with pellagra show extreme sensitivity to sunlight, which likely reflects the inability to mount a normal response to sunlight induced DNA damage.

The possible nutritional consequences of sub-optimal niacin nutrition have been studied in our laboratory using cultured cells. These studies have shown that cells could be readily depleted of NAD by growth in the absence of nicotinate or nicotinamide (17). Further, during nutritional deprivation, a preferential reduction of the NAD pool as compared to the NADP pool occurred. Thus, the mole fraction of the total pyridine nucleotide pool present as NAD (mole % NAD) could be used as a measure of nutritional deprivation. Cells nutritionally depleted of NAD retained many normal properties including normal growth rates and growth control. However, as predicted from the K_m for poly(ADP-ribose) polymerase, the ability to accumulate ADP-ribose polymers in response to carcinogen exposure was greatly reduced as shown in Figure 2. It is particularly noteworthy that only mild depression of NAD content significantly reduced polymer accumulation. Further, Figure 3 shows that nutritionally depleted cells were more sensitive by a factor of 10 to the cytotoxic effects of carcinogenic alkylating agents.

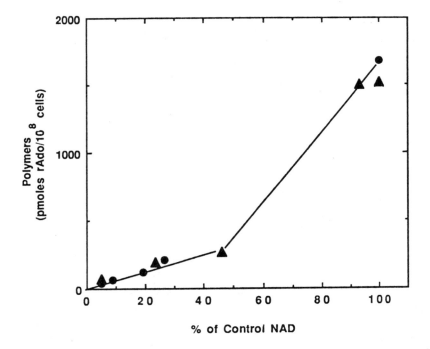

Figure 2. Effect of NAD Deficiency on ADP-Ribose Polymer Accumulation. Cultured C3H10T1/2 cells were deprived of nicotinamide for increasing periods of time to generate cells with varying NAD contents. ADP-ribose polymer synthesis was stimulated by treatment with 34 µM N-methyl-N'-nitro-N-nitrosoguanidine and measured 30 minutes later as described previously (8). Data from two separate experiments are shown.

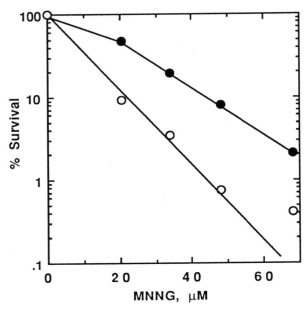

Figure 3. Effect of NAD Deficiency on Survival Following Treatment with a Carcinogenic Alkylating Agent. Cultured C3H10T1/2 cells were grown in nicotinamide free medium until the NAD content was 10% of control. The surviving fraction was measured by colony forming assays following teatment with N-methyl-N'-nitro-N-nitrosoguanidine. Closed symbols, controls; open symbols, NAD depleted cells.

Clinical studies have established the minimum amount of niacin required to prevent the symptoms of pellagra to be approximately 9 milligram equivalents per day. Current recommendations for daily intake, 13 to 18 milligram equivalents per day, have been based on these studies (18-21). In view of the results described above, the question of whether the current niacin nutriture is sufficient to provide optimal tissue levels of NAD for responses to DNA damage raises a possible new horizon for niacin nutriture. Two epidemiological studies support the argument that dietary niacin may modify the action of carcinogens (22, 23), however; we clearly need a better understanding of human niacin metabolism to answer this question. Human niacin nutriture is very complex because niacin equivalents can be obtained from dietary nicotinate, nicotinamide and tryptophan (Figure 4). Thus, nutriture depends upon the amount of these in the diet and upon factors that influence uptake, distribution, and the efficiency of conversion of each to tissue NAD.

In the past, the assessment of niacin nutriture has proven difficult. Until recently, the data available in the literature concerning NAD content of human tissue and

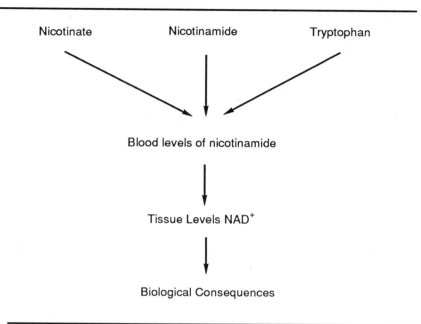

Figure 4. Factors in Human Niacin Nutrition

its relationship to niacin nutriture were few and the early studies were limited by the selectivity and sensitivity of the methodology available. However, a recent study by Fu *et al.* (24) has established clearly that erythrocyte NAD content varies with niacin nutriture within the range of current dietary niacin intake. In this study, human volunteers were switched from a diet containing approximately 19 niacin equivalents per day to a diet containing approximately 10 per day for a period of five weeks. The NAD content of the erythrocyte progressively decreased by 70% while the NADP content decreased only slightly. Thus, the ratio of NAD to NADP progressively declined from 1.8 to 0.5. During restoration of niacin equivalents, both erythrocyte NAD content and the NAD to NADP ratio increased. Fu *et al.* have termed the NAD to NADP ratio "niacin index" and suggested that a "niacin index" below 1.0 may identify subjects at risk for niacin deficiency.

The study of Fu *et al.* has provided the rationale for the development of a biochemical method to assess human niacin status based on the determination of the relative content of NAD and NADP in the erythrocyte. Our laboratory has been studying NAD and NADP content of human cell types and Table 1 shows data for the NAD and NADP content of whole blood and isolated components.

These data demonstrate that over 98% of the NAD and NADP present in whole blood is derived from the erythrocyte fraction. This result suggested to us that whole blood could be used to develop an assay suitable for wide scale screening of NAD and NADP content. Therefore, we developed a procedure for whole blood that allows the determination of the total NAD and NADP content on large numbers of samples at low cost. The general method is described in Figure 5. A key feature of this approach was the adaptation of enzymatic cycling assays (17, 25) to analysis on microtiter plates. Thus, data acquisition can be rapidly achieved using a microtiter plate reader interfaced with a computer. We have expressed the amount of NAD relative to NADP as follows:

$$\text{mole \% NAD} = \frac{\text{NAD}}{\text{NAD} + \text{NADP}} \times 100$$

Table 1.

Nicotinamide Nucleotide Content of Whole Blood and Isolated Fractions

Component	NAD Content		NADP Content		Niacin Number
	pmol/ml	% Total	pmol/ml	% Total	mole % NAD
Whole Blood	34,400	- - -	18,400	- - -	65
Isolated Fractions:					
Erythrocyte	34,400	98	18,400	99	65
Platelet	465	1.3	106	0.5	80
Lymphocyte	218	0.6	62.7	0.3	78
Plasma	< 50	- - -	ND	- - -	- - -
Total	35,100	100	18,600	100	65

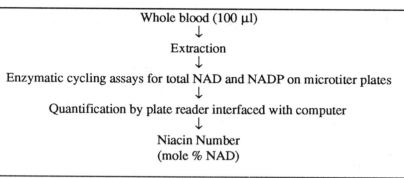

Figure 5. Determination of Niacin Number in Whole Blood

We have termed this value Niacin Number. For example, in the study by Fu *et al.* described above, the Niacin Number declined from 64 to 38 when individuals were transferred from diets containing 19 niacin equivalents per day to 10 per day. We prefer the term Niacin Number over niacin index since it has yet to be proven that this value will be proportional to niacin status in the general population. Table 2 shows the results of repeated analyses of Niacin Number in a single individual. The mean value from seven separate analyses was 64.2 with a percent standard deviation of 2.6. This value was very similar to that calculated for control individuals in the study of Fu *et al.* (24). The immediate goal is to utilize this assay to establish the range of Niacin Number values in a general population. This methodology also provides the opportunity to study the interrelationships outlined in Figure 4 of various dietary precursors of NAD, serum levels of nicotinamide and tissue levels of NAD in humans. The working hypothesis that optimal niacin nutriture is a preventive factor in cancer can be directly tested by determining whether Niacin Number is an intermediate end point in cancer.

Table 2.
Repeated Analyses of Niacin Number in a Single Individual

Analysis	NAD (pmol/ml)	NADP (pmol/ml)	Niacin Number mole % NAD
1	28,120	14,400	66.1
2	31,626	15,948	66.5
3	28,667	15,750	64.5
4	28,280	17,760	61.4
5	31,440	17,920	63.7
6	29,600	16,720	63.9
7	27,680	16,000	63.4
Mean	29,345	16,357	64.2
Std. Dev.	± 1608	± 1227	± 1.7
% Std. Dev.	(5.5%)	(7.5%)	(2.6%)

The authors gratefully acknowledge support for this work by a grant from the National Institutes of Health (CA43894) and a Burroughs Wellcome Research Fellowship to BLW. We also thank Jia-Jen Chen for technical assistance.

References

1. Elvehjem, C. A., Madden, R. J., Strong, F. M., and Woolley, D. W., (1937) *J. Am. Chem. Soc.* 59, 1767.

2. Jacobson, M. K. and Jacobson, E. L., Eds. *ADP-Ribose Transfer Reactions: Mechanisms and Biological Significance.* (1989) Springer-Verlag Publishers, Berlin, Heidelberg, New York.

3. Althaus, F.R. and Richter, C. , Eds. *ADP-Ribosylation of Proteins: Enzymology and Biological Significance.* (1987) Springer-Verlag Publishers, Berlin, Heidelberg, New York.

4. Jacobson, M.K., Aboul-Ela, N., Cervantes-Laurean, D., Loflin, P.T., and Jacobson, E.L. ADP-Ribose Levels in Animal Cells. (1990) *In ADP-ribosylating Toxins and G-Proteins: Insights into Signal Transduction,* J. Moss and M. Vaughan, Eds., American Society for Microbiology, Washington, D.C., pp. 479-492.

5. Doll, R., and Peto, R. The Causes of Cancer,: Quantitative Estimates of Avoidable Risks of Cancer in the United States Today. (1981) *J. Natl. Cancer Institute 66,* 1191-1308.

6. Ames, B.N., and Gold L. S. Dietary Carcinogens, Environmental Pollution, and Cancer: Some Misconceptions. (1990) *Med. Oncol. Tumor Pharmacother. 7,* 69-85.

7. Juarez-Salinas, H., Sims, J.L. and Jacobson, M.K. Poly(ADP-Ribose) Levels in Carcinogen Treated Cells, (1979) *Nature 282,* 740-741.

8. Aboul-Ela, N., Jacobson, E.L. and Jacobson, M.K. Labeling Methods for the Study of Poly- and Mono(ADP-ribose) Metabolism in Cultured Cells. (1988) *Analytical Biochemistry,* 174, 239-250.

9. Alvarez-Gonzalez, R., and Jacobson, M. K. Characterization of polymers of ADP-Ribose Generated *Invitro* and *In vivo.* (1987) *Biochemistry 26,* 3218-3224.

10. Cardenas-Corona, M.E., Jacobson, E.L. and Jacobson, M.K. Endogenous Polymers of ADP-ribose are Associated with the Nuclear Matrix. (1987) *Journal of Biological Chemistry 262,* 14863-14866.

11. Nudka, N., Skidmore, C.J., and Shall, S. The Enhancement of Cytotoxicity of N-methyl-N-nitrosourea and of γ-Radiation by Inhibitors of Poly(ADP-Ribose) Polymerase. (1980) *Eur. J. Biochem., 105,* 525-530.

12. Jacobson, E.L., Smith, J.Y., Mingmuang, M., Meadows, R., Sims, J.L. and Jacobson, M.K. Effect of Nicotinamide Analogues on Recovery from DNA Damage in C3H10T1/2 Cells. (1984) *Cancer Research 44,* 2485-2492.

13. Jacobson, E.L., Smith. J.Y., Wielckens, K., Hilz, H. and Jacobson, M.K. Cellular Recovery of Dividing and Confluent C3H10T1/2 Cells from N-methyl-N'-nitro-N-nitrosoguanidine in the Presence of ADP-Ribosylation Inhibitors. (1985) *Carcinogenesis 6,* 715-718.

14. Jacobson, E.L., Smith, J.Y., Nunbhakdi, V. and Smith, D.G. ADP-Ribosylation Reactions in Biological Responses to DNA Damage. (1985) *In ADP-Ribosylation of Proteins,* F.R. Althaus, H. Hilz and S. Shall, Eds., Springer-Verlag, Berlin, Heidelberg, pp. 277-283.

15.. Lubet, R.A., MacCarvill, J.T., Putman, D.L., Schwartz, J. L. and Schectman, K. M. Effect of 3-Aminobenzamide on the Induction of Toxicity And

Transformation by Ethylmethane-Sulfonate and Methylcholanthrene in Balb/3T3 Cells. (1984) *Carcinogenesis 5*, 459-462.

16. Takahasi, S., Nakae, D., Yokose, Y., Emi, Y., Denda, A., Mikami, T., Ohnishi, T., and Konishi, Y. Enhancement of DEN Initiation of Liver Carcinogenesis by Inhibitors of NAD$^+$ADP-Ribose Transferase in Rats. (1984) *Carcinogenesis 5*, 901-906.

17. Jacobson, E.L., Lange, R.A. and Jacobson, M.K. Pyridine Nucleotide Synthesis in 3T3 Cells. (1979) *Journal of Cellular Physiology 99*, 417-426.

18. Goldsmith, G. A., Sarett, H. P., Register, U. D., and Gibbens, J. Studies of Niacin Requirements in Man I. Experimental Pellagra in Subjects on Corn Diets Low in Niacin and Tryptophan. (1952) *J. Clin. Invest. 31*, 533-542.

19. Goldsmith, G. A., Rosenthal, H. L., Gibbens, J., and Unglaub, G. Studies of Niacin Requirements in Man II. Requirements in Wheat and Corn Diets Low in Tryptophan. (1955) *J. Nutr. 56*, 371-386.

20. Goldsmith, G. A., Gibbens, J., Unglaub, G. and Miller, O. N. Studies of Niacin Requirements in Man III. Comparative Effects of Diets Containing Lime-treated and Untreated Corn in the Production of Experimental Pellagra. (1956) *Am. J. Clin. Nutr. 4*, 151-160.

21. Horwitt, M. K., Harvey, C C., Rothwell, W. S., Cutler, J. L., and Haffron, D. (1956) J. *Nutr. 60 (Suppl. 1)* 43.

22. Prates, M.D., and Torres, F.O. A cancer survey in Lourenco Marques, Portuguese East Africa. (1965) *J. Natl. Cancer Inst., 35*, 729-757.

23. Warwick, G.P. and Harrington, J.S. Some aspects of the epidemology and etiology of esophageal cancer with particular emphasis on the Transkei, South Africa. (1973) *Adv. Cancer Res., 17*, 81-229.

24. Fu, C.S., Swendseid, M.E. Jacob, R.A. and McKee, R. W. Biochemical Markers for Assessment of Niacin Status in Young Men: Levels of Erythrocyte Niacin Coenzymes and Plasma Tryptophan. (1989) *J. Nutr. 119*, 1949-1955.

25. Jacobson, E.L. and Jacobson, M.K. Pyridine Nucleotide Levels as a Function of Growth in Normal and Transformed 3T3 Cells. (1976) *Archives of Biochemistry and Biophysics 175*, 627-634.

Some aspects of nuclear and cytoplasmic ADP-ribosylation. Biological and pharmacological perspectives.

Paul Mandel

Since the discovery of protein phosphorylation by Burnett and Kennedy (1954) protein ADP-ribosylation appears to be the second most important post-translational modification involved in a great variety of cellular processes (see for review Althaus and Richter, 1987; Mandel et al., 1982; Ueda and Hayaishi, 1985). It is noteworthy that the metabolic specificities and biological functions of cells or tissues are often under the control of post-translational modifications and of their effects. Thus it seemed of interest to extend our knowledge on modulation of ADP-ribosylation processes which occur in diverse cell types in biological and pathological conditions. Let us remind that in addition to the nuclear ADPRT (Chambon et al., 1963; Chambon et al., 1966; Doly and Mandel, 1967) mitochondrial (Masmoudi et al., 1988) cytoplasmic (Thomassin et al., 1985) and plasma membrane bound ADP-ribosyl transferases (see for review Althaus and. Richter, 1987) were described. Our interest is focused on three particularly attractive fields : 1) nervous tissue : growth and differentiation; 2) tumoral cell growth and 3) cytoskeleton ADP-ribosylation. In the first part of the present paper we report some aspects of nuclear ADP-ribosylation reaction of two highly differentiated cell types : neurons and astrocytes. The second part concerns nuclear PolyADPR activity and mRNA of mastocytoma P815 tumoral cells. The possibilities offered by manipulation of ADP-ribosylation reaction in antitumoral therapy will also be briefly evoked. In the third part of the paper we shall summarize some aspects of cytoplasmic ADPRT and cytoskeleton ADP-ribosylation which appear of an increasing interest in normal and tumoral cell biology.

1. Nuclear PolyADPR polymerase-transferase activity and gene expression (mRNA) in astrocytes and neurons growing up in cell culture

Nervous tissues provide three major cell types which differ in their structure, metabolism, energy production and pathological alterations : neurons, astrocytes and oligodendrocytes.

The cell culture procedure used provides apparently optimal conditions for cell proliferation and differentiation as well as for growth factors effect analysis. According to the DNA values which provide informations concerning astrocytes proliferation, the number of cells increases until culture day 21 followed by a plateau which corresponds to a period of cell differentiation. It is noteworthy that during the whole period of cell proliferation and of active differentiation that is until about day 30 there is no appreciable change in polyADPRP activity measured in vitro while polyADPRP mRNA undergoes a decrease from day 15 to 25 followed by a short time increase peak and then a plateau (see Chabert et al. in this volume). These observations suggest that striking changes of polyADPRP gene expression may occur without parallel modifications of the measurable enzymatic polyADPRP activity. However, when the effect of bFGF is explored a parallel increase of polyADPRP mRNA and enzymatic activity can be recorded. Thus a growth factor is able to induce a parallel increase of polyADPRP activity and gene expression. The mechanisms involved are under investigation.

Some aspects of ADP-ribosylation were also investigated in primary cultures of rat neurons from cerebral hemispheres of 14 days old rat embryos maintained in the Sato medium. At this period of embryogenesis the rate of cell proliferation is rather low as demonstrated by brain DNA determination. However, a striking neuronal differentiation do occur in these cells in culture (Sensenbrenner et al., 1972).

In our experiments (Chabert et al. in preparation) after 3 days in cell culture brain neurons of 14 days old rat embryos reach an ADPRT activity ($14,4 \pm 2,0$ nmol ADPR^{3}H incorporated/mg DNA) similar to that of 15 days cultured astrocytes derived from new born brains. Since the proliferation rate of neurons at the period of time investigated is rather low compared to that of age corresponding astrocytes it seems likely that the high rate of neuronal ADPRT activity when compared to that of astrocytes correlates a differentiation process. After 2 days in culture the ADPRT activity of embryonic neurons increases by about 50%, and reach after 10 days in culture a value ($23,5 \pm 0,4$ ADPR^{3}H incorportated/mg DNA) similar to that recorded in astrocytes from new born rats. Thus the velocity of neuronal ADPRT increase during early development is much faster than that of astrocytes. Finally from day 5 to day 7 theoretically close to the birth period, neuronal ADPRT activity decreases by about 18% while that of cultured new born astrocytes start to decrease after 12 days in culture. Until this period the proliferation rate of astrocytes is still high while very low in neurons. One might conclude that the high neuronal ADP-ribosylation correlates quite exclusively with cell differentiation (for details, Chabert et al., in preparation). Another interesting aspect is the growth factors effect: bFGF stimulates polyADPR mRNA production of astrocytes during proliferation and NGF that of neurons only during the early culture period. It is noteworthy and somehow unexpected that bFGF appeared more efficient than NGF for a stimulation of neuronal proliferation and ADPRT activity (Chabert et al., in preparation). It might be due to the fact that the embryonic neurons contain already an appreciable amount of NGF while bFGF acts as an exogenous stimulation agent.

In oligodendrocytes the highest polyADPRP activity is reached after 41 days in culture ($14,9 \pm 4,9$ nmol ADPR^{3}H incorporated/mg DNA); it is lower than that maximal recorded in neurons day 11 ($23,4 \pm 1$) but strikingly higher than that optimal of cultured astrocytes (7,5 $\pm$ 1,4) (Sarliève et al., in preparation). PolyADPRP gene expression in culturd oligodendrocytes is under investigation.

Taking altogether one might conclude that cell cultures provide a useful approach for exploration and characterization of ADPRT potential role and gene expression during astrocytes and neuronal proliferation and differentiation. The informations concerning stimulatory effects of NGF and bFGF may be useful in view of the attempts developed in the therapy of degenerative diseases, for instance Alzheimer's disease.

1.1. PolyADPRP activity of neurons and astrocytes in vivo during aging

It is obviously of interest to explore brain ADPRT activity in vivo. In view of the morphological and functional heterogeneity astroctyes and neurons have to be examined separately. Considering the well known decrease of the central nervous system (CNS) capacities during ageing one might expect a drop of nuclear ADPRT activity. Actually more than a four fold increase of polyADPRT activity was recorded in neurons: from 7.6 ± 1.1 to 35.1 ± 2.9 nmol ADPR/mg DNA and about a 60% increase in astrocytes : from 1.33 ± 0.15 to 2.22 ± 0.23 nmol ADPR/mg DNA (Messripour et al., 1991, in preparation). It is highly probable that this increase of ADPRT activity is due to DNA strand breaks, under quantitative evaluation. We observed a similar increase of ADPRT during ageing of lens epithelial cells; it was the only enzymatic activity which increased in these cells during ageing (Bizec et al., 1985). In parallel a significant decrease of double strand DNA was recorded in the nuclei of lens epithelial cells of aged animals, using Birnboim's fluorometric method (Birnboim and Jevcak, 1981).

2. Nuclear polyADPRP in control of tumor growth

Since nuclear polyADPRP appeared to be involved in cell growth and DNA repair (Shall, Hilz and Hayaishi groups; see for review Althaus and Richter, 1987) ADP-ribosylation reaction became a target in cancer research and chemotherapy (Alderson, 1990). Most of the investigations were devoted to tumoral cells in culture. Thus it is of interest to acquire more knowledge about tumors in vivo and to learn how far the investigation model in vitro fits with the biological events which occur in vivo. In order to explore this problem we investigated nuclear PolyADPRP activity and gene expression (mRNA) of the mastocytoma P815 which grow up in cell culture and in vivo as solid tumor and as ascitic cells. Let us remind that the transplantable neoplasm P815 has been used in numerous studies as a reference for cancerologists and immunologists (Dunn and Potter, 1957)

We also investigated some aspects of C6 glioma nuclear polyADPRP activity in cell culture and in solid tumors in vivo (Mandel et al., 1991 in preparation).

Concerning P815 mastocytoma we compared nuclear ADP-ribosylation reaction activity in cultured cells at the confluence in solid tumors at an appreciable size and in ascitic tumoral cells. It appeared that the nuclear ADPRT activity is strikingly higher in confluent cultured cells (15.60 $\pm$ 1.39) than in the solid tumors (3,07 $\pm$ 1,14) or ascitic cells (2.36 $\pm$ 0.17) drawing the attention on the difference between biopsy derived cultured cells and the original solid tumors or ascitic tumoral cells. The relative amount of PolyADPRP mRNA versus 18s rRNA was of 0.93 $\pm$ 0.19 in confluent cultured cells of 1.55 $\pm$ 0.024 in solid tumors and 1.52 $\pm$ 0.20 in ascitic tumoral cells. Thus for PolyADPRP mRNA also striking differences could be recorded between diverse forms of mastocytoma P815 originate from the same cells (Chabert et al., in preparation).

It was of interest to explore whether there are also differences concerning other basic structural components between the P815 mastocytoma cells in culture and in vivo. In this respect we examined the distribution of gangliosides constituents of cell membranes which seem to be involved in cell-cell interaction, morphogenesis and calcium transport (see for review Hakomori and Kannagi, 1983) and thought to act as receptors of some bacterial toxins which induce ADP-ribosylation reactions (see for review Ueda and Hayaishi., 1985).

Basic differences were recorded. The whole amount of gangliosides was in confluent cells 1,9 $\pm$ 0,3 nmol/mg prot. in ascitic cells 0,8 $\pm$ 0,01 and in solid tumors 2,5 $\pm$ 0,3. In addition, when the distribution of diverse gangliosides species was examined the percentage of GM3 was 81,4 + 5,0 in cell culture, of 54,5 $\pm$ 4,5 in ascitic cells and 61.9 $\pm$ 0.7 in solid tumors. Moreover, highly significant differences were recorded for other gangliosides GM2 and GM1. Thus one may conclude that multiple basic differences in gangliosides content do exist between tumoral cells in culture and or in vivo (Dreyfus et al., in preparation).

2.1. Effects of ethanol on mastocytoma P815 cells growth and   on P815 transfected animals survival

Considering the development of a therapeutic approach our attention was focused on the low content of NAD, the polyADPRT substrate in tumoral cells. Several years ago we demonstrated that NAD, NADH as well as NADP NADPH values of ascitic hepatoma cells are less than one sixth of normal hepatocytes (Wintzerith et al., 1961). Similar results were obtained for other tumoral cells when compared to corresponding normals (for review, see Wintzerith et al., 1961). Thus our attention was focussed on the induction in tumoral cells already deficient in NAD of an additional deficit by increasing NAD utilization. Such an effect can be obtained in vivo by treating the animals carrying tumors with alcohol in drinking water or in vitro by addition of ethanol to the culture medium. One might expect that in view of the low NAD level of tumoral cells ADP-ribosylation and thus cell proliferation will be disturbed at a higher extent in tumors when compared to normal cells. Actually in the cell cultures at day 4, that is at confluence, the ADPRT activity is of 15,6 $\pm$ 1,39 and in presence of ethanol 14,94 $\pm$ 1,52 nmol ADPR^{3}H per mg DNA while in the tumors and ascitic tumoral cells of mice consuming alcohol the ADPRT activity measured is significantly reduced : 3.07 $\pm$ 1.14 versus 1.30 $\pm$ 0.20 with alcohol in solid tumors 2,36 $\pm$ 0,17 versus 0,63 $\pm$ 0,05 with alcohol in ascitic cells. Moreover, the survival of mice carrying mastocytoma ascitic tumors is prolonged by ethanol consumption. At day 15 all ethanol-drinking mice are still alive and only 40% in controls (Chabert et al., 1991, in preparation)

A similar experiment was performed with glioma C6 transfected rats (Ledig et al., 1986). When using half of the efficient γ-rays irradiation dosage for tumor suppression, among rats drinking ethanol 80% survive, in the absence of ethanol only 40% survive. The maximum level of ethanol in blood serum, of ethanol/water 10% drinking animals was of 1g/1000 which is within tolerance limit.

Let us remind that ethanol induced sensitization of mouse mammary carcinoma to bleomycin was reported by Mizuno (1981). The author hypothesized that ethanol inhibits DNA repair. A number of reports were devoted to the effects of polyADPR inhibitors on tumoral cell growth (see for review Althaus et al., 1985). Our investigations concerning antimitotic effects potentiation using some metabolic modulators will be reported soon.

3. ADPR-ribosylation of cytoskeleton proteins

We became interested in the ADP-ribosylation of cytoskeleton constituants, in view of their involvement in nerve cells functions (for review see Cold Spring Harbor Symposium, Quantitative Biology, Organization of the cytoskeleton, 1981, 46) and in tumoral cells proliferation (for review see Alderson, 1990 and Fulton, 1984), the two lines of our investigations. Let me remind that three major cytoskeletal structures were identified in the cytoplasm: microtubules, microfilaments and intermediate filaments. Microtubules composed of a protein called tubulin are the largest of the three major filamentous system, and are involved in multiple functions including mitosis. Microtubules bind "microtubule associated proteins": MAPs, which possess sites that are phosphorylated. Inhibition of microtubules subunits association reduces strongly invasion and metastases formation by malignant cells (for review, Alderson, 1990). Intermediate filaments are named considering their size relative to microfilaments and microtubules; their roles appear to be structural rather than dynamic. Neurofilaments are the major filaments of neurons and are believed to provide tensile strength to neuronal axons and dendrites. Microfilaments are thin filaments, composed of actin (for review, see Pollard, 1986). A wide variety of cell functions are possibly modulated via the microfilaments system. ADP-ribosylation of actin by bacterial toxins was already reported (Vandekerckhove et al., 1987). It is conceivable that growth-related cellular functions are regulated by signals which are transmitted through an organized cytoskeleton. Indeed, in recent studies a rapid reorganization of the cytoskeleton was demonstrated in response to certain growth factors and tumor promoters. Most prominent changes are detected in the organization of microfilaments and their associated proteins in tumor cells. There is accumulating evidence suggesting that some of the oncogene proteins which are tyrosine kinases may induce the cytoskeletal changes that occur after transformation. It is already established, that several cytoskeleton proteins undergo a post-translational modification, for instance phosphorylation. Are they also substrates for ADP-ribosylation ?

The first question raised is obviously whether and how ADP-ribosyl transferases can stand in contact with the cytoskeleton. Free ribonucleoprotein particles carrying messenger RNA so called mRNP particles appear to be spacially associated with cytoskeletal network (Bagchi et al., 1987). Actually with Elkaim et al. (1983), we could identify an ADP-ribosyl transferase bound to the mRNP particles. After incubation of particles of different rat and mice organs with labelled NAD, ADP-ribosylated proteins of these particles were obtained. Following mRNP particles isolation by a D_2O saccharose gradient from plasmacytoma cells, containing high amounts of mRNP, the enzyme was purified by affinity chromatography on a 3ABA affigel column. A single band was obtained following SDS gel electrophoresis and western blotting (Chypre et al., 1989).

Only trace amounts of nucleic acids could be detected in the purified enzyme preparation following labeling and polynucleotide kinase and ^{32}PATP. Incubation with DNAse (1 µg per essay) which removes the labelled DNA traces is without effect on the enzymatic activity. Treatment by RNase, slightly increases the ADPRP activity (Weltin et al., in preparation). In view of the high RNA amount present in the mRNP particles it seems likely that RNA traces sticked to the enzyme prepared currently. Thus an RNAse treatment step was added to the former purification procedure (Weltin et al., in preparation). It is noteworthy that in plasmocytoma cells the ADPRT activity amount of the whole cellular mRNP particles is of about 34% of the total cellular ADPRT including that of nuclei (Elkaim et al., 1983). It is an additional support against the hypothesis.that the cytoplasmic ADPRT may be a contaminent of a nuclear origin. Moreover even incubation with 200 µg DNAse per 3 mg of the purified enzyme was without any effect on its ADPRT activity (Thomassin et al., 1985). We can conclude that the cytoplasmic enzyme is DNA independent and differs in this respect from the nuclear ADPRT. The purified cytoplasmic enzyme produces ADP-ribosylation by a dose and time dependent manner of histone H_1 and several other proteins (see below).

Let us now turn to the cytoskeleton proteins ADP-ribosylation by the cytoplasmic ADPRT. The cytoskeleton is composed predominantly of protein filaments that are involved in the structure of the cell and in cellular movements. These filaments can be divided into three classes.

The largest filaments called microtubules are composed of a protein called tubulin. The microtubule constituant tubulin is ADP-ribosylated by the cytoplasmic ADPRT in a time and dose dependent manner. The time course suggests that ADP-ribosylation occurs rather rapidly. The tubulin dimer is a preferential substrate as it was shown by LDS polyacrylamide gel electrophoresis of ADP-ribosylated tubulin (25 µg ^{32}PNAD 1 mM 4 µCi) and autoradiography. Microtubule associated proteins MAPs and Tau also undergo ADP-ribosylation in a time and dose dependent manner. MAP2 is stronger ADP-ribosylated than the Tau proteins. (Jesser et al., in preparation).

The middle-sized filaments are called intermediate filaments; they are composed in mature neurons by neurofilaments. Among the intermediate filaments three neurofilament proteins L, M and H, are the major constituants of the neuronal cytoskeleton. The two higher molecular mass proteins are strongly phosphorylated. It appeared that the M constituant (150 kDa) the most phosphorylated is clearly the most ADP-ribosylated in a time and dose dependent manner : one ADP-ribose per 8 neurofilament subunits has been identified. Some neurofilament associated proteins : c and d are also ADP-ribosylated. Previous phosphorylation, rather potentiates ADP-ribosylation of neurofilaments. Concerning the amino acids binding ADPR - under investigation - we tested the competitive effect of some amino acids. It appears that Arginine, Cysteine, their methyl esters as well as Glutamate, Aspartate and Lysine partially reduce neurofilaments ADP-ribosylation. Moreover, at higher amounts (100 mM) Arginine and Cysteine methylesters may block entirely ADP-ribosylation. The direct determination of the amino acids involved are under investigation. One may conclude that neurofilaments can be ADP-ribosylated by the mRNP cytoplasmic ADPRT, that the M subunit appears to be a privileged substrate, and that previous phosphorylation does not reduce ADP-ribosylation (Jesser et al., in preparation).

The smallest filaments include actin G and F proteins which undergo ADP-ribosylation in a time protein and NAD dependent manner; results confirmed by LDS 7.5 % polyacrylamide gel electrophoresis followed by autoradiography. According our present evaluation, an average of 4 molecules of ADPR are transferred on 10 mol of muscle actin. Among the different amino acids tested for competition with actin ADP-ribosylation: Arginine and Cysteine methyl ester, Cysteine glutamate and Lysine, Arginine methyl ester produces the strongest decrease of actin ADP-ribosylation (Weltin et al., in preparation).

In conclusion, the aim of our presentation was to summarize shortly some aspects of nerve cells ADP-ribosylation, a preliminary step for investigation of its functional role in the multiple brain functions, the effects of pharmacological agents and ageing alterations. The investigations concerning some aspects of tumoral cells ADP-ribosylation provide information which may be useful when approaching the undergoing work on tumoral cells biology and target sites for therapy. It is also highly probable that the investigation of cytoskeleton ADP-ribosylation may be useful for a better understanding of normal and tumoral cells biology.

References

Aktories, K., and Wegner, A. 1989. ADP-ribosylation of actin by clostridial toxins. Mini Review. The J. of Cell Biol., 109:1385-1387.

Alderson, T. 1990. New targets for cancer chemotherapy-poly(ADP-ribosylation) processing and polyisoprene metabolism. Biol. Rev., 65:623-641.

Althaus, F.R., and C. Richter. 1987. ADP-ribosylation of proteins. Enzymology and Biological Significance. Springer Verlag.

Althaus, F.R., Hilz, H., and Shall, S. 1985. ADP-ribosyltion of proteins. Springer Verlag.

Bagchi, T., Larson, D.E., and Sells, B.H. 1987. Cytoskeletal association of muscle-specific mRNAs in differentiating L6 rat myoblasts. Exp. Cell. Res., 168:160-172.

Ben-Ze'ev, A. 1985. The cytoskeleton in cancer cells. Biochim. Biophys. Acta, 780:197-212.

Birnboim, H.C., Jevcak, J.J. 1981. Fluorimetric method for rapid detection of DNA strand breaks in human white blood cells produced by low doses of radiation. Cancer Res., 41:1889-1892.

Bizec , J.C., Klethi, J., and Mandel, P. 1985. Biochimie cellulaire: poly-ADP-ribosyl polymerase (adénosine diphosphateribosyl transférase) du cristallin de bovidés: modulation de son activité au cours du vieillissement. C.R. Acad. Sci. Paris, 300:37-41.

Burnett, G., and Kennedy, E.P. 1954. The enzymatic phosphorylation of proteins. J. Biol. Chem., 211:969-980.

Chabert, M. Niedergang, C., Hog, F., Partisani M., and Mandel, P. 1991. Nuclear poly(ADPR)polymerase expression and activity in rat astrocytes culture: effects of bFGF. in this volume.

Chabert, M., Hu, L., Partisani, M., and Mandel, P. 1991. in preparation.

Chabert, M., Bischoff, P., Kopp, P., and Mandel, P. 1991. in preparation.

Chambon, P., Weill, J.D., and Mandel, P. 1963. Nicotinamide mononucleotide activation of a new DNA-dependent polyadenylic acid synthesizing nuclear enzyme. Biochem. Biophys. Res. Commun., 11:39-43.

Chambon, P. , Weill, J.D., Doly, J., Strosser, M.T., and Mandel, P. 1966. On the formation of a novel adenylic compound by enzymatic extracts of liver nuclei. Biochem. Biophys. Res. Commun., 25:638-643.

Chypre, C., Le Calvez, C., Hog, F., Revel, M.O., Jesser, M., and Mandel, P. 1989. Phosphorylation de la poly(ADP-ribose) polymérase cytoplasmique lié à des particules ribonucléoprotéiques libres par une protéine kinase C associée. C.R. Acad. Sci., Paris, 309:471-476.

Chypre, C., Weltin, D., Le Calvez, C., Hog, F., Kempf, J., Danse, J.M., and Mandel, P., in preparation.

Doly, J., and Mandel, P. 1967. Mise en évidence de la biosynthèse in vivo d'un polymère composé, le polyadénosine diphosphoribose dans les noyaux de foie de poulet. C.R. Acad. Sci., 264:2687-2690.

Dreyfus, H., Guerold, B., Partisani, M., and Mandel, P. 1991. in preparation.

Dunn, T.B. and Potter, M. 1957. A transplantable mast-cell neoplasm in the mouse. J. Nat. Cancer Inst., 18:587-601.

Elkaim, R., Thomassin, H., Niedergang, C. , Egly J.M., Kempf J., and Mandel, P. 1983. Adenosine diphosphate ribosyltransferase and protein acceptors associated with cytoplasmic free messenger ribonucleoprotein particles. Biochimie, 65:653-659.

Fulton, A.B. 1984. The Cytoskeleton. Chapman and Hall. London.

Hakomori, S.I., Kannagi, R. 1975. Structures and organisation of cell surface glycolipids dependency on cell growth and malignant transformation. Biochim. Biophys. Acta, 417:55.

Jesser, M., Hog, F., Rendon, A., Jung, D., and Mandel, P. 1991. in preparation.

Jesser, M., Leterrier, J.F., and Mandel, P. 1991. in preparation.

Ledig, M., Pillement, P., and Mandel P. Alcohol treated glioma cells are sensitive to ionizing radiations. In: Kriegel, H. et al., eds. Radiation risks to the developing nervous system. Stuttgart New York: Gustav Fischer Verlag; 1986:p. 337-350.

Mandel, P., Okazaki H., and Niedergang C. 1982. Poly(adenosine diphosphate ribose). Progr. Nucl. Acid Res. Mol. Biol. 27:1-51.

Mandel, P., Cothenet, V., Chabert, M., Hu, L., and Kopp, P. 1991. PolyADPR polymérase activity and mRNA of C6 glioma. in preparation.

Mandel, P. et al., in preparation.

Masmoudi, A., Islam, F., and Mandel, P. 1988. ADP-ribosylation of highly purified rat brain mitochondria. J. Neurochem., 51:188-193.

Messripour, M., Rastegar A., Chabert, M., Ciesielski L., and Mandel, P. 1991. Age associated changes of rat brain neuronal and glial polyADP-ribose polymerase activity. in preparation.

Mizuno, S. 1981. Ethanol-induced cell sensitization to bleomycin cytotoxicity and the inhibition of recovery from potentially lethal damage. Cancer Res., 41:4111-4114.

Pollard, T.D. 1986. Actin and actin-binding proteins. A critical evaluation of mechanisms and functions. Ann. Rev. Biochem., 55:987-1035.

Sarlième, L.L., Partisani, M., and Mandel, P.. 1991. in preparation

Sensenbrenner, M., Springer, N., Booher, J. and Mandel, P. 1972. Histochemical studies during the differentiation of dissociated nerve cells cultivated in the presence of brain extracts. Neurobiology, 2:282-295.

Thomassin, H., Niedergang, C. and Mandel, P. 1985. Characterization of the poly(ADP-ribose) polymerase associated with free cytoplasmic mRNA-protein particles. Biochem. Biophys. Res. Commun., 133:654-661.

Ueda, K., and Hayaishi, O. 1985. ADP-ribosylation. Annu. Rev. Biochem., 54:73-100.

Vandekerckhove, J., Schering, B., Bärmann M., and Aktories, K. 1987. Clostridium perfringens iota toxin ADP-ribosylates skeletal muscle actin in Arg-177. FEBS Lett., 225:48-52.

Weltin, D., Chypre, C., Hog, F., Kassab, R., and Mandel, P. 1991. in preparation.

Wintzerith, M., Klein, N., Mandel, L., and Mandel, P. 1961. Comparison of pyridine nucleotides in the liver and in an ascitic hepatoma. Nature, 191:467-469.

DNA BASE EXCISION REPAIR STIMULATES POLY(ADP-RIBOSE) SYNTHESIS

Robert J. Boorstein, Joydeep Haldar, Guy Poirier, and Donna Putnam

Department of Pathology, New York University Medical Center, New York, NY and Department of Molecular Endocrinology, Laval University, Quebec, Quebec.

ABSTRACT

3-aminobenzamide (3AB), an inhibitor of poly(ADP-ribose) synthesis, is toxic to cells which incorporate and repair 5-hydroxymethyl-2'-deoxyuridine (HmdUrd). To demonstrate that incorporation and repair of HmdUrd stimulates synthesis of poly(ADP-ribose) from intracellular NAD, V79 hamster cells were treated with HmdUrd and intracellular NAD levels were measured. HmdUrd is incorporated into DNA as a thymidine analogue resulting in extensive substitution of thymine residues with 5-hydroxymethyluracil (HmUra) residues. These HmUra residues are then subject to excision by action of HmUra-DNA glycosylase. Following HmdUrd treatment, NAD levels fell markedly (80-90%) within four hours and remained low for at least 10 hours before partially recovering by 24 hours. The degree of NAD lowering was dose dependent and paralleled net HmdUrd incorporation. The NAD lowering was largely prevented by concurrent treatment with 4 mM 3AB. No effects on NAD levels were seen following treatment with dThd or Brdurd, indicating that the effects on NAD result from incorporation of a nucleoside which puts large amounts of a repairable modification into DNA. To confirm that both incorporation and repair are necessary for the effects of HmdUrd on intracellular NAD, mutant cell strains derived from V79 cells deficient in either the ability to incorporate or to repair HmdUrd were examined. HmdUrd did not produce NAD lowering in either mutant cell strain. These results demonstrate that poly(ADP-ribose) synthesis can result directly and exclusively from repair of DNA base modifications.

RATIONALE

When V79 cells are grown in the presence of the nucleoside 5-hydroxymethyl-2'-deoxyuridine (HmdUrd), a single lesion, 5-hydroxymethyluracil (HmUra), is introduced into DNA. HmdUrd is incorporated into DNA as a thymidine analogue resulting in extensive substitution of thymine residues with 5-hydroxymethyluracil (HmUra) residues. These HmUra residues are then subject to excision by action of

HmUra-DNA glycosylase.

The use of HmdUrd provides a paradigm for further study of the role of poly(ADP-ribose) synthesis in the completion of base excision repair. To date, our studies have provided only inferential evidence that poly(ADP-ribose) synthesis is part of the cellular response to HmdUrd incorporation. 3-aminobenzamide (3AB), an inhibitor of poly(ADP-ribose synthesis, was toxic to cells containing HmUra, but did not interfere with the DNA glycosylase-mediated removal of HmUra from DNA. The degree of sensitization of cells to 3AB was directly dependent on the amount of HmUra in DNA. As compared with equitoxic doses of other agents which are known to stimulate poly(ADP-ribose) synthesis, HmdUrd produced substantially greater sensitization to 3AB. Hela cells, which do not incorporate HmdUrd into their DNA, are not sensitized to 3AB toxicity.

In the current study (Boorstein et al., in preparation) the role of poly(ADP-ribose) synthesis in the repair of HmUra is defined. We demonstrate that HmdUrd induces a specific time and dose dependent lowering of cellular NAD levels in V79 hamster cells that incorporate and repair HmdUrd. In mutant cell strains derived from V79 cells deficient in either the ability to incorporate or to repair HmdUrd, Hmdurd did not produce NAD lowering.

EXPERIMENTAL APPROACH

To test the hypothesis that incorporation and repair of HmdUrd specifically induces poly(ADP-ribose) synthesis, V79 cells were treated with HmdUrd and the cellular NAD levels were determined. To demonstrate that the effects were specific for the nucleoside HmdUrd, which introduces a repairable base modification into DNA, the effects of dThd and BrdUrd were also determined. Inhibition of NAD lowering by 3AB was also measured. Results in wild type V79 cells were then compared with those from mutant cells deficient in their ability to incorporate HmdUrd or to repair HmUra.

RESULTS

1) *Dose dependent lowering of cellular NAD following HmdUrd treatment*: To determine whether treatment of hamster V79 cells affected intracellular NAD levels, cells were treated with HmdUrd for 4 hrs and intracellular NAD was then measured using the non-radioactive cycling method of Jacobson. Low concentrations of HmdUrd (0.1-0.4) resulted in no decrease in NAD levels, but higher doses (1.0-10) induced NAD lowering of as much as 90%.

2) *Correlation of NAD lowering with HmdUrd incorporation*: To determine whether changes in NAD levels paralleled the incorporation of HmdUrd into DNA, V79 cells were treated with increasing concentrations of [^{3}H]HmdUrd and net incorporation into DNA after 4 hrs was measured.

Intracellular NAD decreased with increasing incorporation of HmdUrd.

3) *Correlation of NAD lowering with cytotoxicity*: To determine the effect of HmdUrd on survival, cells were treated with HmdUrd for 4 hrs and then replated to determine colony plating ability. Cells were resistant to low concentrations of HmdUrd (0.1-0.4), but higher doses (1.0-10) caused marked cytotoxicity. The cytotoxic effects of HmdUrd parallel the effects of HmdUrd on intracellular NAD levels.

4) *Time course of NAD lowering*: V79 cells were treated with increasing doses of HmdUrd and NAD levels were measured after 4, 10 and 24 hrs. Decreases in NAD were dose dependent, falling to minimal levels at 4 hrs, remaining low until 10 hrs, and then partially recovering. Net incorporation of HmdUrd into DNA was maximal from 4-10 hrs after addition, but did not markedly increase at later intervals. Cytotoxicity, however, increases continuously over the interval, indicating that the toxic effects of incorporation and repair of HmdUrd are largely cumulative.

5) *Inhibition of NAD lowering by 3AB*: To ascertain whether the HmdUrd induced lowering of NAD levels was due to inhibition of poly(ADP-ribose) synthesis, cells were treated concurrently for four hrs with HmdUrd and 4 mM 3AB, an inhibitor of poly(ADP-ribose) synthesis. Under these conditions, the decrease in NAD was largely inhibited. The effect of lower concentrations of 3AB was less marked with a 4 hr incubation, but even 0.1 mM 3AB inhibited HmdUrd-induced NAD lowering with a 24 hr treatment.

6) *Specificity of HmdUrd as a repairable nucleotide*: To confirm that HmdUrd induced effects on NAD levels result from the incorporation of a nucleoside which produces a repairable lesion in DNA, cells were treated with dThd and BrdUrd which are incorporated into DNA but not repaired. These compounds did not show significant effects on NAD levels.

7) *Effects of HmdUrd on intracellular NAD in mutant V79 cell lines*: We have previously suggested that the toxicity of HmdUrd results from the incorporation of large amounts of the nucleoside into DNA followed by the removal of HmUra residues by the action of HmUra-DNA glycosylase. We have also suggested that the initial steps of the repair of HmUra residues are necessary and sufficient to induce poly(ADP-ribose) synthesis. To prove these hypotheses, we have generated mutant cell strains which are resistant to HmdUrd and which either are deficient in their ability either to repair HmUra (V79mut1) or to incorporate HmdUrd (V79mutIII). The properties of these mutants are summarized below (Boorstein et al., in preparation):

TABLE 1. Comparison of repair capacity and poly(ADP-ribose) response of wild type and HmdUrd resistant cells .

Property	V79wt	V79mut1	V79mutIII
Sensitivity to HmdUrd	Yes	No	No
Incorporation of HmdUrd	Yes	Yes	No
HmUra glycosylase activity	Present	Absent	Present
Repairs HmUra *in vivo*	Yes	No	Not measurable
HmdUrd induced NAD lowering	Yes	No	No
MNNG induced NAD lowering	Yes	Yes	Yes
HmdUrd induced 3AB sensitization	Yes	No	No

CONCLUSIONS

These experiments directly confirm the hypothesis that poly(ADP-ribose) synthesis is stimulated by the generation of DNA strand breaks during the repair of base damage in DNA. V79 cells showed a marked lowering of cellular NAD as HmdUrd was incorporated into DNA and the resulting HmUra residues were subsequently repaired. These effects were blocked by 3AB. Non-repairable nucleosides did not induced NAD lowering. Mutant V79 cell strains deficient in their ability to incorporate HmdUrd or to repair HmUra do not show lower NAD levels in response to Hmdurd, although they respond normally to MNNG.

The heterogeneity of DNA damage produced by ionizing radiation, UV radiation, and alkylating agents, has made it difficult to define the involvement of poly(ADP-ribose) synthesis in base excision repair. With radiation and with alkylating agents, a heterogeneous group of modifications are formed in DNA, and it must always be considered that some undefined lesion is contributing to the overall biological effect. The use of HmdUrd affords the advantage that a single repairable lesion, HmUra, is introduced into DNA. The resulting biological effects can then be ascribed to the consequences of this lesion in DNA. The extent of the effect can be quantified and expressed as a function of the amount of HmUra in DNA, as determined by incorporation of [^{3}H]HmdUrd into DNA. For example, we have found that cells treated with HmdUrd show a loss of about 10 pmoles of NAD for each pmole of HmdUrd incorporated. Assuming that only 5-10% of the HmdUrd is repaired during the interval measured, and that de novo NAD synthesis is negligible, then approximately 100-200 pmoles of NAD are incorporated into polymer for each repaired lesion.

The use of HmdUrd to study the role of poly(ADP-ribose) in base excision repair demonstrates that poly(ADP-ribose) synthesis follows the initial steps of repair of lesions which, by themselves, do not induce poly(ADP-ribose) synthesis. Poly(ADP-ribose) synthesis should be viewed as an integral component of the multistep excision repair process.

It is also imperative to clearly distinguish between the mechanism by which a DNA base modification causes toxicity, and the mechanism by which interference with repair induced poly(ADP-ribose) synthesis causes toxicity. We have shown that high doses of HmdUrd are toxic as a result of repair. In part, as shown here, this toxicity might be due to NAD depletion that results from repair-induced poly(ADP-ribose) synthesis. The mechanism by which 3AB is toxic is qualitatively different. 3AB blocks NAD lowering in this system, without reversing toxicity (Boorstein et al, in preparation). In fact, 3AB is synergistically toxic at doses of HmdUrd which are not by themselves toxic.

ACKNOWLEGEMENTS

This work was supported by the National Institutes of Health, the American Cancer Society and by the Rita and Stanley Kaplan Cancer Center.

Studies on Poly(ADP-Ribosyl)ation in DNA Amplification and Mammalian Longevity

Alexander Bürkle, Karlheinz Grube, and Jan-Heiner Küpper

Poly(ADP-ribosyl)ation is a posttranslational protein modification catalyzed by poly(ADP-ribose) polymerase (PARP[1]), a highly conserved nuclear enzyme which uses nicotinamide-adenine dinucleotide (NAD) as substrate (for review, see Althaus & Richter, 1987). Poly(ADP-ribosyl)ation is thought to play a role in DNA repair and other cellular responses to DNA damage, such as cell cycle perturbations, clastogenesis, mutagenesis, malignant transformation, but also in DNA replication, integration of transfected foreign DNA into the cell genome, signal transduction, differentiation, and aging.

Poly(ADP-Ribosyl)ation and DNA Amplification

DNA amplification is a manifestation of genomic instability and plays a critical role in several steps of tumor development (for review, see Schimke, 1988) as well as in aging processes of fungi (*e.g. Podospora anserina*; Esser et al., 1986) and mammalian cells (Srivastava et al., 1985; Kunisada et al., 1985). Since DNA amplification can be greatly induced in cell culture systems, *e.g.* by carcinogens, we initially have asked the question whether poly(ADP-ribosyl)ation as an immediate response to DNA strand breakage has any influence on the amplification process.

We had initiated our studies on DNA amplification in a Simian virus (SV40)-transformed Chinese hamster cell line (CO60) that amplifies integrated SV40 sequences after carcinogen treatment and served as a short term model system. Our results showed that inhibition of carcinogen-stimulated poly(ADP-ribose) synthesis by 3-aminobenzamide (3AB) was correlated with a two- to sixfold potentiation of inducible DNA amplification in these cells (Bürkle et al., 1987). More recently, we were able to confirm those results in a different cell and amplification system, *i.e.* the development of methotrexate (MTX) resistance associated with dihydrofolate reductase (DHFR) gene amplification in Chinese hamster ovary (CHO) cells (Bürkle et al., 1990). The main finding of this study is shown in Table 1. Treatment with the alkylating carcinogen N-methyl-N'-nitro-N-nitrosoguanidine (MNNG) increased the MTX resistance frequency by about 100-fold (enhancement factor, EF), as expected. Addition of 3AB before MNNG treatment further potentiated the frequency of MTX resi-

---

[1]The abbreviations used are: 3AB, 3-aminobenzamide; AF, amplification factor; DHFR, dihydrofolate reductase; EF, enhancement factor; MNNG, N-methyl-N'-nitro-N-nitrosoguanidine; MTX, methotrexate; PARP, poly(ADP-ribose) polymerase; PE, plating efficiency; SV40, Simian virus 40

Table 1. **Induction of methotrexate (MTX) resistance after treatment of Chinese hamster ovary (CHO) cells with N-methyl-N'-nitro-N-nitrosoguanidine (MNNG) in the presence of increasing concentrations of 3-aminobenzamide (3AB)**

3AB	MNNG	% PE	MTX^r	MTX/PE	EF
-	-	90.2 ± 7.0	10 ± 3.5	11	1
-	2 μM	5.1 ± 0.36	52 ± 6.6	1019	92
0.1 mM	2 μM	4.0 ± 0.33	100 ± 3.5	2500	225
0.4 mM	2 μM	3.0 ± 0.28	112 ± 24.7	3797	342
1.0 mM	2 μM	2.2 ± 0.19	123 ± 14.7	5721	515

Cells were plated in 150-cm^2 culture flasks at 2×10^5 (for untreated controls) or 4×10^5 (for all MNNG treatments). The next day, cells were given 3AB and 1 h later MNNG was added as indicated. Three days later, cultures were trypsinized and replated for the determination of plating efficiency (PE) and MTX resistance. For PE determination, 500 or 2000 (control/MNNG) cells were plated in triplicate onto 10-cm Petri dishes in medium containing 10% dialyzed fetal calf serum. For the parallel determination of MTX resistance 5×10^5 cells were plated in triplicate onto 10-cm dishes in the same medium plus 350 nM MTX.

%PE, number of colonies x 100 / number of cells plated. (Mean $\pm$SD of triplicates);

MTX^r, number of MTX resistant colonies per dish. (Mean $\pm$SD of triplicates);

MTX^r/PE, MTX resistance frequency;

EF, enhancement factor = MTX resistance frequency relative to untreated controls

stance by up to fivefold in a dose-dependent manner, parallel to a potentiation of cytotoxicity (measured as a further reduction of plating efficiency, PE). The same potentiation occurred after cotreatment with benzamide (1 mM), another poly(ADP-ribosyl)ation inhibitor, under conditions which precluded direct drug interactions (data not shown). Benzoic acid, a noninhibitory analogue used as a specificity control, had no effect on the MNNG-induced MTX resistance frequency. Furthermore, 3AB, benzamide, and benzoic acid (each at 1 mM) had no effect on the *spontaneous* frequency of MTX resistance. A number of individual MTX-resistant colonies were expanded to determine their DHFR gene copy number. The relative frequency of DHFR gene amplification events (amplification factor of 2 or more) was similar (14% *vs.* 22%) whether clones were derived from cultures induced with MNNG alone or MNNG in the presence of 1 mM 3AB (data not shown). We thus infer that, along with the potentiation of MTX resistance frequency, 3AB potentiates the frequency of DHFR amplification to the same extent. Recently, Hahn et al. (1990) reported very similar findings with mouse EMT-6 cells, showing that 3AB potentiated X-ray-induced cytotoxicity and MTX resistance.

These results lead us to conclude that poly(ADP-ribosyl)ation should act as a *negative-regulatory factor* in carcinogen-induced DNA amplification, since PARP inhibitors potentiated MNNG-induced SV40 DNA amplification (Bürkle et al., 1987) as well as MNNG-induced MTX resistance associated with DHFR gene amplification (Bürkle et al., 1990; Hahn et al., 1990). Since in all these studies PARP inhibitors were effective only in combination with carcinogen treatment, it is very likely that they acted by an interference with DNA repair, thus blocking efficient damage removal and increasing the cell's responses to damage, *e.g.* cytotoxicity and induction of DNA amplification.

Poly(ADP-ribosyl)ation and Mammalian Species Longevity
Apart from the process of carcinogenesis, it appears that aging of cells and organisms as another complex biological phenomenon is accompanied (if not caused) by genetic instability. In a number of biological systems aging is correlated with DNA excision and/or amplification events, *e.g.* in the aerobic fungus *Podospora anserina* (Esser et al., 1986). Likewise, electron microscopic studies have revealed amplification of extrachromosomal circular DNA molecules in cells of aged rats and humans (Kunisada et al., 1985). On the other hand, during the replicative life span of human diploid fibroblasts genetic instability manifests itself as a continuous loss of certain repetitive DNA sequences that are dispersed over the chromosome length (Shmookler Reis & Goldstein, 1980), but also as a loss of telomeric DNA (located at chromosome ends) (Harley et al., 1990).

The causative factors for aging-associated genetic instabilities have not been elucidated so far. Likely candidates, however, are any kinds of DNA damages generated by ubiquitous endogenous and exogenous agents (*e.g.* oxygen radicals, glucose and other physiological cell metabolites, environmental carcinogens, irradiations). This would also be consistent with the well-known correlation between mammalian life span and certain DNA repair functions which would antagonize the accumulation of such damages more efficiently in longer-lived species (Hart & Setlow, 1974; Hart et al., 1979; Francis et al., 1981). Since poly(ADP-ribosyl)ation is involved in DNA repair and the recovery from DNA damage, it is very interesting that Pero et al. (1985) described a highly significant positive correlation between the life spans of different mammalian species and PARP activity in their nucleotide-permeable mononuclear leukocytes after supralethal gamma-irradiation. Irradiation, however, might not induce the same number of DNA breaks if applied to living cells of different organisms, since many of the breaks are mediated by free-radical mechanisms and/or DNA repair endonucleases whose activities are already known to correlate with the species' life span (Hart & Setlow, 1974;

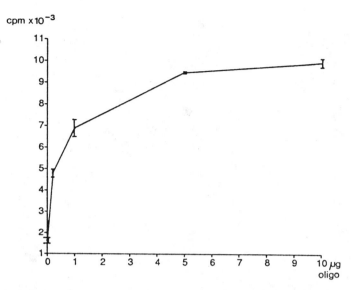

cpm x 10^{-3}

Fig. 1. **Poly(ADP-ribose) polymerase activity in permeabilized Molt-3 cells as a function of double-stranded oligonucleotide concentration.** Molt-3 cells were permeabilized by hypotonic cold shock and incubated in a buffer containing 333 μM NAD, including 0.5 μCi ^{3}H-HAD per 100 μl-enzyme assay, for 10 min at 30°C, followed by trichloroacetic acid precipitation and liquid scintillation counting of acid-insoluble material. Oligonucleotides were included as indicated. Results are given as crude cpm (means of triplicates ± 1 standard deviation). For experimental details, see Grube et al., 1991.

Hart et al., 1979; Francis et al., 1981). Therefore, it is not clear whether the reported correlation between life span and PARP activity is direct (*i.e.* due to a higher enzyme content or a greater specific activity) or indirect (*i.e.* due to other cellular functions). As another potential source of complication, the NAD concentrations used (25 μM) in the quoted study were well below the K_m of PARP.

In order to resolve this interesting question we provided a direct stimulus for PARP in permeabilized cells (Grube et al., 1991). Berger and Petzold (1985) had previously shown that double-stranded deoxyoligonucleotides were excellent activators of *purified* enzyme. As is shown in Fig. 1, the addition of such an oligonucleotide (CGGAATTCCG) to the reaction buffer stimulated PARP activity in permeabilized Molt-3 human lymphoma cells, too. Enzyme activation was concentration-dependent, reaching saturation levels at 5 μg oligonucleotide per 100 μl reaction volume. After background subtraction (zero-time controls without oligonucleotide, yielding 1275 ± 238 cpm) maximal en-

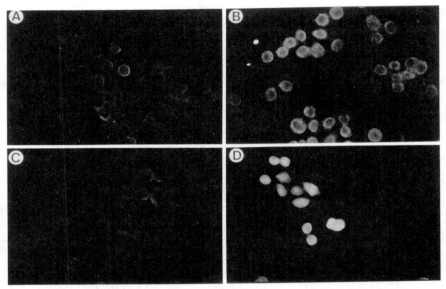

Fig. 2. **Stimulation of poly(ADP-ribose) polymerase activity in ethanol-fixed cells by double-stranded oligonucleotides.** HeLa cells were grown on coverslips, fixed in ethanol for 10 min at -20°C, and postincubated in 100 mM Tris-HCl/pH 8.0, 0.4 mM NAD, 1 mM dithiothreitol, 10 mM $MgCl_2$ for 1 h at 30°C to allow *in-situ* poly(ADP-ribose) synthesis, as described by Ikai et al. (1980). Oligonucleotides or benzamide were included as indicated. Coverslips were then processed for indirect immunofluorescence to visualize polymer synthesis using antibody H10, kindly provided by Dr. H. Kawamitsu, Tokyo, Japan. Panel *A*, postincubation without NAD; *B*, with NAD; *C*, with NAD/benzamide (10 mM); *D*, with NAD/oligonucleotide (100 µg/ml).

zyme stimulation was about 30-fold. The oligonucleotide-stimulated reaction is almost linear during the first 10 min (data not shown).

The activating effect of the same double-stranded oligonucleotide was also evident when HeLa cells were fixed with ethanol and postincubated with NAD to allow poly(ADP-ribose) synthesis to occur *in situ*. Fig. 2 shows immunofluorescences obtained with monoclonal antibody H10 (Kawamitsu et al., 1984) directed against poly(ADP-ribose). In Panel A (control) cells were postincubated in buffer without NAD. Only a nonspecific cytoplasmic background and a very faint nucleolar staining is visible. By contrast, cells subjected to NAD postincubation (Panel B) show an intensification of nuclear fluorescence as compared with controls (in this case mostly in the peripheral region of nuclei), indicating enzyme activity. The presence of the PARP inhibitor benzamide (10 mM) during NAD postincubation prevented poly(ADP-ribose) formation completely (Panel C). However, when double-stranded oligonucleotides (100 µg/ml) were included in the postincubation buffer instead (panel D), a marked further increase in nuclear fluorescence intensity (and hence enzyme activity) was observed as compared with Panel B. Identical results were seen with CV-1 monkey cells (data not shown). Taken together, we could show that double-stranded oligonucleotides can be used as convenient, chemically and stoichiometrically well-defined PARP activators in permeabilized or ethanol-fixed mammalian cells.

Using this modified enzyme activity assay we can thus exclude any influence by other cellular (repair) activities. We are currently re-testing PARP activity as a function of animal life span in Percoll-gradient-purified pheripheral blood mononuclear cells at NAD concentrations within the physiological range. The results obtained so far indeed indicate a significant positive correlation between the life span of 11 mammalian species tested and the maximal PARP activity in their blood mononuclear cells (K. Grube and A. Bürkle, manuscript in preparation).

DNA repair as a protective factor against the constant attacks by endogenous and exogenous DNA-damaging agents is likely to contribute to the maintenance of genome integrity and stability over the lifetime of a species. Therefore, the efficient DNA repair of long-lived species could also be responsible for the delay in tumor development, as compared with short-lived species. We think that the direct correlation of the maximal activity of PARP, a relatively well characterized monomeric enzyme involved in DNA repair, with animal longevity is a good starting point for further studies on the role of poly(ADP-ribosyl)ation in aging processes.

Acknowledgement
We thank Prof. H. zur Hausen for his continuous support and for critical reading of the manuscript and Dr. H. Kawamitsu (Tokyo, Japan) for antibody H10. This work was supported by the Deutsche Forschungsgemeinschaft.

References
1. Althaus, F.R.; Richter, C. ADP-Ribosylation of Proteins. Enzymology and Biological Significance. Molecular Biology, Biochemistry and Biophysics 37. Berlin, Heidelberg: Springer Verlag; 1987.
2. Berger, N.A.; Petzold, S.I. Identification of minimal size requirements of DNA for activation of poly(ADP-ribose) polymerase. Biochemistry 24:4352-4355; 1985.
3. Bürkle, A.; Heilbronn, R.; zur Hausen, H. Potentiation of carcinogen-induced methotrexate resistance and dihydrofolate reductase gene amplification by inhibitors of poly(adenosine diphosphate-ribose) polymerase. Cancer Res.

50:5756-5760; 1990.

4. Bürkle, A.; Meyer, T.; Hilz, H.; zur Hausen, H. Enhancement of N-methyl-N'-nitro-N-nitrosoguanidine-induced DNA amplification in a Simian Virus 40-transformed Chinese hamster cell line by 3-aminobenzamide. Cancer Res. 47:3632-3636; 1987.

5. Esser, K.; Kück, U.; Lang-Hinrichs, C.; Lemke, P.; Osiewacz, H.D.; Stahl, U.; Tudzynski, P. Plasmids of Eucaryotes: Fundamentals and Applications. Berlin, Heidelberg: Springer Verlag; 1986.

6. Francis, A.A.; Lee, W.H.; Regan, J.D. The relationship of DNA excision repair of ultraviolet induced lesions to the maximum life span of mammals. Mech. Ageing Dev. 16:181-189; 1981.

7. Grube, K.; Küpper, J.H.; Bürkle, A. Direct stimulation of poly(ADP ribose) polymerase in permeabilized cells by double-stranded DNA oligomers. Anal. Biochem. 193:236-239; 1991.

8. Hahn, P.; Nevaldine, B.; Morgan, W.F. X-ray induction of methotrexate resistance due to *dhfr* gene amplification. Somat. Cell Mol. Genet. 16:413-423; 1990.

9. Harley, C.B.; Futcher, A.B.; Greider, C.W. Telomeres shorten during ageing of human fibroblasts. Nature 345:458-460; 1990.

10. Hart, R.W.; Sacher, G.A.; Hoskins, T.L. DNA repair in a short- and a long-lived rodent species. J. Gerontol. 34:808-817; 1979.

11. Hart, R.W.; Setlow, R.B. Correlation between deoxyribonucleic acid excision-repair and life-span in a number of mammalian species. Proc. Natl. Acad. Sci. USA 71:2169-2173; 1974.

12. Ikai, K.; Ueda, K.; Hayaishi, O. Immunohistochemical demonstration of poly(adenosine diphosphate-ribose) in nuclei of various rat tissues. J. Histochem. Cytochem. 28:670-676; 1980.

13. Kawamitsu, H.; Hoshino, H.; Okada, H.; Miwa, M.; Momoi, H.; Sugimura, T. Monoclonal antibodies to poly(adenosine diphosphate ribose) recognize different structures. Biochemistry 23:3771-3777; 1984.

14. Kunisada, T.; Yamagishi, H.; Ogita, Z.-I.; Kirakawa, T.; Mitsui, Y. Appearance of extrachromosomal circular DNAs during *in vivo* and *in vitro* ageing of mammalian cells. Mech. Ageing Dev. 29:89-99; 1985.

15. Pero, R.W.; Holmgren, K.; Persson, L. Gamma-radiation induced ADP-ribosyl transferase activity and mammalian longevity. Mutat. Res. 142:69-73; 1985.

16. Schimke, R.T. Gene amplification in cultured cells. J. Biol. Chem. 263:5989-5992; 1988.

17. Shmookler-Reis, R.J.; Goldstein, S. Loss of reiterated DNA sequences during serial passage of human diploid fibroblasts. Cell 21:739-749; 1980.

18. Srivastava, A.; Norris, J.S.; Shmookler-Reis, R.J.; Goldstein, S. c-Ha-*ras*-1 proto-oncogene amplification and overexpression during the limited replicative life span of normal human fibroblasts. J. Biol. Chem. 260:6404-6409; 1985.

DNA TOPOISOMERASE I AND POLY(ADP-RIBOSE) POLYMERASE IN THE EARLY STAGES OF HEPATOCARCINOGENESIS

Giannoni P., Fronza G.[$], Scarabelli L., Orunesu M.,
Abbondandolo A.[$], and Cesarone C.F.

Institute of General Physiology, Corso Europa 26, 16132, Genoa, ITALY;
[$]Laboratory of Mutagenesis, National Institute for Cancer Research, Viale Benedetto
XV, 10, 16132, Genoa, ITALY.

INTRODUCTION

A molecular interpretation of the biological events leading to cellular transformation must take into account the functional role of several catalytic proteins. Chromatin organization, DNA integrity and availability for transcription are modulated by nuclear enzymes. One of these, DNA topoisomerase I (**Topo I**) [E.C. 5.99.1.2.], along with topoisomerase II [1], modulates DNA supercoiling introducing transient strand breaks and thus reducing the superhelical tension which would accumulate in DNA replication and other DNA transitions [2-6]. Modifications of chromatin organization are needed for transcription [7]. DNA topoisomerase I catalytic activity can be inhibited or activated by poly(ADP-ribosyl)ation [8,9] and phosphorylation [10,11], respectively. Poly(ADP-ribosyl)ation-mediated regulation of human DNA topoisomerase I was recently proved [9], suggesting a functional significance for poly(ADP-ribosyl)ation negative-modulation in topoisomerase-related processes (replication, transcription, recombination). Moreover Topo I inhibition via poly(ADP-ribosyl)ation might be responsible for the enhancement of SCE [12].

Interestingly, a drastic depletion of poly(ADP-ribose) polymerase [E.C. 2.5.2.30] (**pADPRP**) in rat liver, using the *in vivo* hepatocarcinogenesis model of Teebor and Becker [13], was recently demonstrated [14]. On the contrary other enzymatic activities, such as DNA polymerase τ [15] and DNA ligase [16] were not modified during the process of transformation.

Since DNA topoisomerase I is a target for pADPRP, we investigated whether this enzyme showed a modified pattern during the Teebor and Becker experimental protocol and if a correlation between pADPRP and DNA topoisomerase I could be found *in vivo*.

MATERIALS and METHODS

Chemicals and Reagents

2-Acetylaminofluorene (2AAF) was purchased from Fluka. Supercoiled plasmid pBR322 and Western blot colorimetric reagents (Nitroblue tetrazolium

(NBT) and 5-Bromo-4-Chloro-3-Indolyl phosphate (BICP)) were from Boehringer. Nitrocellulose membranes were from BioRad. Bicinchoninic acid-based reagents were obtained from Pierce. Human DNA Topoisomerase I monoclonal antibody was a kind gift of Prof. O. Westergaard, University of Aarhus, Denmark. Goat Anti-mouse IgM (μ-chain specific) alkaline phosphatase conjugate antibody was purchased from Sigma.

Animals

Male Wistar rats, weighing 100-120 g, were used. The exposure of animals to the hepato-carcinogen 2AAF was carried out according to Teebor and Becker [13]. Briefly, the 2AAF-treated animals were placed on a four-cycle discontinuous feeding regimen. Each cycle consisted of a 3-week exposure to a standard diet containing 0.05% 2AAF, followed by 1 week of normal diet. Control animals were fed only on normal diet.

DNA Topoisomerase I activity assay

Rats were sacrificed under slight anesthesia. Livers were removed and washed twice in ice-cold physiological saline. All subsequent steps were carried out at 0-4°C. Nuclear and total extracts for Topo I activity assay and Western blot analysis were prepared, with minor modifications, as elsewhere described [17]. Protein content was determined by the bicinchoninic acid method [18].
Extracts were normalized by protein content and assayed for DNA topoisomerase I activity essentially as described by Keller W. [19]. Reaction products were electrophoresed overnight in TPE buffer (30 mM NaH_2PO_4, 1mM EDTA, 36 mM Tris/Cl⁻, pH 7.8) on 1% agarose gel at 1 V/cm at room temperature. Gels were stained with ethidium bromide (1 μg/mL) and photographed. Negatives were processed for densitometric analysis by a GS 300 Hoefer Densitometer, using Hoefer GS 365 Software.
DNA Topoisomerase I activity was determined quantifying the amount of plasmid DNA relaxed in 15 min at 30°C, which showed to be linear in the experimental conditions used. DNA Topo I activity, in 2AAF-exposed animals, is expressed in arbitrary units compared to control rats.

SDS-Electrophoresis and Immunoblot Assay

Frozen samples were lyophilized and resuspended to optimal protein concentration, loading 65 μg protein/lane.
Proteins were separated by means of SDS-polyacrylamide gel electrophoresis as elsewhere described [20]. Immunoblotting procedures were essentially as described by Towbin [21]. Primary antibody incubation (1:500 working dilution in MNT buffer) lasted 3 h, followed by 3x10 min washing with TNT buffer (20 mM Tris/Cl⁻, 0.5 M NaCl, 0.05% Tween, pH 7.5). Secondary alkaline phosphatase conjugated antibody incubation was held for 2 h. Secondary antibody was diluted 1:1000 in TN buffer (20 mM Tris/Cl⁻, 0.5 M NaCl, pH

7.5). Membranes were then washed in TNT buffer for 3x10 min and stained with BICP-NBT standard solutions. Colorimetric reaction was stopped by adding ice-cold 100 mM Tris/EDTA, pH 9.5.

RESULTS

DNA topoisomerase I activity evaluation during 2AAF exposure
According to the experimental protocol, 2AAF-containing diet was supplied, in each cycle, for 3 weeks followed in the last fourth week by normal diet, to allow recovery [13].

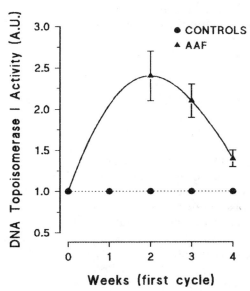

Fig. 1 - Evaluation of DNA topoisomerase I activity in rat liver during the first cycle of 2AAF exposure [13]. The data are the mean ± SE of triplicate assays on 3 different sets of experiments.

Liver samples were obtained at the end of the fourth week of each cycle, as described in Methods. No differences in Topo I activity were noted at the end of each cycle throughout the treatment. Since cell populations are changing during carcinogen exposure [13,14], we have surgically isolated surrounding tissue and preneoplastic nodules (3rd-4th week of the 4th cycle). Also in these samples no significant differences in DNA topoisomerase I activity, with respect to controls, could be detected.

DNA topoisomerase I variations during the first cycle of treatment

Depletion of poly(ADP-ribose) polymerase in rat liver occurred during the first cycle of 2AAF-exposure [14]. We then monitored DNA topoisomerase I activity on extracts obtained from treated and control animals sacrificed at the end of the 2nd, 3rd and 4th week during the first cycle. As shown in Fig. 1, topoisomerase I activity in 2AAF-exposed rats reaches a 2 or 3-fold increase on the 2nd and 3rd week, returning to normal level after the recovery week.
Parallel Western blot analysis (Fig. 2) showed a marked increase in immunoreactive protein content on the 3rd week of treatment, in total extracts obtained from liver of 2AAF-treated animals. Interestingly, DNA topoisomerase I content returned to control values at the end of the cycle (4th week) (Fig. 2, lane 4).

DISCUSSION

DNA topoisomerase I is involved in DNA transcription [6] and gene expression [22], therefore, post-translational modifications which modulate its activity can indirectly influence gene expression, differentiation, and transformation processes. Poly(ADP-ribose) polymerase plays a key role in cellular response to DNA damage and in chromatin structure control [23].
We investigated whether modifications in pADPRP activity, observed early during 2AAF exposure, could account for an altered modulation of Topo I activity. DNA topoisomerase I was determined, in liver samples from 2AAF-treated rats, at the end of each cycle and at the end of the 2nd, 3rd and 4th week of the first one. During the 1st cycle, in fact, we observed a significant increase in both enzymatic activity and catalytic protein level. This phenomenon was concomitant with the reported marked loss of pADPRP activity [14]. No significant variations in Topo I activity were observed at the end of each cycle.

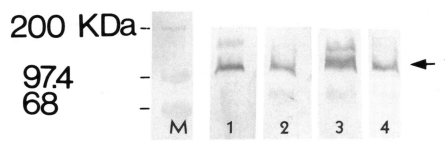

Fig. 2 - Western blot analysis of DNA topoisomerase I in liver extracts from rats exposed to the first cycle of 2AAF treatment according to Teebor and Becker [13]. Lanes: M - Molecular weight markers; 1 - Control extracts; 2-4 - Extracts obtained at the end of the 2nd, 3rd and 4th week of treatment respectively.

Early during 2AAF-feeding different phenomena can occur: activation of Topo I (Fig. 1) and increased enzymatic protein level (Fig. 2). When animals were shifted on standard diet (4th week) both activity and protein content returned in the range of controls.

Being poly(ADP-ribosyl)ation one of the possible mechanisms for nuclear activities control, a minimum in this negative modulator should coincide with a maximum in activation of DNA topo I. This hypothesis was consistent with the increase in activity observed on the 3rd week of the first cycle, which is concurrent with pADPRP loss. Nevertheless it does not apply for the return of Topo I to normal levels on the 4th week of the same cycle.

The loss of chromatin structure control and the accumulation of DNA lesions may lead to the unbalance in Topo I modulation.

It is known, in fact, that DNA distorting lesions, such as UV-induced pyrimidine dimers, can inhibit Topo I catalytic activity both in processive and distributive conditions in vitro [24]. 2AAF-induced bulky adducts could operate similarly, since these lesions distort DNA double helix to a similar extent [25]. This inhibition could also cause, via alteration of DNA supercoiling, increased transcription of Topo I gene accounting for the increased protein level. The return to normal values of both activity and immunoreactive peptide, at the end of the cycle, instead, may depend upon an homeostatic mechanism, which tends to maintain physiological levels of nuclear enzymes. Alterations in this equilibrium may result in dramatic consequences for the cell [26]. In conclusion, even if no direct correlation could be drawn between pADPRP and Topo I activity, during carcinogenic exposure, nevertheless our data suggest a possible link between these activities.

REFERENCES

1. Liu L.F.; Annu. Rev. Biochem.; 58: 351-375; 1989.
2. Romig H. and Richter A.; Biochem. Biophys. Acta; 1048; 274-280; 1990.
3. Yang L., et Al.; Proc. Natl. Acad. Sci. USA; 84; 950-954; 1987.
4. Snapka R.; Mol. Cell. Biol.; 6: 4221-4227; 1987.
5. Richter A., et Al.; Nucl. Acid Res.; 13: 3455-3468; 1987.
6. Gilmour D.S., et Al.; Cell; 44: 401-407; 1986.
7. Aveman K., et Al.; Mol. Cell. Biol.; 8: 3026-2034; 1987.
8. Ferro A.M. and Oliveira B.M.; J. Biol. Chem.; 259: 547-554; 1984.
9. Kasid U.N., et Al.; J. Biol. Chem.; 264: 18687-18692; 1989.
10. Durban E., et Al.; Biochem. Biophys. Res. Comm.; 111: 987-905; 1983.
11. Pommier Y., et Al; J. Biol. Chem.; 265: 9418-9422; 1990.
12. Dillehay, L.E., et Al.; Mutat. Res.; 215: 15-23; 1989.
13. Teebor G.W. and Becker F.F.; Cancer Res.; 31: 3-9; 1971.
14. Cesarone C.F., et Al.; Cancer Res.; 48: 3581-3585; 1988.
15. Chan J.Y.H. and Becker F.F.;Proc. Natl. Acad. Sci.USA; 76:814-818; 1979.

16. Chan J.Y.H. and Becker F.F.; Carcinogenesis (London); 1275-1277; 1985.

17. Cesarone C.F., et Al.; Mutat. Res.; 245: 157-163; 1990.

18. Smith P.K., et Al; Analytical Biochem.; 150: 76-85; 1985.

19. Keller W.; Proc. Natl. Acad. Sci. USA; 72: 4876-4880; 1975.

20. Scovassi A.I., et Al.; J. Biol. Chem.; 259: 10973-10977; 1984.

21. Towbin H., et Al.; Proc. Natl. Acad. Sci. USA; 76: 4350-4354; 1979.

22. Fleischmann G., et Al.; Proc. Natl. Acad. Sci. USA; 81:6958-6962;1984.

23. Althaus F.R., Richter C.; in "ADP-Ribosylation of Proteins"; Springer-Verlag; Berlin; 1987.

24. Pedrini A.M. and Ciarrocchi G.; Proc. Natl. Acad. Sci. USA; 80: 1787-1791; 1983.

25. Lang M.C.E., et Al.; Chem. Biol. Interact.; 28: 171-180; 1979.

26. Giovanella B.C., et Al.; Science; 256: 1046-1048; 1989.

Poly(ADP-ribose) polymerase inhibitors induce murine melanoma cell differentiation by a mechanism independent of alterations in cAMP levels and protein kinase A activity.

B.W. Durkacz, J. Lunec, H. Grindley, S. Griffin, O. Horne & A. Simm

Cancer Research Unit, Medical School, University of Newcastle-upon-Tyne, NE2 4HH, U.K.

Introduction

Murine melanoma cells treated with Melanocyte stimulating hormone (MSH) undergo differentiation characterised by enhanced melanogenesis and altered morphology (1). The effects of MSH are mediated via the adenylate cyclase-cAMP pathway leading to activation of protein kinase A (2). How this signal induces alterations in gene expression in this system is not known. By analogy with other cAMP-responsive systems, it is likely to involve phosphorylation of proteins such as cAMP-regulatory element binding protein (CREB).

Other known effectors of murine melanoma cell differentiation also act through the adenylate cyclase-cAMP pathway. For example, inhibitors of cyclic nucleotide phosphodiesterase and bacterial toxins (e.g. cholera, pertussis) which mono-ADP-ribosylate G proteins and lead to persistent activation of adenylate cyclase.

Two lines of evidence for an involvement of poly(ADP-ribose) polymerase (PADPRP) in gene expression and hence differentiation are:- 1) 3-aminobenzamide (3AB) treatment of intact cells increases the expression of certain genes, e.g. metallothioneins, cytochrome p450 and murine mammary tumour virus (MMTV) (e.g. 4, 5); 2) rapid de-ADP-ribosylation of HMG proteins correlates with glucocorticoid induction of MMTV and metallothionein gene expression (4,5). Because of these previous observations, we investigated the effects of PADPRP inhibitors on the differentiation of murine melanoma cells.

Results

Published evidence has suggested that 3AB has other metabolic effects independent of inhibition of PADPRP relating to disturbance of *de novo* purine synthesis (6). This has led to criticism of work proposing a role for poly(ADP-ribose) in cellular processes based exclusively on the use of 3AB. Therefore we utilised a range of PADPRP inhibitors (benzamide and its 3-position substituted derivatives), for which we estimated the *in vitro* IC50 values for this enzyme (see Table 1). We also examined the effect of these compounds on steps in the cAMP-mediated signal transduction pathway to exclude as far as possible other known metabolic effects which could

TABLE 1

IN VITRO IC50 VALUES FOR
FOR POLY(ADP-RIBOSE)POLMERASE

3-Acetamidobenzamide (3AAB)	2.8μM
Benzamide (B)	4.0μM
3-Methoxybenzamide (3MB)	5.2μM
3-Aminobenzamide (3AB)	10.0μM
4-Aminobenzamide (4AB)	164.0μM

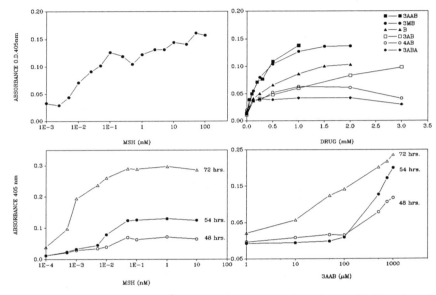

Figure 1. Cells were seeded at $2x10^4$ per well in 96-well microtitre plates in Eagle's minimal essential medium containing 10% foetal calf serum. 24h. later, MSH or benzamides were added at the concentrations specified. 48h. later (or at times indicated), the plates were scanned at 405nM in a Perkin-Elmer plate reader. Each data point represents the mean of 3 readings.

induce differentiation in this system.

B16 F1 murine melanoma cells were treated in culture with MSH or 0.5mM 3AAB. Compared to control cells, the treated cells became more dendritic and more highly pigmented. The appearance of the 3AAB-treated cells was identical to the MSH-treated cells (results not shown). Induction of melanogenesis was confirmed by monitoring the increase in absorbance at 405nM. MSH induced melanogenesis in the 0.01-10nM concentration range. All the benzamides were effective inducers of melanogenesis in the 0.2-2mM range (Fig.1). In general agreement with the estimated IC50 values for PADPRP (Table 1), the potency of these compounds as inhibitors of PADPRP correlated with their effectiveness as inducers of melanogenesis. By contrast, the non-inhibitory compound, 3-aminobenzoic acid (3ABA), was ineffective. The second part of Fig.1 shows the time course of melanogenesis in response to increasing doses of MSH and 3AAB. In both cases, increased melanin production became apparent between 48-72 hours. Note that at late times (72h), 3AAB can induce melanogenesis at doses as low as 10-50 μM. It was also possible to demonstrate synergism following coincubation of MSH and 3AAB for 48h. (results not shown). 3AAB was selected for all subsequent experiments as it proved the most effective inducer of differentiation.

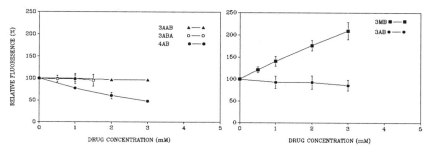

Figure 2. The effect of the benzamides on cyclic nucleotide phosphodiesterase activity was assayed by using a fluorescent analogue of cAMP, 2-etheno-cAMP, according to the method described in (7).

Inhibition of cyclic nucleotide phosphodiesterase by methyl xanthines (e.g. theophylline) increases cAMP levels and induces B16 F1 melanogenesis (results not shown). There are some structural similarities between the 3-position substituted benzamides and theophylline (exemplified by the fact that theophylline is also a weak inhibitor of PADPRP, IC50 125μM, our results). Therefore it was important to establish the effect of the PADPRP inhibitors on this enzyme. The results are shown in Fig.2. It can be seen that only one of the compounds, 4AB, exerted any inhibitory effect on this enzyme. Note that 4AB was a poor inhibitor of PADPRP and induced melanogenesis very weakly.

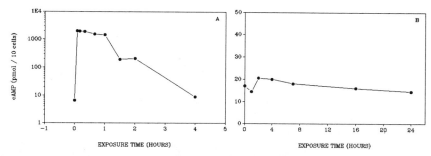

Figure3. The effect of increasing exposure time to MSH (20nM) and 3AAB (1mM) on cAMP levels in cells was assayed using a radioimmunoassay kit obtained from New England Nuclear. Each data point represents the mean of triplicate samples. Results are expressed as pmol/10^6 cells.

Activation of adenylate cyclase in the signal transduction pathway leads to a rise in cAMP levels. We wished to exclude the possibility that the benzamides raised cAMP levels. Therefore the effect of MSH and 3AAB (at doses which induce maximal melanogenesis) on cAMP levels was measured. Fig. 3a shows that 20nM MSH caused a very rapid rise in cAMP levels, up to 200-fold by 5 minutes. cAMP levels had dropped by 90% after about 90 minutes and returned to basal levels by 4 hours. In contrast, 1mM 3AAB (Fig. 3b) had no effect on cAMP levels, which remained at basal values (6-16pmol/10^6 cells) over a 24 hour time period.

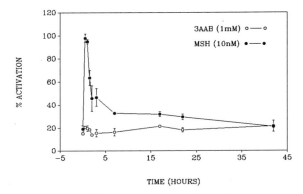

TIME (HOURS)

Figure 4. Protein kinase A activity was measured in crude cell extracts by following the phosphorylation of a peptide based on a PKA substrate consensus sequence. The assay system was obtained from GIBCO BRL.

cAMP activates PKA by binding to the holoenzyme and causing dissociation of the regulatory and catalytic subunits. The PKA activity in crude cell extracts reflects endogenous cAMP levels. Total available PKA activity is revealed by the addition of saturating amounts of cAMP to the assay tube. Fig. 4 shows a time course response to fixed doses of MSH and 3AAB (again selected for maximal induction of melanogenesis). MSH caused a rapid activation of PKA, reaching a peak by 10 minutes and returning to basal levels by 5 hours. These kinetics mirrored very closely the cAMP response (see Fig.3). In comparison, 3AAB had no effect over the same time course (up to 45 hours).

Conclusions

We have made the novel observation that a family of compounds which are inhibitors of PADPRP are effective inducers of melanogenesis and differentiation in murine melanoma cells. Any effects of these compounds on components of the adenylate cyclase-cAMP signal transduction pathway up to and including PKA activation have been excluded. These data implicate poly(ADP-ribosylation) reactions in the regulation of the differentiation status of melanoma cells. Furthermore, by excluding effects on cAMP levels and PKA activity, they suggest a role for PADPRP in this system in the regulation of gene transcription.

References

1) Wong, G. & Pawelek, J. (1973) Nature New Biol. 241 : 213
2) Bitensky, M.W., Demopoulos, H.B. & Russel, V. (1973) in "Pigmentation, its genesis and biologic control" Ed. Riley, V. (Appleton-Century Crofts, N.Y.) : 247
3) Althaus, F.R. & Richter, C. (1987) ADP-Ribosylation of proteins. (Springer-Verlag, Berlin, Heidelberg, N.Y.)
4) Tanuma, S., Johnson, L.D. & Johnson, G.S. (1983) J. Biol. Chem. 258 : 15371
5) Tanuma, S., Kawashima, K., & Endo, H. (1987) Biochim Biophys Acta 910 : 197
6) Moses, K., Willmore, E. & Durkacz B.W. (1990) Cancer Res. 50 : 1992
7) Kincaid, R.L. and Manganiello, V.C.(1988) Methods in Enzymol. 159 : 457

Enhancement of antimetabolite cytotoxicity by 3-aminobenzamide in Chinese hamster ovary cells is independent of poly(ADP-ribose) polymerase inhibition.

Elaine Willmore & Barbara W. Durkacz

Cancer Research Unit, Medical School, University of Newcastle-upon-Tyne, NE2 4HH U.K.

Introduction

Although the biochemical mechanisms of poly(ADP-ribose) polymerase (PADPRP) have been largely elucidated, its biological roles are far from understood. The majority of work relating to function has relied on the use of competitive inhibitors of PADPRP, the 3-position substituted benzamides. This type of work has implicated poly(ADP-ribosylation) reactions in cellular processes as diverse as gene transcription (e.g. 1) and DNA repair (for a review, see 2). However, the benzamides may have other metabolic effects unrelated to an inhibition of PADPRP.

The best validated role for PADPRP is in the process of DNA repair. PADPRP is activated by DNA strand breaks, and inhibitors potentiate the cytotoxicity of a variety of DNA damaging agents by increasing net DNA strand break frequencies. It has been postulated that poly(ADP-ribosylation) modulates repair by altering local chromatin structure at damaged sites, or by affecting the activity of repair enzymes or endonucleases.

Since most antimetabolites used in cancer chemotherapy are cytotoxic because of their effect on DNA, we postulated that PADPRP inhibitors may potentiate their cytotoxicity. 5-Fluorouracil (FU) is incorporated into RNA and DNA. It also inhibits thymidylate synthase (3) and this effect leads to misincorporation of FdUTP and dUTP into DNA which are subject to excision repair. 6-Mercaptopurine (MP) inhibits *de novo* purine synthesis and is incorporated into DNA as 6-Thioguanine (4).

We investigated the effects of 3-aminobenzamide (3AB) on the cytotoxicity of MP and FU in Chinese hamster ovary cells (CHO-K1). 3AB enhanced the cytotoxicity of both MP and FU, but the data obtained lead us to conclude that PADPRP inhibition does not contribute to the cytotoxic mechanism. Although the benzamides are powerful tools for elucidating PADPRP function, these results emphasise the necessity of excluding possible effects on other metabolic processes.

Results and Discussion

Using clonogenic survival assays, it was found that 3AB (3mM) enhanced the cytotoxicity of FU with a dose enhancement factor (DEF) at 10% survival of 2 (Fig. 1). We measured NAD levels in FU-treated cells (Fig. 1). In contrast to cells treated with monofunctional alkylating agents, in which NAD is depleted rapidly by PADPRP activation (within 1-2 hours), there was no immediate effect on NAD levels. NAD did decrease in FU-treated cells

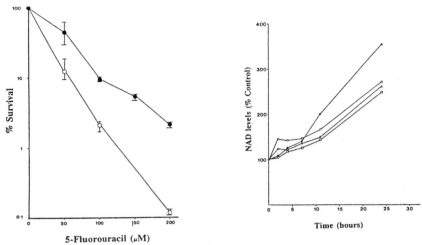

Figure 1 (Left) Survival of CHO-K1 cells following a 16 hr exposure to FU in the presence (O) and absence (●) of 3mM 3AB.
(Right) CHO-K1 cells were harvested at intervals after FU addition and assayed for NAD. (●) Untreated; (O) 50μM ; (△) 150μM ; (□) 200μM.

relative to controls over a 24 hour period, but this decrease was not prevented by 3AB (results not shown) and therefore cannot be attributed to increased PADPRP activity. Alkaline elution studies were used to study the effect of FU on DNA strand breakage. Although there was no effect in mature (high molecular weight) DNA (results not shown), we observed a dose-dependent increase in nascent DNA strand breaks, presumably due to excision of uracil and FU residues **(Fig. 2)**. 3AB treatment did not increase the number of FU-induced DNA strand breaks, but in some cases decreased the amount of breaks observed. These results indicate that modulation of FU-induced DNA repair processes by 3AB is not responsible for the enhanced cytoxicity involved.

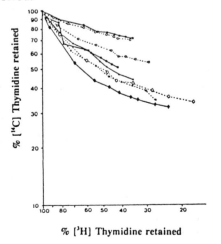

Figure 2. Effect of FU on nascent DNA. CHO-K1 cells were exposed to FU for 24 hrs and labelled during the last 4 hrs of drug exposure with $[^{14}C]$ thymidine prior to elution so that only nascent DNA was labelled. Internal standard (control) cells were labelled with $[^{3}H]$ thymidine for 4 hours then X-irradiated (300cGy) prior to elution. For FU alone (●) Control; (■) 50μM; (▲) 100μM; (◆) 150μM. For 3mM 3AB + FU; (O) 3AB alone; (□) +50μM FU; (△) +100μM; (◇) +150μM.

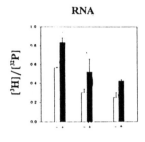

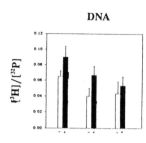

Figure 3. Effect of 3AB on the incorporation of [³H] FU into nucleic acids. Cells were exposed simultaneously to [³H]FU and H₃[³²P]O₄ for 16 hrs before harvesting for Caesium Sulphate density gradient centrifugation. Histograms show ratios of [³H]/[³²P] in the presence (■) and absence (□) of 3mM 3AB.

Caesium sulphate density gradient centrifugation was used to study the amount of [³H]FU incorporated into nucleic acids both in the absence and presence of 3AB. There was a significant increase in the amount of [³H]FU incorporated into RNA (at least 50%) and DNA (45%) in the presence of 3AB (**Fig. 3**), suggesting that the enhanced cytotoxicity of FU by 3AB is due to increased salvage of the base.

3AB potentiated the cytotoxicity of MP with a DEF of 30 at 10% survival (**Fig. 4 (a)**). We also used other inhibitors of PADPRP to see if these had a similar effect. **Fig. 4 (b) (c) & (d)** show that neither 3-acetamidobenzamide (3AAB, 1mM), 3-methoxybenzamide (3MB, 1mM) or Benzamide (B, 1mM) enhanced the cytotoxicity of MP as effectively as 3AB, the maximum DEF obtained being 2.5 (for 3AAB) at 10% survival. 3AAB, 3MB & B all have IC50 values for PADPRP which are lower than that of 3AB (see legend for **Fig. 4**), indicating that they are more potent inhibitors of the enzyme. This result indicates that the degree of inhibition of poly(ADP-ribosylation) which

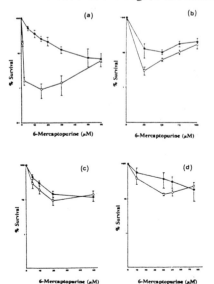

a benzamide affords does not correlate with the degree of enhancement of cytotoxicity of MP. It is apparent, then, that 3AB enhances MP cytotoxicity by a mechanism independent of inhibition of poly(ADP-ribosylation). This was confirmed by the result that MP does not significantly reduce NAD levels (results not shown), indicating that there is little or no activation of PADPRP.

Figure 4. Survival of CHO-K1 cells following a 16 hr exposure to MP in the presence (O) and absence (●) of (a) 3mM 3AB, IC50 value 10μM (b) 1mM 3AAB, IC50 2.8μM (c) 1mM B, IC50 4μM and (d) 1mM 3MB, IC50 5.5μM.

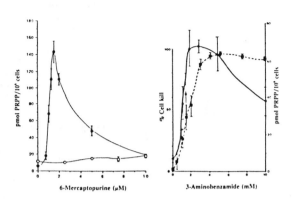

Figure 5. (Left) CHO-K1 cells were treated for 16 hrs with increasing doses of MP in the presence (●) and absence (O) of 3mM 3AB and assayed for PRPP. (Right) Cells were treated for 16 hrs with 1.5μM MP in the presence of increasing doses of 3AB. (■) Cell kill; (◆) PRPP levels.

Since our results ruled out an involvement of 3AB in inhibition of DNA repair, we speculated that 3AB could be having a synergistic effect on the inhibition of *de novo* purine synthesis by MP. The first step in *de novo* purine synthesis is rate-limited by phosphoribosylpyrophosphate (PRPP). The first enzyme which converts MP into its active metabolites (hypoxanthine-guanine phosphoribosyl transferase, HGPRT) is also rate-limited by PRPP. We measured PRPP levels in MP and MP + 3AB treated cells (**Fig. 5**). There was a 14-fold increase in PRPP levels when cells were treated with MP + 3AB ($\sim$ 142 pmol/ 10^6 cells) as compared with MP alone ($\sim$ 10 pmol/ 10^6 cells) at a value of 2μM MP (**Fig. 5, left**). This rise in PRPP levles correlated well with the increase in cytotoxicity (**Fig. 5, right**) and indicates a target for 3AB (currently under investigation) in the *de novo* purine synthesis pathway.

Summary

We have demonstrated that 3AB potentiates the cytotoxicity of both FU and MP by mechanisms independent of an effect on DNA repair. In the case of FU, 3AB increases misincorporation of fraudulent nucleotides into RNA and DNA. In MP-treated cells, 3AB causes a synergistic rise in PRPP levels which presumably leads to augmented salvage of MP by HGPRT. Both these novel effects cause enhancement of antimetabolite cytotoxicity (which may be of clinical significance) and verify that the metabolic effects of 3AB are not specific to an inhibition of poly(ADP-ribosylation). Therefore, it is clear that when these inhibitors are used to ascribe roles for poly(ADP-ribosylation) in cellular functions, it is important to exclude other possible metabolic targets.

References

1. Tanuma, S., Kawashima, K. & Endo, H. (1987) Biochim. Biophys. Acta 910: 197-201.
2. Cleaver, J.E. & Morgan, W.F. (1991) Mutat. Res. 257: 1-18.
3. Piper, A.A. & Fox, R.M. (1982) Cancer Research 42: 3753-3760
4. Liliemark, J., Pettersson, B., Engberg, B., Lafolie, P., Masquelier, M., & Peterson, C. (1990) Cancer Research 50:108-112.
5. Milam, K.M. & Cleaver, J.E. (1984) Science 223:589-591

Evidence for the Participation of Poly(ADP-ribosyl)ation in Collagenase Gene Expression in Rabbit Synovial Fibroblasts after Treatment with Active Oxygen Released by Xanthin/Xanthinoxidase

Wilhelm Ehrlich, Hans Huser, and Hans Kröger

Robert Koch-Institut, Abt. Biochemie,Nordufer 20, D-1000 Berlin 65, Germany

Introduction

Rheumatoid arthritis is an autoimmune disease characterized by chronic inflammation. In this disease the membrane (synovium) that lines the joint proliferates into a huge mass that invades and destroys the connective tissues, i.e. articular cartilage, tendon and bone (1). It is suggested that this is a tumor like process which is induced by the combined action of autoantigen stimulated polymorphonuclear leukocytes, T and B cells, and monocytes/macrophages (1, 2, 3). The proliferating tissue is compromised primarily of the fibroblast-like synovial cells that line the joint and that are derived from mesenchymal stem cells (1). The neutral proteinase collagenase is a principle gene product of the synovial fibroblasts and is responsible for the extensive amount of connective tissue destruction seen in rheumatoid disease (1, 4 - 7).
Stimulated inflammatory polymorphonuclear leukocytes and monocytes/macrophages release large amounts of active oxygen (AO), in particular superoxide and hydrogenperoxide that are converted to the hydroxyl radical in the presence of free metall ions, i.e. iron and copper (8, 9).
There is strong evidence that AO can act as tumor promotors and contribute to cell transformation and that poly(ADP-ribose) is involved in this process by modulating the expression of specific genes (10, 11, 12). The gene coding for collagenase belongs to a family of genes that are inducible by phorbolesters (13, 14, 15) which are well known tumor promoters (10, 13).
Here we report on experiments which indicate that similar events take place in the genesis of rheumatoid arthritis. For our studies we use primary cultures of synovial fibroblasts isolated from the joints of rabbits. We determined the accumulation of poly(ADP-ribose) and the induction of collagenase in these cells after treatment with xan-

thin/xanthinoxidase (x/xo). This system mimics the release of AO from polymorphonuclear leukocytes and monocytes/macrophages during inflammation (11, 16, 17).

Results and Discussion

Monolayer cultures of rabbit synovial fibroblasts were established from the knee synovium of young rabbits as described previously (18, 19, 20). They were treated with increasing amounts of AO released by xanthin/xanthinoxidase for 30 min in medium 199 supplemented with 8 percent fetal calf serum. Cells were washed and supplied with Dulbecco's modified Eagle's medium (DMEM) plus 0.2 percent lactalbumin hydrolysate. The culture supernatants taken after 0, 24 and 48 hrs respectively were monitored for collagenase activity, which represents collagenase secreted by the cells, with a spectrophotometric assay using rat tail collagen coated microtiter plates (21). For examining the accumulation of poly(ADP-ribose) cells were harvested after treatment and the content of poly(ADP-ribose) was determined according to the method of Jacobson (22) which avoids cell permeabilization (ε-RAdo, $1,N^6$-ethenoribosyladenosine). DNA content was quantified as described by Labarca and Paigen (23).

From Fig. 1 a and 1 b it is evident that AO generated by x/xo cause a strong induction of collagenase in rabbit synovial fibroblasts. There is a large almost linear dose dependent increase in collagenase activity found in culture supernatants 48 hrs after exposure of cells to increasing concentrations of x/xo from 12.5 µg/ml x plus 0.0025 U/ml xo to 50 µg/ml x plus 0.01 U/ml xo. Moreover, there is a coincident accumulation of poly(ADP-ribose) in intact synovial fibroblasts following exposure to increasing concentrations of x/xo as shown in Fig. 2. From the good correlation between increase in collagenase and poly(ADP-ribose) levels it appears that poly(ADP-ribose) may indeed play a role in collagenase gene expression. This is confirmed by other experiments from us which show that the induction of collagenase by interleukin-1ß in synovial fibroblasts is suppressed by the poly(ADP-ribose) polymerase inhibitor 3-aminobenzamide (data not shown). Support for a role of poly(ADP-ribose) in this process and in the genesis of rheumatoid arthritis in general comes from the fact that autoantibodies against poly(ADP-ribose) and poly(ADP-ribose) polymerase occur in patients with rheumatoid diseases (24, 25). In further studies we will investigate whether the effects of AO on synovial cells are exerted through DNA breakage and chromosome alterrations. Future work will also include

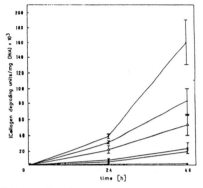
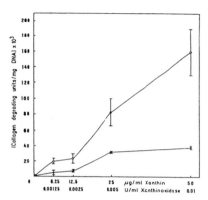

Fig. 1 a. Time dependent induction of collagenase in rabbit synovial fibroblasts after 30 min treatment of monolayer cultures with increasing amounts of active oxygen generated by (x) 6.25 μg/ml xanthin (x) + 0.00125 u/ml xanthin oxidase (xo), (■) 12.5 μg/ml x + 0.0025 U/ml xo, (▲) 25 μg/ml x + 0.005 U/ml xo, and (●) 50 μg/ml x + 0.01 U/ml xo. (O) untreated control and (△) cells treated with 1.67 U/ml interleukin-1β. The culture supernatants taken after 0, 24 and 48 hrs following exposure were monitored for collagenase activity with a spectrophotometric assay using rat tail collagen coated microtiter plates.

Fig. 1 b. Induction of collagenase in rabbit synovial fibroblasts in response to increasing amounts of active oxygen. Monolayer cultures were exposed for 30 min to increasing concentrations of AO generated by xanthin/xanthinoxidase as indicated. Following treatment cells were cultured for (●) 24 and (O) 48 hrs respectively and culture supernatants were monitored for collagenase activity with a spectrophotometric assay using rat tail collagen coated microtiter plates.

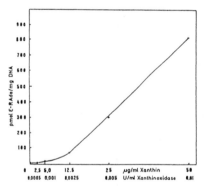

Fig. 2. Poly(ADP-ribose) accumulation in rabbit synovial fibroblasts in response to increasing concentrations of active oxygen. Monolayer cultures were exposed for 30 min to increasing concentrations of AO generated by xanthin/xanthinoxidase as indicated. The cells were harvested and the poly(ADP-ribose) concentrations determined according to Jacobson et al. (22).

inhibitors of poly(ADP-ribosyl)ation and their potential to suppress collagenase induction. This will be done in view of their use as agents in rheumatoid arthritis therapy.

Acknowledgements. We would like to thank Dr. F.R. Althaus for his helpful advice concerning pADPR determination. We also thank Ms. H. Wohlert, Ms. M. Klewer, Ms. R. Grätz and Ms. K. Krohn for excellent technical assistance. This work was supported by the Völkl-Stiftung.

References

1) Harris, E.D. (1985) Textbook of Rheumatology. W.N. Kelly, E.D. Harris, Jr., S. Ruddy and C.B. Sledge(eds.) C.B. Saunders & Co., Philadelphia, pp. 886-914.
2) Harris, E.D. and Sporn, M.B. (1981) American J Medicine 70:1231-1236
3) Brinckerhof, C.E. and Harris, E.D. (1981) Am J Pathology 103:411-419
4) Murphy, G., Nagase, H., and Brinckerhoff, C.E. (1988) Coll Rel Res 8:389-391
5) Gross, R.H., Sheldon, L.A., Fletcher, C.F., and Brinckerhoff, C.E. (1984) Proc Natl Acad Sci 81:1981-1985
6) Fini, M.E., Karmilowicz, M.J., Ruby, P.L., Beeman, A.M. Borges, K.A., and Brinckerhoff, C.E. (1987) Arthritis and Rheumatism 30:1254-1264
7) Frisch, S.M., Clark, E.J. and Werb, Z. (1987) Proc Natl Acad Sci 84:2600-2604
8) Weiss, S.J. (1989) N Engl J Med 320:365-376
9) Segal, A.W. (1989) Human Monocytes. M. Zembala, G.L. Asherson (eds.), Academic Press, New York, pp. 89-100.
10) Cerutti, P. (1985) Science 227:375-381
11) Mühlematter, D., Larsson, R., and Cerutti, P. (1988) Carcinogenesis 9:239-245
12) Cerutti, P., Krupitza, G., and Mühlematter, D. (1989) ADP-Ribose Transfer Reactions. M.K. Jacobson and E.L. Jacobson (eds.), Springer Verlag, New York, pp. 225-234.
13) Karin, M. and Herrlich, P. (1988) Genes and Signal Transduction in Multistage Carcinogenesis, N.H. Colbuan (ed.) Marcel Dekker, Inc., New York.
14) Angel, P., Baumann, I., Stein, B., Delius, H., Rahmsdorf, H.J., and Herrlich, P. (1987) Mol Cell Biol 7:2256-2266
15) Brinckerhoff, C.E., McMillan, R.M., Fahey, J.V. and Harris, E.D. (1979) Arthritis and Rheumatism 22:1109-1116
16) Kellogg, E. and Fridovich, I. (1975) J Biol Chem 250:8812-8817
17) Lautier, D., Poirier, D., Boudreau, A., Alaoui Jamali, M.A., Castonguay, A., Poirier, G. (1990) Biochem Cell Biol 68:602-608
18) Kieval, R.I., Young, C.T., Prohazka, D., Brinckerhoff, C.E. and Trentham, D.E. (1989) J Rheum 16:67-74
19) Dayer, J.M., Krane, S.M., Russel, R.G.G., and Robinson, D.R. (1976) Proc Natl Acad Sci 73:945-949
20) Brinckerhoff, C.E., Benoit, M.C., and Culp, W.J. (1985) Proc Natl Acad Sci 82:1916-1920
21) Nethery, A., Lyons, J.G., and O'Grady, R.L. (1986) Anal Biochem 159:390-395
22) Jacobson, M.K., Payne, M., Juarez-Salinas, H., Sims, J., and Jacobson, E.L. (1984) Methods Enzymol 106:483-494
23) Labarca, C. and Paigen, K. (1980) Anal Biochem 102:344-352
24) Yamanaka, H., Willis, E.H., Penning, C.A., Peebles, C.L., Tan, E.M., and Carson, D.A. (1989) ADP-Ribose Transfer Reactions. M.K. Jacobson, E.L. Jacobson (eds.), Springer Verlag, New York, pp. 139-144
25) Althaus, F.R. and Richter, C. (1987) ADP-Ribosylation of Proteins, Springer Verlag Berlin, pp. 115-116

Enhancement of Oncogene-Mediated Transformation in Cloned Rat Embryo Fibroblast (CREF) Cells by 3-Aminobenzamide

Zao-zhong Su and Paul B. Fisher

Departments of Pathology and Urology, Cancer Center/Institute of Cancer Research, Columbia University, College of Physicians and Surgeons, New York, New York 10032, U.S.A.

The carcinogenic process involves a complex series of genetic alterations in the evolving tumor cell resulting in changes in genes which induce both positive (oncogene) and negative (tumor suppressor) control of cellular proliferation (1-3). In vitro studies employing rat embryo fibroblast cells and more recently in vivo studies using transgenic mice have provided convincing evidence that specific oncogenes can cooperate in inducing both cellular transformation and the development of specific neoplasms (1-3). Based primarily on data obtained using early passage and established rodent cells it appears that oncogenes which act in the nucleus can cooperate most efficiently with oncogenes which act in the cytoplasm (1,3). Cooperative interactions between nuclear oncogenes, such as myc, N-myc, mutant p53, fos, jun, adenovirus (Ad) E1A, polyomavirus and SV40 large T antigens, papillomavirus E7 and tax of human T cell leukemia virus type 1, and cytoplasmic oncogenes, such as Ha-ras, Ki-ras, N-ras, src and polyomavirus middle T antigen, have been described (reviewed in 1,3). However, there are examples in which two nuclear oncoproteins (4) or two cytoplasmic oncoproteins (5) can cooperate in inducing transformation of rat embryo fibroblasts. Recent studies have also indicated that overexpression of the signal-transducing molecule protein kinase C can cooperate with both a cytoplasmic oncogene Ha-ras (6) and a nuclear acting oncogene E1A (7). Evidence has also been presented indicating that transformed cells expressing both nuclear and cytoplasmic oncogenes can be further progressed or inhibited in their expression of the transformed phenotype by the transfer and expression of a second cytoplasmic or nuclear acting oncogene (8-10). The mechanisms underlying these complex interactions between different classes of oncogenes in mediating cellular transformation remain to be defined. One hypothesis is that cooperating oncogenes induce distinct signal pathway alterations in cells resulting in the induction and suppression of specific genes as a consequence of modulation in transcription (reviewed in 3).

Our laboratory has studied the role of cellular and viral genes in mediating the initiation and the maintenance of stable expression of cellular transformation by adenovirus type 5 (Ad5) (11-22). Employing a specific clone of Fischer rat embryo fibroblast (CREF) cells (23), which is transformed >150- to >200-fold more efficiently by Ad5 than uncloned early passage rat embryo cells, we have demonstrated that specific carcinogenic and DNA damaging agents can enhance the frequency of viral transformation (15,16,18-20). In the CREF model system, the poly(ADP-ribose) synthesis inhibitor 3-aminobenzamide (3AB) has been shown to: (a) enhance de novo and methyl methanesulfonate (MMS) and γ irradiation enhancement of viral transformation of CREF cells by a cold-sensitive host-range mutant of type 5 adenovirus, H5hr1; (b) enhance transformation of CREF cells by Ad5 E1A or the Ha-ras (T24) oncogene; and (c) enhance the frequency of antibiotic resistant colonies after transfection with a bacterial neomycin or hygromycin resistance gene (20). Enhancement of viral transformation in CREF cells was not associated with a selective effect on the growth of Ad5 transformed cells in either monolayer or agar culture (20). Similarly, 3AB did not alter the percentage of MMS- or γ-irradiated-pretreated H5hr1-infected cells retaining free Ad5 DNA or the random pattern or amount of viral DNA integrated in control or carcinogen-treated H5hr1-transformed cells (20). As indicated in Fig. 1, 3AB also did not alter the steady-state levels of Ad5 E1A or glyceraldehyde phosphate dehydrogenase (GAPDH) mRNA in the H5hr1-transformed CREF clone, A2. These observations suggest that cellular processes influenced by the nuclear enzyme ADPRT, or additional processes affected by 3AB, may be important mediators of stable transformation induced by transfected DNA and both de novo and carcinogen-enhanced viral transformation of specific target cells, such as CREF.

Growth of Ad5 E1A, Ha-ras (T24) and v-src transfected CREF cells in 3AB resulted in enhanced transformation (Table 1). Although transformation frequencies varied from experiment to experiment and the degree of oncogene-enhancement of transformation induced by 3AB was often small (between 1.3- and 6.4-fold), additional experiments have resulted in similar statistically significant increases in transformation frequencies induced by 3AB (20 and data not shown). These results indicate that 3AB can modify trans-

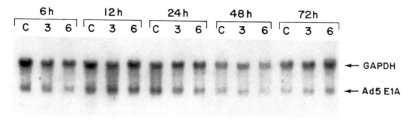

Fig. 1. Effect of 3AB on steady-state Ad5 E1A and GAPDH mRNA levels in the H5hr1-transformed CREF clone, A2. Logarithmically growing cells were grown without (C) or with 3AB (3 or 6 mM) for 6, 12, 24, 48 or 72 hr, total cytoplasmic RNA was isolated, mRNAs were separated by electrophoresis through 0.6% agarose gels, transferred to nitrocellulose filters and probed with a combination of a multiprimed [^{32}P]-labeled Ad5 E1A and a GAPDH probe (20,21).

formation induced by genes acting either in the nucleus (Ad5 E1A) or in the cytoplasm (Ha-ras and v-src). In contrast, 3AB inhibits the ability of Ha-ras and bacterial resistance genes to transform NIH 3T3 cells (20,24,25). Studies by Diamond et al. (24), also indicate that 3AB can inhibit v-raf-, v-mos- and v-src-induced transformation of NIH 3T3 cells, whereas 3AB did not alter v-fos transformation and enhanced SV40 transformation of NIH 3T3 cells. These results indicate that the ability of 3AB to modify transformation is both cell type and oncogene specific.

The combination of the Ad E1A and the Ha-ras oncogene results in enhanced transformation in both early passage rat embryo cells and CREF cells (26,27). To determine if 3AB could enhance the ability of the Ad5 E1A and the Ha-ras gene to cooperate in inducing transformation we have exploited the differences in morphologies of transformants induced by wild-type Ad5 E1A (wt-E1A) versus Ha-ras. Ad5 E1A results in transformed CREF cells with an epithelioid appearance, whereas Ha-ras results in transformed CREF cells with a distinctive fibroblastic morphology. CREF cells containing both Ad5 E1A and Ha-ras display the fibroblastic morphology characteristic of Ha-ras transformants. Transfection of Ad5 E1A-transformed CREF cells with Ha-ras results in fibroblastic double-transformants which can be easily identified on an epithelioid Ad5 E1A-transformed CREF background. By employing CREF cells transformed by an Ad5 E1A gene under the transcriptional control of an MMTV promoter [inducible by dexamethasone (DEX)] (20,28) (MMTV-

Table 1. Effect of 3AB on Morphological Transformation of CREF Cells Transfected with an Ad5 E1A, a Ha-ras (T24) or a v-src Oncogene.

Transforming Gene	DNA (μg)	Experimental Conditions [a]		
		No Additions	5 mM 3AB	10 mM 3AB
Ad5 E1A	5	26	44 (1.7)	72 (2.8)
	10	58	96 (1.7)	109 (1.9)
	20	79	127 (1.6)	140 (1.8)
	30	104	154 (1.5)	210 (2.0)
Ha-ras (T24)	5	16	20 (1.3)	30 (1.9)
	10	48	67 (1.4)	78 (1.6)
	20	63	91 (1.4)	131 (2.1)
v-src	5	8	14 (1.6)	N.T.
	10	15	25 (1.7)	N.T.
	20	40	77 (1.9)	N.T.

[a]CREF cells were seeded at 1×10^6/10 cm plate, 24 hr later cultures were transfected with 5 to 30 μg of oncogene DNA, 6 hr later cells were briefly glycerol shocked and then incubated for an additional 2 days at 37°C in complete growth medium (DMEM supplemented with 5% FBS) (20). Cultures were then subcultured at 1 or 3×10^5 cells/6 cm plate in DMEM-5 with and without 5 or 10 mM 3AB. Cultures were fed 2X per week with and without 3AB and transformed colonies were counted 21 days later.

E1A) we were able to directly monitor the effect of Ad5 E1A expression and Ha-ras, alone and in combination with 3AB, on Ha-ras-induced transformation. As shown in Fig. 2A, Ha-ras results in approximately a 2-fold higher frequency of transformation when transfected into wt-E1A CREF or MMTV-E1A CREF grown in 10^{-7} DEX, as opposed to CREF, CREF grown in 10^{-7} DEX or MMTV-E1A CREF grown in the absence of DEX. This enhancement effect was not simply related to a DEX modification of Ha-ras transformation, since wt-E1A CREF cells were transformed by Ha-ras at the same efficiency in the presence or absence of 10^{-7} DEX. When grown in 3AB, the frequency of transformation induced by Ha-ras was increased under all experimental conditions, i.e. CREF alone, CREF + 10^{-7} DEX, wt-E1A CREF, wt-E1A CREF + 10^{-7} DEX, MMTV-E1A CREF and MMTV-E1A CREF + 10^{-7} DEX. As shown in Fig. 2B, cotransfection of CREF cells with wt-E1A and v-src in the presence of 3AB also enhances transformation above that observed with the combination of oncogenes in the absence of

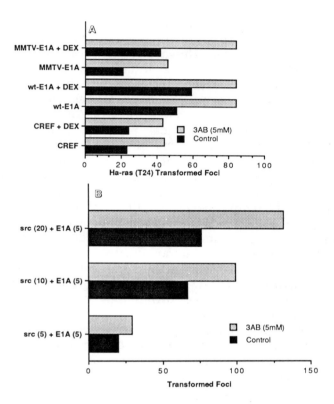

Fig. 2. Effect of 3AB on transformation induced by single and multiple oncogenes. (A) CREF, wt-E1A-transformed CREF and MMTV-E1A-transformed CREF cells were transfected with 10 μg of Ha-ras (T24) DNA, reseeded at 1 X 10⁵ cells/6 cm plate in media lacking or containing 10^{-7} M DEX and grown in the presence or absence of 3AB and DEX for 3 weeks with media changes 2X per week (20). Values indicate the average number of Ha-ras transformed foci per plate from 5 to 7 plates per condition. Replicate plates varied by ≤ 15%. (B) CREF cells were co-transfected with v-src + Ad5 E1A DNA (amount of DNA in μg given in brackets), reseeded at 1 X 10⁵ cells/6 cm plate in media lacking or containing 5 mM 3AB. and refed 2X per week ± 3AB. Transformed foci were enumerated after three weeks growth. Values indicate the average number of transformed foci per plate from 5 to 7 plates per condition. Replicate plates varied by ≤ 15%.

3AB. These results indicate that 3AB can enhance transformation induced by a single nuclear acting or cytoplasmic acting oncogene as well as the combination of both types of oncogenes.

At present, we do not know the mechanism by which 3AB enhances single- and multiple-oncogene-mediated-transformation of CREF cells. However, since the mechanism by which Ad5 E1A induces cellular transformation is apparently different than that involved in transformation induced by cytoplasmic acting oncogenes, such as Ha-ras and v-src, it is reasonable to assume that 3AB can alter transformation occurring by different biochemical pathways. This hypothesis is supported by studies indicating that 3AB can also enhance transformation in specific target cells by both physical and chemical carcinogens or the combination of carcinogens and viruses (20,29,30). The complexity of 3AB effects on transformation is emphasized by the observations that this agent can also inhibit transformation induced by carcinogens and specific oncogenes in certain target cells, such as C3H 10T1/2, NIH 3T3 and human fibroblasts (20,25,31-33). Although the reason for the differential response to 3AB in inducing transformation in not known, current evidence suggests that it may involve differences in the experimental protocols utilized and/or the specific cell types employed. In addition, since poly(ADP-ribose) polymerase is involved in a number of important cellular processes (reviewed in 34), it may be difficult to identify specific biochemical changes involved in mediating the effects of 3AB on carcinogen and oncogene induced transformation. In the case of CREF cells, enhanced transformation does not appear to involve a direct effect of 3AB on viral DNA integration or expression of the Ad5 E1A transforming gene at a mRNA level. Further studies are required, however, to determine if 3AB can alter Ad5 E1A, and or Ha-ras or v-src, at a post-transcriptional level. It also remains to be determined if 3AB can alter the interaction between Ad5 E1A and cellular proteins which function as either promoters or suppressors of transformation (1-3,22,27,35). Elucidation of the mechanism by which 3AB enhances viral and oncogene-mediated transformation of CREF cells should provide new insights into common biochemical pathways which may underlie transformation induced by diverse oncogenic agents.

Acknowledgements

This research was supported by National Cancer Institute Grants CA35675 and CA43208. P.B.F. is a Chernow Research Scientist in the Departments of Pathology and Urology.

References:

1. Weinberg RA. Science, 230:770-776, 1985
2. Bishop JM. Cell, 64:235-248, 1991.
3. Hunter T. Cell, 64:249-270, 1991.
4. Ruppert JM, Vogelstein B & Kinzler KW. Mol Cell Biol 11:1724-1728, 1991.
5. Reed JC, Haldar S, Croce CM & Cuddy MP. Mol Cell Biol 10: 4370-4374, 1990.
6. Hsiao WL, Housey GM, Johnson MD & Weinstein IB. Mol Cell Biol 9: 2641-2647, 1989.
7. Su ZZ, Duigou GJ & Fisher PB. Mol. Carcinogenesis, in press, 1991.
8. Duigou GJ, Babiss LE & Fisher PB. NY Acad. Sci. 567:302-306, 1988.
9. Yu D, Scorsone K & Hung M-C. Mol. Cell. Biol. 11:1745-1750, 1991.
10. Su ZZ, Leon JA, Austin V, Zimmer SG & Fisher PB. In preparation, 1991.
11. Babiss LE, Ginsberg HS & Fisher PB. Proc. Natl. Acad. Sci. USA 80:1352-1356, 1983.
12. Babiss LE, Fisher PB & Ginsberg HS. J Virol 49:731-740, 1984.
13. Babiss LE, Zimmer SG & Fisher PB. Science 228:1099-1101, 1985.
14. Babiss LE, Liaw WS, Zimmer SG, Godman GC, Ginsberg HS & Fisher PB. Proc. Natl. Acad Sci USA 83: 2167-2171, 1986.
15. Hermo H Jr, Duigou GJ, Babiss LE & Fisher PB. Carcinogenesis 8:967-975, 1987.
16. Hermo H Jr, Duigou GJ, Zimmer SG & Fisher PB. Cancer Res.48:3050-3057,1988.
17. Herbst RS, Hermo H Jr, Fisher PB & Babiss LE. J Virol 6:4634-4643, 1988.
18. Su, ZZ & Fisher PB. Mol Carcinogenesis 2:252-260, 1989.
19. Su ZZ, P Zhang, C Geard & Fisher PB. Mol. Carcinogenesis 3:141-149, 1990
20. Su ZZ, P Zhang & Fisher PB. Mol. Carcinogenesis 3:309-318, 1990.
21. Su ZZ, D Grunberger & Fisher PB. Mol. Carcinogenesis, in press, 1991.
22. Duigou GJ, Su ZZ, Babiss LE, Driscoll B, Fung YKT & Fisher PB. Oncogene, in press, 1991.

23. Fisher PB, Babiss LE, Weinstein IB & Ginsberg HS. Proc Natl Acad Sci USA 79;3527-3531, 1982.
24. Diamond AM, Der CJ & Schwartz JL. Carcinogenesis 10:383-385,1989.
25. Fazaneh F, Panayotou GN, Bowler LD, Hardas BD, Broom T, Walther C & Shall S. Nucleic Acids Res 16:11319-11326, 1988.
26. Ruley HE. Nature 304:602-606, 1983.
27. Lillie JW, Lowenstein PM, Green MR & Green M. Cell 50:1091-1100, 1987.
28. Brunett LM & Berk AJ. Mol Cell Biol 8:4799-4807, 1988.
29. Lubet RA, McCarvill JT, Putman DL, Schwartz JL & Schectman LM. Carcinogenesis 5:459-462, 1984.
30. Kasid UN, Stefanik DF, Lubet RA, Dritschilo A & Smulson ME. Carcinogenesis 7:327-330,1986.
31. Borek C, Morgan WF, Ong A & Cleaver JE. Proc Natl Acad Sci USA 81:243-247, 1984.
32. Borek C, Ong A, Morgan WF & Cleaver JE. Radiat Res 99:219-227, 1984.
33. Milo GE, Kurian P, Kirstein E & Kun E. FEBS Lett 179:332-336, 1985.
34. Cleaver JE, Borek C, Milam K & Morgan WF. Pharmacol Ther 31:269-293, 1985.
35. Green MR. Cell 56:1-3, 1989.

CONTROL OF PROCOLLAGEN GENE TRANSCRIPTION AND PROLYL HYDROXYLASE ACTIVITY BY POLY(ADP-RIBOSE)

Q. Perveen Ghani, M. Zamir Hussain,
Jincai Zhang, and Thomas K. Hunt

INTRODUCTION

The synthesis of collagen increases during induced injury, growth and wound healing. There is an emerging consensus that collagen synthesis is controlled at the level of collagen gene transcription. The synthesis of specific collagen types in keloids (Bauer et al, 1986), virus transformed fibroblasts (Adams et al, 1982; Parker et al, 1979) and transforming growth factor-β (TGF-β) or interleukin-1 (IL-1) treated fibroblasts (Ignotz et al, 1987; Postlethwaite et al, 1988) is accompanied by an increase in the levels of the respective procollagen mRNAs. However, molecular events leading to this stimulation is not clear. Poly(ADP-ribose) is closely associated with the differentiation of several eukaryotic cells (Farzaneh et al, 1982; Ohashi et al, 1985) and is also implicated in the expression of individual genes (Kun et al, 1986; Poirer et al, 1982). Tanuma et al (Tanuma et al, 1983) showed a clear relationship between ADP-ribosylation of HMG 14/HMG 17 and glucocorticoid regulated expression of MMTV gene. The use of 3-aminobenzamide (3-AB), lowered poly(ADP-ribose) levels on these nuclear proteins and enhanced the level of MMTV mRNA. Similarly, the expressions of c-myc and c-fos proto-oncogenes were stimulated by the treatment of fibroblasts with 3-methoxybenzamide (McNerney et al, 1987). Earlier, we found that exposure of fibroblasts to 20 mM lactate lowered [NAD^+], total ADP-ribosylation and activated prolyl hydroxylase causing increased collagen synthesis (Hussain et al, 1989). The activity of prolyl hydroxylase usually correlates with the rate of collagen synthesis and generally increases when elevated amounts of collagen are produced (Hussain et al, 1989). We report here the effects of changing the level of poly(ADP-ribose) synthesis on $\alpha 1$ (I) and $\alpha 1$ (III) procollagen mRNA levels and the activity of prolyl hydroxylase. Results of these studies indicated that ADP-ribosylation may be involved in collagen gene transcription as well as the post-translational prolyl hydroxylation and thus regulate overall synthesis of collagen.

RESULTS AND DISCUSSION

In this study poly(ADP-ribose) synthesis was modulated in two ways. First, confluent cultures of human skin fibroblasts were treated with non-toxic concentration of 5 mM 3-AB or 10 mM nicotinamide, inhibitors of nuclear

poly(ADP-ribose) synthetase for 18 hours. Second, ADP-ribosylation was reduced by depleting intracellular NAD^+ through lactate (20 mM) treatment (Hussain et al, 1989) and was increased by exposing fibroblasts to exogenous NAD^+ (Ghani et al, 1990). Earlier experiments had showed that an 18 hour exposure to 1.0 mM NAD^+ increased the cellular NAD^+ pool by more than 3 X fold and poly(ADP-ribose) synthesis by 6 X fold and that exogenous NAD^+ was translocated intact into fibroblasts (Ghani et al, 1990, Loetscher et al, 1987). Procollagen mRNAs were measured by slot-blot hybridization using pro-α1 (I) and pro-α1 (III) collagen cDNA clones.

We found that the levels of both procollagen mRNAs as well as prolyl hydroxylase activity correlated with cellular NAD^+ pool. As seen in Figure 1, when the intracellular NAD^+ was lowered by lactate or poly(ADP-ribose) synthesis was blocked by 3-AB/nicotinamide, procollagen mRNA levels were markedly stimulated (up to 4 x fold). There was no significant change in actin mRNA level under these conditions. Preliminary nuclear transcription assay using isolated nuclei from fibroblasts treated with nicotinamide also showed dose-dependent enhancement in the rate of synthesis of procollagen mRNAs. It is noteworthy that the activity of prolyl hydroxylase was not changed by poly(ADP-ribose) synthetase inhibitors but was increased by lactate as well. However, both treatments caused significant increases in collagen synthesis (Table 1). This difference can be explained if one assumes that the activating effect of lactate on prolyl hydroxylase is mediated through lowering the level of cytoplasmic ADP-ribose that normally is inhibitory to prolyl hydroxylase (Hussain et al, 1989). Our recent findings strongly support this hypothesis. Exposure of fibroblasts to non-toxic doses of oleic acid, linoleic acid, novobiocin or vitamin K_1, all potent inhibitors of cytoplasmic arginine-mono-ADP-ribosyl transferase (Sabir et al, 1991) stimulated the activity of prolyl hydroxylase 2-3 x fold. When intracellular $[NAD^+]$ was elevated by exposing cultures to 1.0 mM NAD^+, procollagen mRNA levels as well as prolyl hydroxylase activity decreased by 70% and 48% respectively (Figure 1 and Table 1).

Figure 1. Levels of collagen mRNAs in fibroblasts exposed to NAD^+ and inhibitors of poly(ADP-ribose)

	Autoradiograph		**Relative Level**	
	$\alpha1(I)$	$\alpha1(III)$	$\alpha1(I)$	$\alpha1(III)$
Control (A)			1	1
Control (B)			1	1
Nm, 10 mM			3.6	2.1
3-AB, 5 mM			4.2	2.9
NAD^+, 1 mM			0.3	0.3
Lactate, 20 mM			3.9	3.5

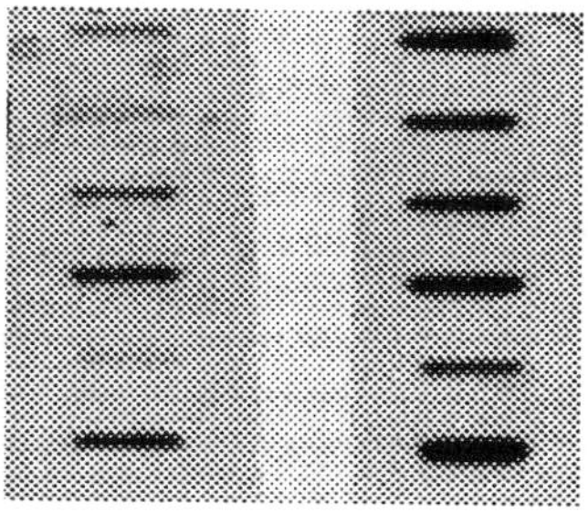

mRNA was quantitated by densitometric analysis of hybridized autoradiographes. Nicotinamide (Nm) was compared to control (B): 3-AB, NAD^+ and lactate were compared to control (A).

Table 1. Effect of changing the level of ADP-ribosylation on prolyl hydroxylase activity and collagen synthesis in fibroblasts.

Treatment	Prolyl Hydroxylase Activity dpm/mg protein $x10^{-3}$	Collagenase-Sensitive radioactivity dpm/μg DNA $x10^{-3}$
None	7.01 ± 0.68	95.6 ± 10
Lactate, 20 mM	14.79 ± 1.30	176.7 ± 18
Nicotinamide, 10 mM	6.80 ± 0.71	153.3 ± 16
3-AB, 5 mM	7.11 ± 0.69	198.1 ± 18
NAD^+, 1mM	3.65 ± 0.33	58.3 ± 6

Prolyl hydroxylase was assayed by following the release of ^3H_2O from $[4-^3H]$-proline-labeled unhydroxylated collagen from tibiae(Hussain et al, 1989). Collagen synthesis was determined by measuring collagenase susceptible radioactivity in fibroblasts labeled with 10 μCi/ml of 2,3,4,5, ^3H-proline.

Results of our studies demonstrated that ADP-ribosylation reactions may control collagen synthesis in fibroblasts. Nuclear poly(ADP-ribose) down regulated collagen gene transcription and cytoplasmic ADP-ribose down regulated the activation of prolyl hydroxylase which hydroxylates specific prolyl residues in polypeptide precursors of collagen (Prockop et al, 1979). It is known that levels of the products of both types of ADP-ribosylation reactions are sensitive to a change in cellular NAD^+ pool but the respective ADP-ribosylating enzymes have distinct inhibitor specificity. Based on these considerations, a possible scheme of regulation of collagen synthesis is presented in Figure 2. According to this scheme, a high concentration of a metabolite or an exogenous compound capable of NAD^+ reducing potential (e.g. lactate, ethanol), lowers the normal NAD^+ pool, decreasing the level of ADP-ribosylation products. Thus, lower levels of poly(ADP-ribose) in the nucleus results in an enhanced production of procollagen mRNAs that are translated into an increased level of unhydroxylated collagen peptides. Concurrently, lower levels of ADP-ribose in the cytoplasm relieve prolyl hydroxylase from ADPR suppression thereby ensuring full hydroxylation of these unhydroxylated collagen peptides. Hydroxyproline imparts characteristic stability to collagen triple helix and confers resistance to intracellular protease degradation (Kivirikko and Myllyla, 1982). Thus, by depleting both ADP-ribosylation reaction products, a compound like lactate, can act as a dual signal to 'turn on' collagen synthesis. Theoretically, a closely similar stimulating effect on collagen synthesis can be obtained by exposing fibroblasts simultaneously to inhibitors of both nuclear and cytoplasmic ADP-ribosyl transferases. Furthermore, it is logical to expect that an increase in the cellular NAD^+ pool should repress collagen gene transcription and diminish prolyl hydroxylase activity causing a reduced level of collagen synthesis. Our data confirmed this.

A possible regulatory role of ADP-ribosylation in collagen synthesis has wide significance. Further studies will provide insight into molecular mechanisms of collagen gene expression. In addition, it should help in selecting potential compounds to promote wound healing and also to reduce fibrosis.

Figure 2.   Regulation of collagen synthesis by ADP-ribosylation

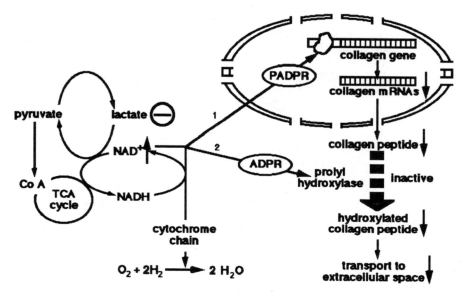

We hypothesize that normally poly (ADP-ribose) enables a putative nuclear protein to bind to collagen transcription component and represses collagen gene. 1= inhibited by nicotinamide/3-AB. Concurrently, cytoplasmic ADP-ribose inhibits prolyl hydroxylase. 2= inhibited by oleic acid, lenoleic acid, novobiocin and vitamin K_1. A compound like lactate lowers both nuclear poly (ADP-ribose) and cytoplasmic ADP-ribose levels, releases these inhibitions and thus allows stimulated transcription and prolyl hydroxylase activation.

Acknowledgements: This work was supported in part by the National Institute of Health Grants GM-27345 and DE-6622

REFERENCES

1. Baurer, E. A. , Uitto, J., Cruz, D. S., Turner, M. L. Progressive nodular fibrosis of the skin: Altered procollagen and collagenase expression by cultured fibroblasts. J Invest. Dermatol. 87:210-216; 1986.

2. Parker, M. I., Judge, K., Gevers, W. Loss of type I procollagen gene expression in SV40 transformed human fibroblasts is accompanied by hypermethylation of these genes. Nucleic Acid Res. 10:5879-5891; 1982.

3. Adams, S., Alwine, J. C., Crombrugghe, B., Pastan, I. Use of recombinant plasmids to characterize collagen RNAs in normal and transformed chick embryo fibroblasts. J. Biol. Chem. 254:4935-4938; 1979.

4. Ignotz, R. A., Endo, T., Massague, J. Regulation of fibronectin and type I collagen mRNA levels by transforming growth factor-ß. J. Biol. Chem. 262:6443-6446; 1987.

5. Postlethwaite, A. E., Raghow, R., Stricklin, G. P., Poppleton, H., Seyer, J. M., Kang, A. H. Modulation of fibroblast-functions by interleukin 1: Increased steady-state accumulation of type I procollagen messenger RNAs and stimulation of other functions but not chemotaxis by human recombinant interleukin 1 α and β. J. Cell Biol. 106:311-318; 1988.

6. Farzaneh, F., Zalin, R., Brill, D., Shall, S. DNA strand breaks and ADP-ribosyl transferase activation during cell differentiation. Nature 300:362-366; 1982.

7. Ohashi, Y., Ueda, K., Hayaishi, O., Ikai, K., Niwa, O. Induction of murine terato-carcinoma cell differentiation by suppression of poly(ADP-ribose) synthesis. Proceedings of the National Academy of Science, USA. 81:7132-7136; 1985.

8. Kun, E., Minaga, T., Kirsten, E., Hakam, A., Jackowski, G., Tseng, A., Jr., Brooks, M. Possible participation of nuclear poly(ADP-ribosylation) in hormonal mechanisms. In: Litwack, ed. Chemical Actions of Hormones, Vol. 13, New York: Academic Press; 1986:p. 33-53.

9. Poirer, G. G., Murcia, G. D., Jongstra-Bilen, J., Niedergang, C., Mandel, P. Poly(ADP-ribosyl)ation of polynucleosomes causes relaxation of chromatin structure. Proceedings of the National Academy of Science, USA. 79:3423-3427; 1982.

10. Tanuma, S., Johnson, L. D., Johnson, G. S. ADP-ribosylation of chromosomal proteins and mouse mammary tumor virus gene expression. J. Biol. Chem. 258:15371-15375; 1983.

11. McNerney, R., Darling, D., Johnstone, A. Differential control of proto-oncongene c-myc and c-fos expression in lymphocytes and fibroblasts. Biochem. J. 245:605-608; 1987.

12. Hussain, M. Z., Ghani, Q. P., Hunt, T. K. Inhibition of prolyl hydroxylase by poly(ADP-ribose) and phosphoribosyl-AMP. J. Biol. Chem. 264:7850-7855; 1989.

13. Ghani, Q. P., Enriquez, B., Hunt, T. K., Hussain, M. Z. Prolyl hydroxylase activity in fibroblasts exposed to NAD$^+$. FASEB J. 4:A2120, 1990 (abst.)

14. Loetscher, P., Alvares-Gonzales, R., Althaus, F. R. Poly(ADP-ribose) may signal changing metabolic conditions to the chromatin of mammalian cells. Proceedings of the National Academy of Science, USA. 84:1286-1289; 1987.

15. Sabir, J., Tavassoli, M., Shall, S. Purification, characterization of NAD:Arginine mono(ADP-ribosyl) transferases and study of inhibitors. This volume.

16. Prockop, D. J., Kivirikko, K. I., Tuderman, L. Biosynthesis of collagen and its disorders. New Engl. J. Med. 301:13-23; 1979.

17. Kivirikko, K. I., Myllyla, R. The hydroxylation of prolyl and lysyl residues. In: Weiss, J. B., Jayson, M. I. V., eds. Collagen in Health and Disease. New York: Churchill-Livingstone, Inc.; 1982;p. 101-120.

Poly(ADP-ribose) Synthesis in Lymphocytes of Systemic Lupus Erythematosus Patients

Hai-Ying Chen, Raymond M. Pertusi, Bernard Rubin, and Elaine L. Jacobson

Several studies have implicated altered poly(ADP-ribose) metabolism in the disease systemic lupus erythematosus (SLE). For example, circulating antibodies to poly(ADP-ribose) (1, 2) and poly(ADP-ribose) polymerase (3) have been reported in patients with SLE. In addition, Sibley et al. (4) have reported a 70 percent decrease in the accumulation of ADP-ribose polymers of peripheral blood lymphocytes of patients with SLE. We have confirmed this finding and report here studies on the mechanism of this defective metabolism.

The study of Sibley et al. (4) used an assay for poly(ADP-ribose) polymerase with a substrate concentration of 0.03 μM NAD, which is 3000 to 5000 fold below the Km. In the study described here, assay conditions were optimized to measure maximal enzymatic activity in lymphocytes from SLE and control subjects by varying DNase, NAD, and time of incubation (5). We established that maximal stimulation of poly(ADP-ribose) polymerase occurred at 30 μg/ml DNase and net product formation was nearly linear for at least 15 minutes (data not shown). The data of Figure 1 show the effect of varying NAD concentration on net formation of ADP-ribose polymers under these assay conditions. Polymer accumulation increased up to a maximum at 150 μM NAD. Therefore, 150 μM NAD was chosen as a standard assay condition. These assay conditions yielded maximal activity and thus provided increased sensitivity important for characterizing this metabolism in the limiting volume of blood available from SLE patients.

The results of Figure 1 compare two control and two SLE subjects. These data are consistent with the earlier observation of Sibley et al. (4) in that net synthesis of poly(ADP-ribose) was diminished in SLE subjects. We then studied a population of 15 SLE subjects who fulfilled the American Rheumatology Association Criteria (1982) for SLE and 13 age and sex-matched controls. The results are shown in Figure 2. The mean specific activity in the control group was 392 ± 128 units. The mean specific activity for the SLE group was 222 ± 137 units (range = 20 to 440). All controls tested, with one exception, showed a minimum activity of 280 units. Thus, 60% of the SLE group demonstrated activities below the minimum value of the control group. A one way analysis of variance (ANOVA) ($0.0001 < p < 0.005$) and a student's unpaired t test (0.0005

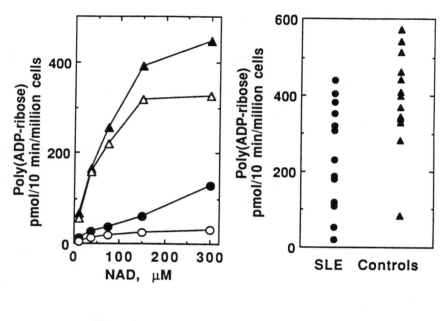

Figure 1

Figure 2

Figure 1. Accumulation of Poly(ADP-ribose) in Permeable Lymphocytes from SLE and Control Subjects. Triangles, two different controls; Circles, two different SLE subjects.

Figure 2. Distribution of Poly(ADP-ribose) Accumulation in Lymphocytes of SLE and Control Subjects.

< p < 0.005) showed the two groups to be significantly different. The heterogeneity demonstrated in the SLE population suggests that there may be at least two different populations within the SLE group with regard to ADP-ribose polymer accumulation.

Because of the complexity of poly(ADP-ribose) metabolism, several possible mechanisms could account for the diminished accumulation of ADP-ribose polymers observed in lymphocytes from SLE subjects. This metabolism is a dynamic system in which the synthesis of ADP-ribose polymers catalyzed by poly(ADP-ribose) polymerase is in concert with the degradation of these polymers catalyzed by poly(ADP-ribose) glycohydrolase (6). A third enzyme, ADP-ribose protein lyase, completes the metabolic cycle by removing the protein proximal ADP-ribose moiety. The assay for poly(ADP-ribose) described above and that used by Sibley et al. (4) does not measure rates of synthesis and degradation, but net accumulation. Therefore, we investigated several possible mechanisms which could account for the diminished accumulation of ADP-ribose polymers in SLE lymphocytes.

A primary concern was that there might be increased competition for the radiolabelled substrate, ^{32}P-NAD, in the SLE samples. NADases are known to be activated by disruption of plasma membranes and can rapidly deplete NAD in cell extracts (6). Furthermore, we observed that platelet contamination was significantly greater in SLE lymphocyte preparations than in control samples. Platelet membranes could contribute NADase activity in the assay. HPLC analyses showed that the concentration of ^{32}P-NAD was not significantly changed during the assay in either the control or SLE samples (data not shown). Thus, the differences observed in the SLE population were not due to increased competition for the substrate by other enzymes.

A second possible mechanism which could explain the observed decrease in polymer accumulation in SLE could be an increased turnover of the polymers due to elevated poly(ADP-ribose) glycohydrolase activity in the assay. We have examined this possibility by determining polymer stability. This was accomplished by adding 500 μM benzamide after 10 minutes of incubation. This compound has been shown to inhibit 99% of poly(ADP-ribose) polymerase activity at this concentration (7). The polymers formed prior to addition of benzamide were stable for up to 20 minutes in both control and SLE samples. Thus, poly(ADP-ribose) glycohydrolase was not active under these assay conditions. This is most likely due to the fact that poly(ADP-ribose) glycohydrolase activity was released from the nucleus and thus diluted many fold during hypotonic shock of the lymphocytes. Thus, the altered ADP-ribose metabolism of SLE lymphocytes was not due to differential turnover of polymers during the assay.

We next examined the possibility that the chromatin of SLE lymphocytes may be more resistant to further modification by ADP-ribose polymers during the *in vitro* assay than controls. This could result from a decreased turnover of ADP-ribose polymers *in situ* if poly(ADP-ribose) glycohydrolase activity was lower than controls. To test this hypothesis, we measured the poly(ADP-ribose) glycohydrolase activity in lymphocytes of four control and five SLE subjects. SLE subjects with abnormally low ADP-ribose polymer accumulation as well as subjects with normal range values were studied. These data are shown in Table 1. No significant difference was observed between the control and SLE groups nor was there a difference between SLE samples with normal range versus diminished poly(ADP-ribose polymerase activity.

The above studies established that the decreased accumulation of ADP-ribose polymers observed in Figures 1 and 2 is the result of decreased ADP-ribose polymer synthesis and not due to alterations in turnover or availability of substrate. Thus, poly(ADP-ribose) polymerase activity is decreased in SLE.

Table 1. Poly(ADP-ribose) Glycohydrolase Activity in Lymphocytes of Control and SLE Subjects

Subjects	Glycohydrolase Activity[1]	Poly(ADP-ribose) Accumulation[2]
Control		
1	118	426
2	130	534
3	120	250
4	116	444
Mean	121 ±6	414 ± 118
SLE		
1	152	440
2	128	90
3	106	108
4	115	420
5	100	110
Mean	120 ±2	234 ± 180

[1] pmol of ADP-ribose/10^6 cells/10 min

[2] pmol of ADP-ribose polymerized/10^6 cells/10 min.

No significant difference in poly(ADP-ribose) glycohydrolase activity was observed in the SLE group or in those subgroups of SLE with low or normal polymer synthesis.

We have initiated studies to characterize the decreased poly(ADP-ribose) polymerase activity in SLE lymphocytes. The $K_{m\ app}$ for NAD for poly(ADP-ribose) polymerase in human lymphocytes was analyzed in a subpopulation of SLE patients and controls. The mean $K_{m\ app}$ for NAD for poly(ADP-ribose) polymerase in controls was 62 μM (data not shown). SLE samples with low or normal range poly(ADP-ribose) polymerase activities showed no significant difference in $K_{m\ app}$ from controls. Thus, the observed differences between control and SLE samples was not due to an altered affinity of the enzyme for the substrate, NAD. In order to learn more about the decrease in poly(ADP-ribose) polymerase activity, the size and and degree of branch complexity of the polymers formed in SLE lymphocytes were compared to that in controls. These experiments showed that the product of the poly(ADP-ribose) polymerase reaction in SLE, even though it was only 5% of the amount in controls, distributed over similar polymer sizes ranging from a 2-mer to 80-mer and contained a similar pattern of branched polymers as controls (data not shown). These data suggest that the mechanism of the enzyme molecules in SLE is not altered, but that fewer active enzyme molecules are present.

Several possible alterations could lead to fewer active poly(ADP-ribose) polymerase molecules in SLE. One possibility is that the expression of the enzyme may be decreased or the rate of turnover of the protein may be increased in SLE. It is also possible that SLE lymphocytes contain a normal number of enzyme molecules, but that they are inactive due to modification or a lack thereof. Experiments to quantify the number of poly(ADP-ribose) polymerase molecules in SLE lymphocytes by Western blot analyses using an antibody to poly(ADP-ribose) polymerase are in progress and should distinguish between these two possibilities.

The relationship of decreased poly(ADP-ribose) polymerase activity to the clinical manifestations of SLE are unclear. It is interesting that defective DNA repair has been reported in patients with this disease (8). The sensitivity that many SLE patients show to sunlight could reflect the inability to complete DNA repair following UV exposure. It is possible that the diminished poly(ADP-ribose) polymerase activity may be related to this defect. However, whether altered ADP-ribose polymer metabolism is a primary or secondary phenomenon is yet to be determined. Certainly, the clinical symptoms of this disease are heterogeneous. This proved to be true also in the pattern of decreased poly(ADP-ribose) polymerase activity within the population studied here. With regard to clinical correlations, we observed that four patients in this small population had renal insufficiency or failure. The poly(ADP-ribose) polymerase activities in lymphocytes from these subjects were among the lowest observed. Whether altered ADP-ribose metabolism is related to this clinical manifestation is not yet established but warrants further investigation. Decreased poly(ADP-ribose) polymerase activity may serve as a marker for major organ system involvement or as an indicator of disease activity. Hopefully, testing such possibilities will lead to better understanding and treatment of this disease.

We gratefully acknowlegde support by grants from the National Institutes of Health (CA43894) and the TCOM Faculty Research Fund. We also thank Charles Kiehlbauch for technical assistance in analysis of ADP-ribose polymer size distributions.

References

1. Clayton, A. L., Berstein, R.M. Tavassoli, M., Shall, S.,Bunn, C., Hughs, G. R. V., Chantler, S. M. Measurement of Antiboty to Poly(adenosine diphosphate-ribose): Its Diagnostic Value in Systemic Lupus Erythematosus. (1984) *Clin. Exp. Immunol. 56*, 263-271.

2. Kanai, Y., Kawaminami, Y., Miwa, M., Matsushima, T., Sugimura, T., Moroi, Y., and Yokohari, R. Naturally Occurring Antibodies to Poly(ADP-ribose) in Patients with Systemic Lupus Erythematosus. (1977) *Nature 265*, 175.

3. Muller, S., personal communication.

4. Sibley, J. T., Haug, B. L., and Lee, J. S. Altered Metabolism of Poly(ADP-ribose) in the Peripheral Blood Lymphocytes of Patients with Systemic Lupus Erythematosus. (1989) *Arthritis and Rheumatism 32*, 1045-1049.

5. Berger, N. A., Sikorski, G. W., Petzold, S. J., and Kurohara, K. K. Association of Poly(adenosine diphosphate ribose) Synthesis with DNA Damage and Repair in Normal Human Lymphocytes. (1979) *J. Clin. Invest. 63*, 1164-1171.

6. Jacobson, M. K., Aboul-Ela, N., Cervantes-Laurean, D., Loflin, P. T., and Jacobson, E. L. ADP-Ribose Levels in Animal Cells. (1990) *In ADP-ribosylating Toxins and G-Proteins: Insights into Signal Transduction*, J. Moss and M. Vaughan, Eds., American Society for Microbiology, Washington, D.C., pp. 479-492.

7. Rankin, P. W., Jacobson, E. L., Benjamin, R. C., Moss, J. and Jacobson, M. K. Quantitative Studies of Inhibitors of ADP-Ribosylation *In Vitro* and *In Vivo*. (1989) *Journal of Biological Chemistry, 264*, 4312-4317.

8. Harris, G., Asbery, L., Lawley, P. D., Denman, A. M., Hylton, W. Defective Repair of O^6 Methylguanine in Autoimmune Diseases. (1982) *Lancet II*, 952-956.

Possible involvement of poly(ADP-ribosyl)ation in phenobarbital promotion activity in rat hepatocarcinogenesis

Toshifumi Tsujiuchi, Masahiro Tsutsumi, Ayumi Denda and Yoichi Konishi

Department of Oncological Pathology, Cancer Center, Nara Medical College, 840 Shijo-cho, Kashihara, Nara 634, Japan

Introduction

Poly(ADP-ribosyl)ation has been suggested to be involved in various biological processes, including DNA repair(Shall, 1984), cell differentiation (Farzaneh et al., 1982), cell proliferation(Menegazzi et al., 1988) and malignant transformation(Borek et al., 1984). Previously, we reported that inhibitors of poly(ADP-ribose) polymerase enhanced liver carcinogenesis which was given in the initiation phase of diethylnitrosamine(DEN) (Takahashi et al., 1982; Takahashi et al., 1984). Recently, 12-O-tetradecanoylphorbol-13-acetate(TPA), a mouse skin tumor promoter, was found to increase poly(ADP-ribosyl)ation of nuclear proteins in human monocytes, mouse fibloblasts and rat hepatocytes(Singh et al., 1985; Singh and Cerutti, 1985; Romano et al., 1988), and suggested the involvement of poly(ADP-ribosyl)ation on tumor promotion processes. In the present experiment, we studied the effects of inhibitors of poly(ADP-ribose)polymerase on the promotion by phenobarbital (PB) in liver carcinogenesis initiated by DEN in rats.

Materials and Methods

Male Fischer 344 rats(Shizuoka Laboratory Animal Center, Shizuoka, Japan), 6 weeks old, were given a single intraperitoneal injection of DEN(Wako Pure Chemicals Co. Ltd., Osaka, Japan) at a dose of 200mg/kg body weights. After 2 week recovery period, the rats were placed on either basal diet(Oriental M. Oriental Yeast Co. Tokyo, Japan), diet containing 0.05% PB(Maruishi Pharmaceutical Co. Inc. Ltd., Osaka, Japan), diet containing various doses of inhibitors alone or diet containing 0.05% PB plus various doses of inhibitors for 10 weeks, and then sacrificed 12 weeks after the beginning of the experiment. The inhibitors of poly(ADP-ribose)polymerase used in the present experiments were 3-aminobenzamide(ABA)(Tokyo Kasei Co. Ltd., Tokyo, Japan), luminol(Nakarai Chemical Co. Ltd., Osaka, Japan) and 3-methoxybenzamide(MBA)(Wako Pure Chemicals Co. Ltd., Osaka, Japan). The enzymatic phenotypical altered foci expressed positive glutathione S-transferase placental form(GST-P) were quantitatively analyzed by IMAGELYZER MODEL HTB-c995(Hamamatsu Television Co. Ltd., Shizuoka, Japan) connected to Desktop Computer System-45(Hewlett-Packard, Co. USA).

Results and Discussion

The results of this study were shown in Table 1. The dose of 1% ABA exerted no effects, but both 1.5 and 2% of ABA given with PB significantly inhibited the induction of GST-P-positive foci in the liver initiated with DEN, with the numbers of foci per cm^2, average sizes of the foci and the percentage of liver area occupied by the foci being decreased compared with PB alone group. Luminol at doses of 3 and 6% inhibited dose dependently the PB promotion, but not 1 and 2% luminol, with the numbers of foci per cm^2 and the percentage of liver area occupied by the foci. All the doses of luminol significantly decreased the average sizes of the foci developing by PB. The doses of 1 and 2% MBA dose-dependently inhibited the promotion by PB, with the numbers of foci per cm^2 and the percentage of liver area occupied by the foci.

The present results clearly indicated that the inhibitors of poly(ADP-ribose)polymerase, ABA, luminol and MBA, inhibited the induction of GST-P-positive foci by PB. It has been reported that the inhibitors include enhancement of the pancreas islet cell carcinogenesis induced by streptozotocin and alloxan in rats(Yamagami et al., 1985) and oral carcinogenesis by diethylbenz(a)-antracene in hamster cheek porch(Miller et al., 1989). In contrast, the inhibitors exerted the inhibitory effects on the colon carcinogenesis by methylazoxymethanol acetate in rats(Nakagawa et al., 1988). In the present experiment, we studied the effect of ABA on active hepatocyte cycle induced by partial hepatectomy in rats and ABA delayed hepatocyte cycle with delayed appearous of poly(ADP-ribose)polymerase activity(data not shown). Therefore, these results suggest that ABA might inhibited hepatocyte proliferation during PB-induced promotion in liver carcinogenesis.

Table 1. Effects of PB and/or ABA, luminol and MBA on the development of GST-P-positive liver in rats initiated with DEN

Experimental group	No. of rats	GST-P-positive foci		
		No. of foci/cm^2	Average size (mm^{2}x10^{-2})	% area occupied by foci
A				
1. BD	10	8.0±5.8	5.5±3.3	0.27±0.18
2. PB	10	27.5±9.9	5.7±6.7	1.55±0.81
3. 2%ABA	9	12.1±4.4	3.4±5.5	0.41±0.14
4. PB+1%ABA	10	33.9±7.2	6.7±8.6**	2.24±0.59
5. PB+2%ABA	12	15.9±6.1**	3.2±4.1***	0.51±0.25***
6. 2%luminol	9	8.7±4.6	2.9±3.3	0.25±0.15
7. PB+1% luminol	10	25.6±9.8	4.2±5.6***	1.06±0.51
8. PB+2% luminol	12	31.3±10.1	4.2±5.7***	1.29±0.52
B				
1. BD	10	3.8±1.2	5.3±4.8	0.21±0.08
2. PB	9	13.4±0.3	6.6±7.9	0.90±0.31
3. 1.5%ABA	10	4.3±1.4	6.5±7.8	0.28±0.11
4. PB+1.5%ABA	11	4.1±1.2***	4.2±3.7***	0.17±0.06***
5. 6% luminol	11	3.0±1.0	6.1±6.6	0.19±0.10
6. PB+3% luminol	10	6.2±1.9***	5.1±6.3***	0.32±0.13***
7. PB+6% luminol	10	2.9±0.8***	2.9±2.7***	0.08±0.03***
8. 2%MBA	10	4.2±1.3	5.9±5.6	0.25±0.09
9. PB+1%MBA	9	9.3±1.6***	6.0±6.1	0.56±0.12*
10. PB+2%MBA	11	7.3±1.2***	5.5±7.5**	0.39±0.12***

*p<0.02, **p<0.01, ***p<0.001 compared with group 2 in experiments A and B.

Acknowledgements

This study was supported in part by a Grant-in-Aid for Cancer Research from the Ministry of Education. Science and Culture, and by a Grant-in-Aid from the Ministry of Health and Welfare for the Comprehensive 10-Year Strategy For Cancer Control, Japan.

References

1. Shall, S. (1984) Adv Radiat Biol 11:1-69
2. Farzaneh, F., Zalin, R., Brill, D. and Shall, S. (1982) Nature 300:362-366
3. Menegazzi, M., Gelosa, F., Tommasi, M., Uchida, K., Miwa, M., Sugimura, T. and Suzuki, H. (1988) Biochem Biophys Res Commun 156:995-999
4. Borek, C., Morgan, W.F., Ong, A. and Cleaver, J.E. (1984) Proc Natl Acad Sci USA 81:243-247
5. Takahashi, S., Ohnishi, T., Denda, A. and Konishi, Y. (1982) Chem-Biol Interactions 39:363-368
6. Takahashi, S., Nakae, D., Yokose, Y., Emi, Y., Denda, A., Mikami, S., Ohnishi, T. and Konishi, Y. (1984) Carcinogenesis 5:901-906
7. Singh, N., Poirier, G., and Cerutti, P. (1985) Biochem Biophys Res Commun 126:1208-1214
8. Singh, N. and Cerutti, P. (1985) Biochem Biophys Res Commun 132:811-819
9. Romano, F., Menapace, L. and Armato, U. (1988) Carcinogenesis 9:2147-2154
10. Yamagami, T., Miwa, A., Takasawa, S., Yamamoto, H. and Okamoto, H. (1985) Cancer Res 45:1845-1849
11. Miller, E.G., Rivera-Hidalgo, F. and Binnie, W.H. (1989) ADP-Ribose Transfer Reactions. Mechanisms and Biological Significance, Jacobson, M.K. and Jacobson, E.L. (eds.), Springer-Verlag, New York, pp.287-290
12. Nakagawa, K., Utsunomiya, J. and Ishikawa, T. (1988) Carcinogenesis 9:1167-1171

1-Methylnicotinamide stimulates cell growth, causes DNA hypermethylation and inhibits induced differentiation of Friend erythroleukemia cells.

Jim R. Kuykendall and Ray Cox.

Cancer Research Laboratory, Veterans Administration Medical Center, 1030 Jefferson Avenue, Memphis TN 38104 and Department of Biochemistry, 800 Dunlop Avenue, University of Tennessee, Memphis TN 38163.

Formation of 1-methylnicotinamide (1-MN) occurs via enzymatic methylation of excess nicotinamide (NA) by nicotinamide methyltransferase, a phase II detoxification enzyme which is found predominantly in the liver and kidney (1, 2). Although 1-MN is known to be the primary metabolite of excess NA, the cellular functions of this compound are not well understood. 1-MN was found to stimulate DNA synthesis and cell growth of cultured hepatocytes (2). We have previously noted (3) that 1-MN has an unique effect of causing DNA hypermethylation in Friend erythroleukemia cells (FELCs). Induced erythroid differentiation of FELCs, as determined by increased hemoglobin expression, has been correlated with a transient genome-wide DNA hypomethylation using a variety of chemical inducers including NA (4-9). However, direct evidence for a causal involvement of DNA hypomethylation in the process of FELC differentiation is lacking. In this study, 1-MN was combined in FELC culture with several chemical inducers which are known to cause DNA hypomethylation, in an effort to reverse the induction of differentiation by antagonism of the DNA hypomethylation.

FELCs were cultured as described (10), by diluting log phase cells to 2 x 10^5 cells/ml prior to drug addition. 1-MN has growth stimulatory properties in cultured FELCs at all doses tested (1-10 mM), with up to 1.5-fold increase in cell density at 96 hrs in cultures exposed to the optimum concentration of 5 mM 1-MN (Figure 1A). Cell viability was two times higher in 1-MN treated cells at doses up to 5 mM (Figure 1B). 1-MN became toxic at 10 mM, with about three times higher cell death than controls.

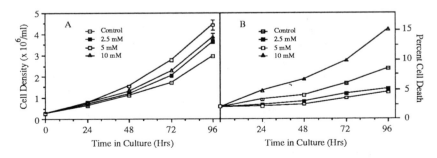

Figure 1. Effect of increasing concentrations of 1-MN on cell growth (A) and viability (B) of FELCs over 96 hrs. All data represent the mean ± S. E., n=4.

FELCs were exposed to either 20 mM NA or 2.5 mM 1-MN continuously for up to 96 hrs in culture. Crude nuclear extracts were prepared from FELCs and DNA methylase assays were performed according to the published procedures (11, 12). DNA isolated from NA-treated FELCs was used as substrate for DNA methylation assays, and found to accept more radioactively labeled methyl-groups than control cell DNA, showing that it is hypomethylated (Figure 2). This hypomethylation is transient, peaking at 24 hrs after NA exposure. DNA isolated from cells cultured with 1-MN became progressively hypermethylated with up to 75 % less methyl-incorporation at 72 hrs (Figure 2), while methyl-accepting ability was constant in untreated cells over this time period. This shows that 1-MN and NA have opposing effects on DNA methylation in FELCs.

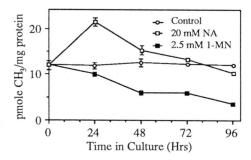

Figure 2. Methyl-accepting ability of DNA from FELCs cultured with 1-MN. DNA isolated from 1-MN treated or untreated cells was used as substrate in assays for methyl-accepting ability (11,12) using crude nuclear extracts as enzyme source. Data represent the mean ± S. E., n=3.

Induction of differentiation of FELCs is correlated with DNA hypomethylation, a process which has recently been found to involve removal of 5-methylcytosine and replacement with cytosine (8, 9). In an effort to determine if this hypomethylation is directly involved in the differentiation process, we have attempted to reverse or override the DNA hypomethylation in cells exposed to NA by addition of 1-MN. Log phase FELCs were exposed to drugs for 24 hrs in culture and DNA purified for use as a methyl-accepting substrate in DNA methylase assays, as before. While control cell DNA accepted 15.7 ± 0.7 pmoles CH_3/mg protein, DNA from cells treated with 2.5 mM 1-MN was hypermethylated with 13.5 ± 0.2 pmoles CH_3/mg protein (14 % decrease). DNA from 10 mM NA treated cells was hypomethylated accepting 19.2 ± 0.3 pmoles CH_3/mg protein (22 % higher incorporation). DNA from cells cultured with NA and 1-MN was able to accept 17.2 ± 0.4 pmoles CH_3/mg protein. This value was 8 % higher than control DNA, but represents a 64 % decrease in DNA hypomethylation from that cells cultured with NA alone. This data shows that 1-MN is not only able to cause DNA hypermethylation in cultured FELCs (Figure 2), but can also partially prevent DNA hypomethylation in NA treated cells.

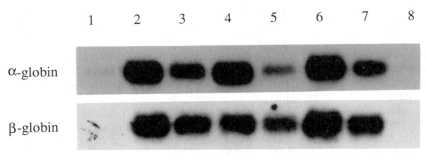

Figure 3. Inhibition of globin mRNA synthesis in differentiating FELCs by 1-MN. Northern blot analysis of globin-specific mRNA from cells culture with: 1) no drugs, 2) 5 mM N'-MN, 3) 5 mM N'-MN + 2.5 mM 1-MN, 4) 5 mM HMBA, 5) 5 mM HMBA + 2.5 mM 1-MN, 6) 270 mM DMSO, 7) 270 mM DMSO + 2.5 mM 1-MN, 8) 2.5 mM 1-MN.

When combined with several inducers in FELC culture, 2.5 mM 1-MN was able to antagonize induction of hemoglobin synthesis. Heme accumulation was determined as previously described (13) based on reduction of 2,7-diamino-fluorene (DAF) to fluorene blue. There was a reduction in accumulation of heme in cells treated for 96 hrs with either 5 mM N'-methylnicotinamide (N'-MN), 5 mM hexamethylene bisacetamide (HMBA) or 270 mM dimethylsulfoxide (DMSO) by 40-50 % (Table 1). Detection of globin mRNA synthesis was determined by northern blotting analysis (14, 15). FELCs cultured with 5 mM N'-MN, 5 mM HMBA or 270 mM DMSO exhibited high levels of globin mRNA at 48 hrs after exposure (Figure 3). The addition of 2.5 mM 1-MN to the inducers treated-cultures caused a significant decline in the appearance of both α- and β-globin mRNA at 48 hrs (Figure 3). Densitometric scanning analysis (Table 1) of the autoradiographs from these blots (Figure 3) allowed relative quantitation of the globin mRNA levels. The accumulation of α-globin mRNA in inducer treated cells was inhibited by 1-MN by about 60-80 %, while β-globin mRNA levels were decreased by about 35-65 % (Table 1). These results lead us to conclude that 1-MN can act as an inhibitor of erythroid differentiation at the level of globin gene transcription. Time course experiments (data not shown) show that 1-MN needs to be present only during the first 24-48 hrs after exposure of the inducer to achieve near optimum inhibition.

NA and N'-MN act as inducers of FELC differentiation, causing DNA hypomethylation 24-48 hrs after exposure (3). 1-MN acts in an opposing manner causing progressive DNA hypermethylation and antagonizing induced DNA hypomethylation in FELCs by NA. 1-MN is the only reported DNA hypermethylating agent which is non-toxic to cell growth. At this time, the mechanism responsible for DNA hypermethylation by 1-MN, and its possible relationship to growth stimulation in cultured FELCs are not understood. It is tempting to speculate that the hypermethylating activity of 1-MN may be responsible for its ability to antagonize induction of differentiation by several inducers known to cause DNA hypomethylation. It is not known if the growth stimulating activity of 1-MN on FELCs is related to the DNA hypermethylating activity. Work is currently in progress to determine if 1-MN causes DNA hypermethylation at specific sequences in the DNA of differentiating FELCs, particularly in the upstream regulatory sequences of α- and β-globin genes.

Table 1. Effect of 2.5 mM 1-MN on heme accumulation and globin-specific mRNA synthesis in FELCs cultured with several chemically unrelated inducers.

Compounds	ng Heme per 10^6 Cells at 96 hrs of Culture	Peak Area (AU x mm) α_1-globin	β_{maj}-globin
Control	41.2 ± 0.8	0.316	0.363
2.5 mM 1-MN	36.0 ± 0.8 (-9.1)	0.188 (-40.5)	0.320 (-11.8)
5 mM N'-MN	775.2 ± 20.0	7.588	8.360
+ 2.5 mM 1-MN	369.6 ± 6.8 (-52.3)	4.163 (-45.1)	2.990 (-64.2)
4 mM HMBA	453.5 ± 7.5	3.986	7.178
+ 2.5 mM 1-MN	225.3 ± 9.5 (-43.7)	2.599 (-79.3)	1.485 (-34.8)
270 mM DMSO	339.5 ± 9.0	7.081	8.435
+ 2.5 mM 1-MN	168.0 ± 2.3 (-50.5)	4.142 (-41.5)	3.533 (-58.1)

Band intensities from autoradiographs (24 hr exposure) shown from Figure 3 were determined from analogous autoradiographs (8 hr exposure) and peak area above background was calculated. Values in parentheses represent the percent change in values from inducer-treated and inducer plus inhibitor-treated samples. Heme determinations represent the mean $\pm$ S. E., n=4.

Acknowledgements
These studies were supported by the United States Veterans Administration (4323-01) and the USPHS Research Grant CA-15189 from the National Cancer Institute. This work was used in partial fulfillment of the requirements for Ph.D. thesis (JRK).

References
1. Johnson GS and Chaing PK (1981) 1-Methylnicotinamide and NAD metabolism in normal and transformed rat kidney cells. Arch Biochem Biophys 210: 263-269.
2. Hoshino J, Kuhne U and Kroger H (1982) Enhancement of DNA synthesis and cell proliferation by 1-methylnicotinamide in rat liver cells in culture: implication for its in vivo role. Biochem Biophys Res Commun 105: 1446-1452.
3. Kuykendall JR and Cox R (1989) Niacin analogs that induce differentation of Friend erythroleukemia cells are able to cause DNA hypomethylation. In, ADP-Ribose Transfer Reactions: Mechanisms and Biological

Significance (MK Jacobson and EL Jacobson, eds). Springer-Verlag, New York, pp. 338-344.

4. Christman JK, Price P, Pedriman L and Acs G (1977) Correlation between hypomethylation of DNA and expression of globin genes in Friend erythroleukemia cells. Eur J Biochem 81: 53-61.

5. Christman JK, Weich N, Schonenbrum B, Schneiderman N and Acs G (1980) Hypomethylation of DNA during differentiation of Friend erythroleukemia cells. J Cell Biol 83: 366-370.

6. Christman JK (1984) DNA methylation in Friend erythroleukemia cells: the effects of chemically induced differentiation and of treatment with inhibitors of DNA methylation. Current Topics Micro and Immun 108: 49-73.

7. Razin A, Levine A, Kafri T, Agostini S, Gomi T and Cantoni GL (1988) Relationship between transient DNA hypomethylation and erythroid differentiation of murine erythroleukemia cells. Proc Natl Acad Sci USA 85: 9003-9006.

8. Razin A, Moshe S, Kafri T, Roll M, Giloh H, Scarpa S, Carotti D and Cantoni G (1986) Replacement of 5-methylcytosine by cytosine: a possible mechanism for transient DNA demethylation during differentiation. Proc Natl Acad Sci USA 83: 2827-2831.

9. Kuykendall JR and Cox R (1990) Detection of 5-methylcytosine removal from DNA of Friend erythroleukemia cells following exposure to several differentiating agents. Cancers Letters 47: 149-152.

10. Friend C, Scher RC, Holland JG and Sato T (1971) Hemoglobin synthesis in murine erythroleukemia cells in vitro: stimulation of erythroid differentiation by dimethylsulfoxide. Proc Natl Acad Sci USA 68: 378-382.

11. Cox R and Goorha S (1986) A study of the mechanism of selenite-induced hypomethylated DNA and differentiation of Friend erythroleukemia cells. Carcinogenesis 7: 2015-2018.

12. Cox R (1985) Selenite, a good inhibitor of rat-liver DNA methylase. Biochem Internatl 10: 63-69).

13. Kaiho S and Mizuno K (1985) Sensitive assay systems for the detection of hemoglobin with 2,7-diaminofluorene: histochemical and colorimetry for erythrodifferentiation. Anal Biochem 149: 117-120.

14. Thomas PS (1980) Hybridization of denatured RNA or small DNA fragments. Proc Natl Acad Sci 77: 5201-5205.

15. Thomas PS (1983) Hybridization of denatured RNA transferred or dotted to nitrocellulose paper. Methods Enzymol 100: 255-266.

2-Aminobenzamide antagonizes DNA de-methylation which precedes induced differentiation of Friend erythroleukemia cells by N'-methylnicotinamide.

Jim R. Kuykendall and Ray Cox.

Cancer Research Laboratory, Veterans Administration Medical Center, 1030 Jefferson Avenue, Memphis TN 38104 and Department of Biochemistry, 800 Dunlop Avenue, University of Tennessee, Memphis TN 38163.

Induction of Friend erythroleukemia cell (FELC) differentiation has been shown to be preceded by a genome wide hypomethylation of the DNA (1-3). Recent work suggests that DNA becomes actively de-methylated by removal of 5-methylcytosine (5-mC) and replacement with cytosine during chemically induced FELC differentiation (4, 5). Nicotinamide (NA) and several of its analogs induce FELC differentiation (6-12), but inhibition of ADP-ribosylation is not required for induction. Aminobenzamides (-AB) are potent inhibitors of ADP-ribosylation with very low inducer activity which were found to inhibit induction of differentiation (10, 12) by hexamethylene bisacetamide (HMBA) and dimethylsulfoxide (DMSO). In previous studies, we have shown that NA and its inducer-capable analogs cause DNA hypomethylation in cultured FELCs (11), with N'-methylnicotinamide (N'-MN) being the most potent. Subsequent work determined that up to 6 % of 5-mC in the existing DNA of differentiating FELCs was removed during the first 24 hrs of N'-MN exposure (5). N'-MN has no inhibitory effect on ADP-ribose transferase activity, but causes the highest level of DNA de-methylation of any compound tested. We designed experiments using 2-AB in combination with N'-MN in cultured FELCs in order to determine if ADP-ribosylation may be involved in the process of DNA de-methylation.

Table 1. Effect of 5 mM benzamide and its amino-derivatives on induction of hemoglobin synthesis and DNA methylation in cultured FELCs.

Compounds [5 mM]	% Hb$^+$ Cells	ng Heme per 10^6 Cells	Methyl-accepting Ability of DNA[a]
None	8 ± 1	49.5 ± 2.4	18.8 ± 0.5
BAm	65 ± 3	530.4 ± 8.4	36.1 ± 1.6
2-AB	16 ± 2	60.4 ± 2.4	17.5 ± 1.4
3-AB	19 ± 2	67.2 ± 4.4	18.8 ± 2.9
4-AB	37 ± 3	174.4 ± 4.4	28.2 ± 2.1

[a]Data is expressed as pmol CH_3 incorporated per mg protein present in each assay. All data represent the mean $\pm$ S.E., n=4.

In this study, benzamide (BAm) and its amino-derivatives were tested for inducer activity in FELCs after 96 hrs exposure in culture, and for effects on

DNA methylation after 24 hrs. Crude nuclear extracts were prepared from FELCs and DNA methylase assays were performed according to the published procedure from our laboratory (13, 14). DNA isolated from drug-treated FELCs was used as substrate for DNA methylation assays, where hypomethylated DNA was able to incorporate more radioactively labeled methyl-groups than control cell DNA. Hemoglobin production (Hb$^+$) was determined using previously described procedure (15) based on reduction of 2,7-diaminofluorene (DAF) to fluorene blue in the presence of hydrogen peroxide. Only BAm and 4-AB were able to cause significant induction of heme-synthesis (Table 1) or DNA hypomethylation. Most importantly, 2-AB and 3-AB both had only weak inducer activity and do not affect DNA methylation in cultured FELCs (Table 1) compared to untreated cells. The benzoate analogs (-ABA) of these compounds were found to be weak inducers with no effect on DNA methylation or heme synthesis (data not included).

Table 2. Inhibition of induction of heme accumulation in FELCs by aminobenzamides.

Compounds	Viable Cells/ml (x 10^6)	% Cell Death	% Hb$^+$ Cells	ng Heme per 10^6 Cells
None	2.64 ± 0.12	7.8 ± 0.6	6 ± 1	66 ± 1
5 mM N'-MN	2.50 ± 0.05	7.3 ± 0.9	98 ± 1	796 ± 27
+ 5 mM 2-AB	1.11 ± 0.04	16.3 ± 0.9	68 ± 1	237 ± 4
+ 5 mM 2-ABA	2.44 ± 0.03	9.5 ± 0.6	95 ± 1	686 ± 12
+ 5 mM 3-AB	1.76 ± 0.02	13.5 ± 0.3	81 ± 1	619 ± 35
+ 5 mM 3-ABA	2.17 ± 0.09	11.5 ± 0.5	95 ± 1	866 ± 8
5 mm N'-MN	2.43 ± 0.09	7.3 ± 0.5	95 ± 1	792 ± 10
+ 0.1 mM 2-AB	2.48 ± 0.06	7.6 ± 1.3	95 ± 1	692 ± 14
+ 0.5 mM 2-AB	2.43 ± 0.10	8.0 ± 0.9	85 ± 1	652 ± 20
+ 1.0 mM 2-AB	2.24 ± 0.06	7.1 ± 0.8	77 ± 1	584 ± 25
+ 2.5 mM 2-AB	1.98 ± 0.08	9.9 ± 1.2	72 ± 2	486 ± 13
+ 5.0 mM 2-AB	1.23 ± 0.04	17.3 ± 2.1	70 ± 1	282 ± 5
+ 7.5 mM 2-AB	1.17 ± 0.03	18.1 ± 2.8	61 ± 2	212 ± 4

Cell counts represent at least 300 each. All data represent the mean ± S. E., n=4.

Since 2-AB and 3-AB are potent inhibitors of ADP-ribosylation, without having significant inducer activity, we combined them in culture with N'-MN, a potent inducer with no ADP-ribose transferase inhibitory activity. We have confirmed the previous report (10) that 3-AB can inhibit induction of differentiation (Table 2), but at 5 mM concentration found 2-AB to be a better inhibitor (Table 2). 2-ABA had a slight inhibitory effect on heme accumulation, while 3-ABA stimulated heme accumulation. Neither aminobenzoate had a significant effect on the appearance of Hb$^+$ cells (Table 2). However, 2-AB was more toxic to cell growth in combination with N'-MN than was 3-AB. Combining increasing concentrations of 2-AB with 5 mM N'-MN gave a dose-dependent inhibition of cell growth and differentiation with up to 35 % fewer Hb$^+$ cells and up to 75 % reduction heme accumulation at 7.5 mM (Table 2). We used 2-AB as the inhibitor of ADP-ribosylation and induced differentiation in subsequent experiments.

Table 3. Commitment to differentiation and induced hypomethylation of DNA in FELCs cultured for 24 hrs with 5 mM N'-MN are antagonized by 5 mM of 2-AB.

[5 mM]	% DAF-reactive Colonies[a]	Cells/ml (x 10^5)		Methyl-accepting Ability of DNA[b]	
None	7.3 ± 0.6	5.96 ± 0.63		12.99 ± 1.02	
2-AB	11.3 ± 0.6	3.68 ± 0.21		12.81 ± 1.68	
N'-MN	92.5 ± 1.6	4.96 ± 0.40	(-17)	50.50 ± 5.5	(+389)
N'-MN + 2-AB	61.3 ± 0.9 (-34)	4.56 ± 0.33	(-23)	23.04 ± 2.06	(+177)

[a]Data represents the percentage of colonies containing cells which are synthesizing heme. Cells were exposed for 24 hrs to drugs, harvested and replated in drug-free semi-solid medium containing 0.8 % methylcellulose. After an additional 72 hrs in culture, cells were stained with DAF (15).

[b]Data expressed as pmol CH$_3$ incorporated per mg protein in each assay.

All data represent the mean ± S.E. of triplicate assays.

The commitment of FELCs to differentiate in the presence of N'-MN was determined by colony cloning in semi-solid medium, by a modification of a published procedure (16). Briefly, cells were cultured for 24 hrs in the presence of 5 mM N'-MN and/or 5 mM 2-AB, then washed and replated in drug-free medium containing 0.8 % methylcellulose. Plates were stained 72 hrs later with DAF to determine the presence of hemoglobin-producing (Hb$^+$) cells in isolated colonies. We found that less than 8 % of colonies of control cells contained spontaneously generated Hb$^+$ cells, while 92 % of the colonies in N'-MN treated cultures contained Hb$^+$ cells (Table 3). 2-AB is a weak inducer, with 11 % Hb$^+$ cells, but there was a 34 % decrease in commitment to differentiate in N'-MN treated cells by the addition of equimolar 2-AB (Table 3). In parallel experiments, the methyl-accepting ability of the DNA was examined from similarly treated cells 24 hr exposure to these drugs. The methyl-accepting ability of DNA was increased by almost four fold in N'-MN treated cells (389 % over controls), but was essentially unaffected by 2-AB (Table 2). However, methyl-accepting ability of DNA from cells cultured with 2-AB and N'-MN was reduced to 177 % that of control cells (Table 3), representing a 54 % reduction of N'-MN induced DNA hypomethylation by 2-AB. While all compounds caused a decrease in cell growth, there was only a 5 % decrease in growth of N'-MN treated cultures (17 % decrease) by addition of 2-AB (23 % decrease). It seems unlikely that this small decrease in cell growth is responsible for the effect on DNA hypomethylation. From these data we conclude that 2-AB is unable to inhibit DNA methylation in growing FELCs, but can partially antagonize the induced de-methylation preceding differentiation by N'-MN.

The molecular mechanisms involved in the removal 5-mC from the DNA of differentiating FELCs are not understood, but thought to involve removal of the entire nucleoside, rather than cleavage of the methyl-group from the cytosine (4). This process may involve an endonuclease activity or strand nicking followed by an exonuclease activity. In either case, one would expect DNA repair mechanism to be involved. The ability of 2-AB to inhibit this process suggests a positive modulatory role for ADP-ribosylation in DNA de-methylation, probably at the level of 5-mC removal. Since 2-AB is also able to inhibit induced differentiation, it is tempting to speculate that ADP-ribose transferase inhibitors, such as 2-AB, may derive at least part of their inhibitory activity via their antagonistic effect on DNA hypomethylation. The higher inducer activity of N'-MN than either NA or BAm may be due to its lack of ADP-ribose transferase inhibitory activity. One must also realize that ADP-ribosylation probably affects additional steps.

Acknowledgements

These studies were supported by the United States Veterans Administration (4323-01) and the USPHS Research Grant CA-15189 from the National Cancer Institute. This work was used in partial fulfillment of the requirements for the Ph.D. thesis of JRK. Preliminary work in this study was presented at the 8th International Symposium for ADP-Ribosylation Reactions in Fort Worth, TX in 1986.

References

1. Christman JK, Price P, Pedriman L, Acs G (1977) Correlation between hypomethylation of DNA and expression of globin genes in Friend erythroleukemia cells. Eur J Biochem 81: 53-61.
2. Christman JK (1984) DNA methylation in Friend erythroleukemia cells: the effects of chemically induced differentiation and of treatment with inhibitors of DNA methylation. Current Topics Micro and Immun 108: 49-73.
3. Razin A, Levine A, Kafri T, Agostini S, Gomi T, Cantoni GL (1988) Relationship between transient DNA hypomethylation and erythroid differentiation of murine erythroleukemia cells. Proc Natl Acad Sci USA 85: 9003-9006.
4. Razin A, Moshe S, Kafri T, Roll M, Giloh H, Scarpa S, Carotti D and Cantoni G (1986) Replacement of 5-methylcytosine by cytosine: a possible mechanism for transient DNA demethylation during differentiation. Proc Natl Acad Sci USA 83: 2827-2831.
5. Kuykendall JR and Cox R (1990) Detection of 5-methylcytosine removal from DNA of Friend erythroleukemia cells following exposure to several differentiating agents. Cancers Letters 47: 149-152.

6. Tereda M, Fujiki H, Marks P and Sugimura T (1979) Induction of erythroid differentiation of murine erythroleukemia cells by nicotinamide and related compounds. Proc. Natl. Acad. Sci. USA 76, 6411-6414.

7. Morioka K, Tanaka K, Nokuo T, Ishizawa M, Ono T (1979) Erythroid differentiation and poly(ADP-ribose) synthesis in Friend erythroleukemia cells. Gann 70: 37-46.

8. Zlatanova JS, Swetly P (1980) Poly-ADP-ribosylation of nuclear proteins in differentiating Friend cells. Biochem Biophys Res Commun 92: 1100-1116.

9. Morioka K, Tanaka K, Tetsuo O (1980) Poly(ADP-ribose) and differentiation of Friend erythroleukemia cells. Restriction of chain length. J Biochem 88: 517-524.

10. Brac T, Ebisuzaki, K (1987) ADP-ribosylation and Friend erythroleukemic-cell differentiation: action of poly(ADP-ribose) polymerase inhibitors. Differentiation 34: 139-143.

11. Kuykendall, JR and Cox, R. (1989) Niacin analogs that induce differentiation of Friend erythroleukemia cells are able to cause DNA hypomethylation. In, ADP-Ribose Transfer Reactions: Mechanisms and Biological Significance (MK Jacobson and EL Jacobson, eds). Springer-Verlag, New York, pp. 338-344.

12. Kuykendall, JR and Cox, R. (1990) 2-Aminobenzamide, an inhibitor of ADP-ribosylation, antagonizes induced DNA hypomethylation during differentiation of murine Friend erythroleukemia cells by N'-methyl-nicotinamide. Differentiation 44: 69-73.

13. Cox R and Goorha S (1986) A study of the mechanism of selenite-induced hypomethylated DNA and differentiation of Friend erythroleukemia cells. Carcinogenesis 7: 2015-2018.

14. Cox, R (1985) Selenite, a good inhibitor of rat-liver DNA methylase. Biochem Internatl 10: 2015-2018.

15. Kaiho S and Mizuno K (1985) Sensitive assay systems for the detection of hemoglobin with 2,7-diaminofluorene: histochemical and colorimetry for erythrodifferentiation. Anal Biochem 149: 117-120.

16. Fibach E, Reuben RC, Rifkind RA and Marks PA (1977) Effect of hexamethylene bisacetamide on the commitment to differentiation of murine erythroleukemia. Cancer Res 37: 440-444.

ADP-RIBOSYLATION OF RUBISCO IN LETTUCE CHLOROPLASTS.

Pearson, C. K., Wilson, S. B., Dutnall, R. N. & Gault, W. G.
Department of Molecular & Cell Biology, University of Aberdeen, Marischal College, Aberdeen, Scotland, U.K. AB9 1AS.

Protein ADP-ribosylation in plants

There are only a few published studies on ADP-ribosylation reactions in plants and these deal with reactions associated with the cell nucleus (1-4). No studies are available regarding endogenous reactions in extranuclear organelles. Millner & Robinson (5) reported ADP-ribosylation of a pea chloroplast thylakoid membrane polypeptide of about 60kDa, thought to be a Gs protein, but this was totally dependent on the presence of cholera toxin.

We describe here our preliminary findings of endogenous protein ADP-ribosylation in isolated lettuce chloroplasts, showing for the first time that the major chloroplast enzyme Ribulose bisphosphate carboxylase-oxygenase (Rubisco) is ADP-ribosylated when chloroplasts are incubated with radiolabelled NAD^+.

Rubisco can catalyses two reactions, their relative rates being governed by the concentrations of available oxygen and carbon dioxide:

(1) Ribulose 1,5-bisphosphate + CO_2 ⟶ 2 Phosphoglycerate,
the initial CO_2-fixing reaction of photosynthesis and

(2) R 1,5-BP + O_2 ⟶ 3 PGA + Phosphoglycolate,
the initial oxygenase reaction of photorespiration.

Understanding the factor(s) controlling Rubisco is crucial since the selective perturbation of these two competing reactions to increase CO_2 fixation might lead to an improved rate of photosynthesis and thus increased productivity in many commercially important crops. ADP-ribosylation may be one of the controlling factors.

Optimal conditions for incorporation of radiolabel from NAD into acid insoluble material in islolated chloroplasts

Lysed chloroplasts incorporate up to 50pmoles of radiolabelled NAD per mg protein over a 60min incubation.

Plant tissues may contain high levels of endogenous phenolic compounds. These can be rapidly oxidised to quinones and brown pigments by the enzyme polyphenoloxidase. Such compounds inhibit the activity of many enzymes and subcellular organelles. The presence of polyvinylpyrrolidine (at 1%), an inhibitor of the oxidase, in the assay improved NAD incorporation by over 30%.

A single pH optimum of 6.5 and a single temperature optimum of about 28^{0}C were observed. The reaction was found to be cold-labile; freezing the chloroplasts for even a few minutes considerably decreased the subsequent incorporation of radiolabel from NAD.

We have yet to determine the apparent Km and Vmax for NAD. Any addition of non-radioactive NAD in the range 10μM to 500μM reduced the incorporation of radiolabel to background values. The apparent Km is therefore likely to be in the micromolar range.

Identification of incorporated radiolabel as ADP-ribose

Although characterisation is not complete, nevertheless we think that the radiolabelled material is probably ADP-ribose because: (a) the substrate NAD was labelled in the adenine moiety (b) The precipitable material is rendered acid-soluble by alkali within 20-30 minutes (c) PEI-Cellulose tlc of the material released by the alkali showed only ADP-ribose and 5`-AMP and (c) HAP column chromatography revealed a single radioactive peak of material eluting with 1mM phosphate buffer. These data also indicate that the modification is monomeric ADP-ribose.

The reaction is influenced by light intensity

A series of neutral density filters was used to examine the effect of different light intensities on the reaction. Maximum incorporation of NAD occurred in the dark. An increase in light intensity resulted in a decrease in incorporation until a photon flux density of about 50 quanta/mmol m^{-2}s^{-1} (this is about bright daylight intensity) was achieved. Further increasing the light intensity had no additional effect on the reaction.

240

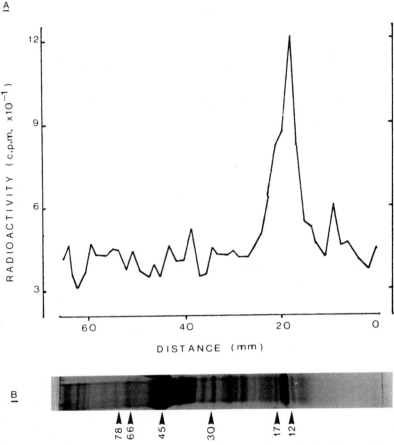

Figure 1. A chloroplast preparation purified on Percoll gradients (6) was incubated with 10μCi of ^{3}H-NAD$^+$ for 1h and then precipitated with TCA. The TCA precipitate was washed repeatedly with TCA and then ethanol-ether (1:1) and solubilized in 0.625M Tris buffer, pH 6.8, containing 2% (w/v) SDS, 20% (v/v) glycerol, 5% (v/v) mercaptoethanol and Bromophenol Blue and boiled for 3-5min before electrophoresis. This was carried out in a SDS-discontinous buffer system (7) using a linear 10-20% (w/v) acrylamide gradient.

SDS-PAGE of radiolabelled chloroplast proteins

Figure 1 (A) shows a single major radioactive peak migrating with a protein of mol. mass about 13kDa. Sequence analysis of some 20 N-terminal residues of this band eluted from the gel (B) revealed it to be the small subunit of the major chloroplast enzyme Rubisco (Ribulose bisphosphate carboxylase-oxygenase).

CONCLUDING REMARKS

Our data show that isolated lettuce chloroplasts can incorporate radiolabel from NAD into a molecule which is probably monomeric ADP-ribose and that the the crucially important chloroplast enzyme Rubisco appears to be the major, if not the only, acceptor. We don`t know at this stage the extent to which the reaction is enzymic, catalysed by an ADP-ribosyltransferase, or results from the non-enzymic attachment to protein of free ADP-ribose produced by the action of an NAD glycohydrolase.

The light dependence of the ADP-ribosylation reaction is of interest since the activity of Rubisco is influenced by light intensity. It is therefore possible that ADP-ribosylation plays a role in this scenario.

It must be emphasised that these data are only preliminary. We have yet to determine with greater certainty that Rubisco is indeed an acceptor and then establish the extent to which the enzyme may be modified. It is to be expected that if ADP-ribosylation is an important regulatory event for this enzyme a considerable proportion of the molecules would become modified under appropriate physiological cirmumstances. Present estimates based only on the specific activity of the radiolabelled NAD added to the assay suggest that up to 10% of the holoenzyme may be modified. This is likely to be an underestimate since we still have to establish the endogenous NAD concentration in the assay.

REFERENCES

1. Whitby, A. J. Sonte, P. R. & Whish, W. J. D. (1979) Biochem. Biophys. Res. Commun. 90: 1295-1304.
2. Willmitzer, L. & Wagner, K. G. (1982) in `ADP-ribosylation Reactions Biology and Medicine . Ueda, K. and Hayaishi, O. (eds) pp. 241-252, Academic Press, New York.
3. Tramontano, W. A., Phillips, D. A., Carman, C. A. & Massaro, A. M. (1990) Phytochemistry 29: 31-34.
4. Tramontano, W. A. & Decostanzo, D. C. (1990) Phytochemistry 29: 2797-2800.
5. Millner, P. A. & Robinson, P. S. (1989) Cell Signaling 1: 421-433.
6. Robinson, S. P., Edwards, G. E. & Walker, D. A. (1979) in `Plant Organelles`. E. Reid (ed), pp. 13-24, Ellis Horwood, Chichester.
7. Laemmli, U. K. (1970) Nature 277: 680-685.

IN VIVO EVIDENCE IN HUMANS OF AN ASSOCIATION BETWEEN ADP-RIBOSYLATION LEVELS IN MONONUCLEAR LEUKOCYTES AND IMMUNE FUNCTION

Ronald W. Pero, Hakan Olsson, Leif Salford, and Walter Troll

Historically, mononuclear leukocytes (MNL) have been the principal cell source used for estimation of cancer risk in humans. Despite repeated demonstrations of effects from genotoxic exposures in vivo on MNL, evidenced by elevated levels of DNA damage (e.g., cytogenetic abnormalities, and DNA adducts, strand breaks and repair inhibition), it only recently has been appreciated what DNA damage does to the function of MNL (1,2). This is an important switch in emphasis, from one where MNL were viewed as a surrogate cell system, reflecting genetic factors or carcinogenic exposures present in epithelial cells (i.e., target cells for cancer), to one where MNL are viewed more as target cells for cancer development because of their role in the immune function (3).

Recently our laboratory focused on the functional importance of DNA damage accumulation via inhibition of DNA repair processes in MNL. The naturally occurring reactive oxygen species, H_2O_2, an important promoter of the carcinogenic process (4), is a potent activator of poly(ADP)ribosyl transferase (ADPRT) (5), an enzyme known to be involved in DNA repair, cell proliferation and differentiation (6). MNL from individuals with cancer, including breast cancer, which were exposed in vitro to H_2O_2 have suppressed ADPRT values (5,7). Moreover, suppressed ADP-ribosylation can occur at least in part from a reduction/oxidation (redox) imbalance, due to the fact that ADPRT is a sulfhydryl-containing enzyme that is up- and down-regulated by reduced and oxidized cellular glutathione, respectively (8). Because T-lymphocyte responses to polyclonal mitogenic stimulation are likewise known to be modulated by the redox status in a similar way (9-11), a relationship between ADP-ribosylation and immune function is strongly supported by several studies involving direct estimation of ADPRT (12,13), and by using inhibitors of ADPRT (14). Because of the well-documented in vivo immunosuppressive properties of steroidal drugs in humans (15,16), there is ample jus-

tification to test the hypothesis that ADP-ribosylation levels in MNL reflect immune function by examining if individuals receiving steroidal drugs have suppressed MNL ADP-ribosylation.

For this purpose we studied two clinically well-defined populations. The first study involved 50 patients with supratentorial astrocytomas, grades II,
II-III, III, III-IV and IV. After diagnosis of a malignant brain tumor by computerized tomography (CT-
Scan), corticosteroids were prescribed (32 out of 50
patients). Corticosteroid treatment was 16 mg/day betamethasone divided into 2 doses 12 hr apart. The
dose-tapered regime of betamethasone was 8 mg/day for
2-3 days, followed by 2 mg/day for 2-3 days and ending
with 1 mg/day for 2-3 days. The dose was tapered following surgery, or if the period before surgery ex-
ceeded 1 wk the dose was tapered to the level needed
to relieve symptoms. The number of days the patients
received betamethasone, including the dose-tapering
period, was used to estimate duration.

The second study involved 83 consecutive female individuals, who had surgery to remove invasive breast
cancer, and were randomized during the primary care
period before initiation of any chemo- and radio-ther-
apeutic programs into two groups: one (n=40) received
no drug treatment and the other (n=42) was scheduled
to receive 20 mg/day of tamoxifen for a 2 yr period
but were sampled between 7-368 days. A 20 ml heparin-
ized blood sample was taken from each individual studied. The MNL fraction was isolated from the blood by
density gradient centrifugation using an Isopaque-Fi-
coll cushion (1.077 gm/ml). The procedure for analyzing ADPRT activity was adapted from the permeabilization technique of Berger with modifications as described (5). Data were recorded as cpm trichloroacetic
acid precipitable ^{3}H-adenine labeled NAD^+ per 5×10^5
cells in the absence (constitutive level) and presence
of 100 μM H_2O_2 (activated level). ADPRT values were
converted to Ln values for statistical analysis to
minimize influences from nonparametric distributions.

In the study concerned with the analysis of corticosteroid treatment, other clinical parameters considered were tumor grade, age (15-81 yr), sex (20 females, 30 males), smoking (10/50), alcohol use
(16/50), and antiepileptic medications (24/50). After
adjustment by multiple regression analysis only beta-

methasone treatment had any significant influence on ADP-ribosylation levels in MNL. Data in Fig. 1 show highly significant suppression of ADPRT activity in MNL in addition to the steroid treatment. Betamethasone treatment known to cause immunosuppression *in*

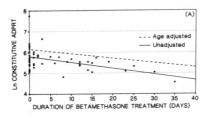 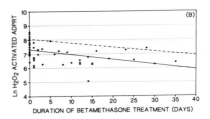

Fig. 1. ADP-ribosylation levels in the MNL of patients with malignant gliomas (astrocytomas II-IV) being treated for peritumoral edema with betamethasone (16 mg/day). Age-adjusted Ln constitutive ADPRT, $Y=6.16-0.022(x)$, $r=0.55$, $p<0.01$; unadjusted Ln constitutive ADPRT, $Y=5.78-0.027(x)$, $r=0.51$, $p<0.0001$; age-adjusted Ln H_2O_2-activated ADPRT, $Y=8.08-0.29(x)$, $r=0.57$, $p<0.006$; unadjusted Ln H_2O_2-activated ADPRT, $Y=0.32-0.036(x)$, $r=0.43$, $p<0.001$. Panel A has $n=49$ with 3 hidden values and Panel B has $n=50$ with 4 hidden values.

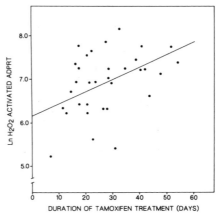

Fig. 2. A linear correlation between duration of tamoxifen (20 mg/day) and ADP-ribosylation levels in MNL of nonsmoking women with breast cancer. Data shown are unadjusted values giving $Y=0.025(x) + 6.19$, $r=0.42$, $p<0.01$.

vivo in humans also causes dose-dependent suppression of ADPRT activity. Tamoxifen is well-known to block the effects of estrogens by binding to the estrogen receptor (17). Because estrogens are known to be immunosuppressive, tamoxifen treatment would be expected to upgrade immune function and thereby ADPRT activity. The data presented in Fig. 2 demonstrate this. After adjustment for age, smoking habits, estrogen use, menstruation and tumor size at surgery, there still remained a significant dose-dependent positive association to duration of tamoxifen treatment.

A steroid effect on immune function would mandate the presence of receptors, and MNL have been shown to have both corticosteroid and estrogen receptors (16, 18,19). Breast cancer patients are known to have immune suppression (20), high serum levels of estradiol (21) and suppressed ADPRT activity (22). Hence, our data are consistent with the interpretation that an improved immune function is being mediated by tamoxifen *via* blockage of the estrogen receptor and is detectable by elevated levels of MNL ADPRT. This point is further supported by *in vitro* data, showing tamoxifen stimulation of natural killer cell activity (23), and it could also explain why estrogen receptor-negative breast cancers respond to tamoxifen treatment (24). Another action of tamoxifen that does not depend on estrogen receptors was recently observed. When 5 μM of tamoxifen is added to human neutrophils, the formation of oxygen radicals such as H_2O_2, $\cdot O_2^-$ and HOCl caused by an inflammatory phorbol ester tumor promoter is completely blocked. HOCl is also a potent inhibitor of ADPRT. Suppressed oxygen radical production is a characteristic of chemopreventive agents, like retinoids and protease inhibitors, and identifies tamoxifen as a chemopreventive agent (25) with immunomodulating properties.

This work was supported by the Swedish Cancer Society (1728-890-0IXA) and NIH (CA 53003).

References

1. Luster M & Blank JA. *Ann. Rev. Pharmacol. Toxicol.* 27:23-49, 1987.
2. Luster M. *Health Environ. Digest 3(1)*:1-3, 1989.

3. Strigini P, Carobbi S, Sansone R, Lombardo C, & Santi L. *Cancer Detect. Prevent.* *15(2)*:115-26, 1991.
4. Troll W & Wiesner R. *Ann. Rev. Pharmacol. Toxicol.* *25*:509-28, 1985.
5. Pero RW, Johnson DB, Miller DG, *et al.* *Carcinogenesis* *10(9)*:1657-64, 1989.
6. Cleaver JE, Borek C, Milam K, *et al.* *Pharmac. Ther.* *31*:269-93, 1987.
7. Roush G, Pero RW, Osborne M, *et al.* *Preventive Oncol.* *18*:780, 1989.
8. Pero RW, Anderson MW, Doyle GA, *et al.* *Cancer Res.* *50*:4619-25, 1990.
9. Fidelus RK, Ginouves P, Lawerance D, *et al.* *Expt. Cell Res.* *170*:269-75, 1987.
10. Fischman CM, Udey MC, Kurtz M, *et al.* *J. Immunol.* *127(6)*:2257-62, 1981.
11. Hamilow DL & Wedner HJ *J. Immunol.* *135(4)*:2740-47, 1985.
12. Johnson DB, Markowitz MM, Joseph PE, *et al.* *Int. J. Biochem.* *22*:67-73, 1990.
13. Scouvassi AI, Stefanini M, Lagomarsini P, *et al.* *Carcinogenesis* *8*:1295-300, 1987.
14. Johnstone AP & Williams GT. *Nature* *300*:368-79, 1982.
15. Ahmed SA & Talal N. *Scand. J. Rheumatol.* *18*:69-76, 1989.
16. Capps T & Fauci A. *Immunology Rev.* *65*:133-55, 1982.
17. Jordan VC. *Pharmacol. Rev.* *36(4)*:245-76, 1984.
18. Cohen JHM, Daniel L, Cordier G, *et al.* *J. Immunol.* *131*:2767-71, 1983.
19. Daniel L, Savrueine G, Monier JC, *et al.* *J. Steroid. Biochem.* *18*:559-63, 1983.
20. Hakin AA. *Cancer* *61*:689-701, 1988.
21. Henderson BE, Ross RK, Pike MC, *et al.* *Cancer Res.,* *42*:3232-9, 1982.
22. Roush GC, Pero RW, Osborne M, *et al.* *Preventive Oncol.* *18*:780, 1989.
23. Modeville R, Chali SS & Chasseau JP. *Eur. J. Cancer Clin. Oncol.* *20(7)*:983-5, 1984.
24. Fair BJA. (Ed) Hormone therapy. *Clinics in Oncology, Vol. 1* (1982).
25. Troll W & Lim J. *Proc. Am. Assoc. Cancer Res.,* *32*:149, 1991.

Evaluation of the DNA damage and inhibition of DNA repair caused by the benzamide derivative Metoclopramide in human peripheral mononuclear leukocytes.

Lybak, Stein and Pero, Ronald W.

Metoclopramide (MCA) is a N-carboxamide substituted benzamide that has been used in the clinic for more than 20 years. It is used worldwide for preventing nausea and vomiting induced by chemotherapeutic drugs, and it has also been used in the treatment of esophageal reflux, dyspepsia, and a variety of other gastrointestinal disorders (1). Doses of 1 - 3 mg/kg repeated every 2 hr to a total of 3 - 5 doses have been given as a single drug treatment or in combination with steroids and / or diphenhydramine to prevent nausea and vomiting induced by chemotherapeutic drugs (2).

Our interest in MCA as a possible potentiator of cisplatin and ionizing radiation stems from the structural resemblence of MCA to some of the known inhibitors of poly-adenosine-diphospho-ribosyl-transferase (poly-ADPRT) such as benzamide and 3-aminobenzamide. Inhibitors of poly-ADPRT have been shown to potentiate ionizing radiation and cytotoxic drug treatment. However, N-carboxamide substituted benzamides were not shown to inhibit poly-ADPRT when assessed directly on enzymatic preparations (3).

Because N-carboxamide substituted benzamides have been shown to be pharmacologically active, our laboratory has investigated the possibility of whether these agents might possess cytotoxic enhancing properties using animal tumor models. We have shown that MCA enhances the effect of cisplatin (4) and ionizing radiation (5) treatment of human squamous cell carcinoma from the head and neck region xenografted to nude mice. The MCA dose used in these experiments was 2 mg/kg given at the same time, 24 hr and 48 hr after the cisplatin dose or the radiation treatment. We have later on found that the optimal time for MCA administration is 8 hr after cisplatin administration and 1 hr before ionizing radiation treatment. These effects were seen without increased toxicities measured as weight loss and mortality. Radiation of skin and bone marrow in vivo has also been carried out, and we could not see any potentiation by MCA on these normal tissues. There is a growing possibility that MCA could be used as a clinical useful potentiator, and we have started a clinical phase 1 study at the Oncology Clinic in Lund to evaluate the possibility of using MCA as a potensiator in patients with inoperable squamous cell carcinoma of the lung.

We have used quantitative estimation of nucleoid sedimentation , ADPRT and NA-AAF induced UDS as biochemical endpoints to indicate DNA damage and repair. The data presented in these studies support the contention that MCA can induce DNA damage and inhibit DNA repair in resting viable HMLs.

HML's treated with MCA for 30 minutes and subsequently studied using the nucleoid sedimentation technique, show a dose response decrease in nucleoid sedimentation, and this decrease in nucleoid sedimentation is independent of temperature (Fig. 1). A similar effect can be seen with chlorpromasin and the calmodulin inhibitor W-7 (N-(6-aminohexyl)-5--chloro-1-naphthalenesulfonamide), although the effect of W-7 is not temprature independent (Fig.1). We interpret the decrease in nucleoid sedimentation caused by MCA as a direct evidence of DNA damage.

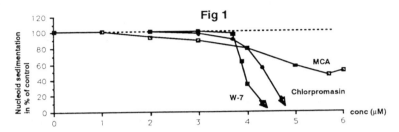

Fig 1

ADPRT measurements were also used to evaluate DNA breaks caused by MCA. HML's exposed to increasing doses of MCA showed a dose response increase in ADPRT activity for doses above 100 μM (Fig 2). If we treated the cells with 15 Gy radiation to activate the ADPRT level, and simultaneously treated them with MCA, an activation of the ADPRT level could be seen with doses as low as 100 nM (Fig 2). The increase in ADPRT activity, being due mostly to poly-ADPRT, is taken as another indicator that MCA causes DNA damage.

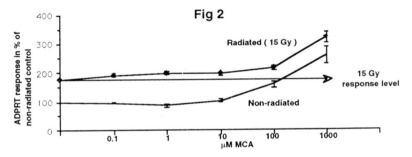

Fig 2

No direct inhibition of ADPRT was seen with MCA in either radiation treated cells or in non-treated cells or in permeabilized cells. Sims et al (3) have previously ascertained that ADP-ribosylation, is not inhibited by N-carboxamide substituted benzamides, whereas other benzamide analogs were potent inhibitors of ADP-ribosylation. The study by Sims and co-workers was carried out on subcellular preparations of poly-ADPRT, and our data, using HML cytoskeletons treated with MCA, are consistant with their observations. However, although a direct inhibition of poly-ADPRT activity by MCA has been ruled out as a possible molecular mechanism of radiosensitization, an indirect effect on poly-ADPRT has not been excluded. This point is a current subject of our research effort.

The role of MCA on the repair of DNA damage induced by ionizing radiation and NA-AAF in HMLs has also been studied. MCA showed a dose response inhibition of the repair of the bulky DNA lesions induced by NA-AAF when estimated by unscheduled DNA synthesis (UDS) (Fig 3a). When MCA and radiation were combined, we observed an additive effect measured as a decrease in nucleoid sedimentation (Fig. 3b). In the time study of MCA + radiation shown in Fig. 3b, we find that the DNA damage induced by the combination of MCA + radiation is repaired slower than the DNA damage induced by the radiation only. The DNA damage induced by MCA alone or in combination with radiation is primarily repaired with poor kinetics, indicating an inhibited state, or the presence of difficultly repairable DNA lesions, or some combination thereof. Nevertheless, these data do not exclude a MCA effect on inhibiting the repair of some types of lesions induced by gamma radiation.

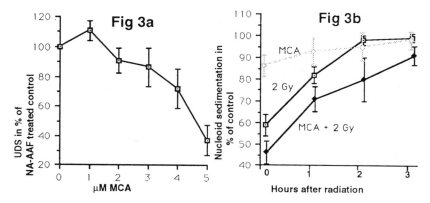

Cantoni et al (6) have shown that the induction of DNA damage by H_2O_2 can be totally

blocked by chelating the intracellular Ca^{++} pool by Quin-2. We have used this technique to examine if MCA might induce the DNA damge by production of H_2O_2. The DNA damage induced by H_2O_2 is totally reversed in a dose dependent manner by Quin-2, but the MCA induced DNA damage is not effected by chelating the intracellular Ca^{++}, nor is the DNA damage induced by chlorpromazine or W-7 which are well known calmodulin inhibitors (Fig 4).

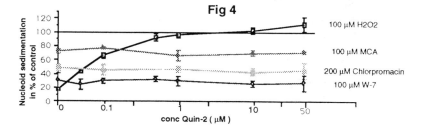

To examine if MCA works by the production of radicals, we have also examined the DNA damage induced by MCA and H_2O_2 with increasing doses of the electrophilic scavanger, nordihydroguaiaretic acid (NHGAA). The DNA damaging effect of can be totally reversed by NHGAA in a dose dependent manner, but the damage caused by MCA is not effected by NHGAA (Fig. 5). In conclusion, we find that the DNA damaging effect of MCA is most likely not caused by a radical mechanism.

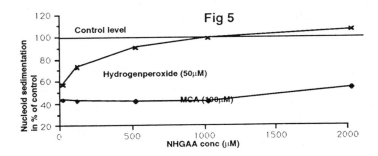

Fig 5

References:

(1) Harrington, R.A.; Hamilton, C.W.; Brogben, R.N.; Linkewich, J.A.; Romankiewicz, J.A.; Heel, R.C. Metoclopramide. An updated rewiew of its pharmacological properties and clinical use. *Drugs* 1983; 25:451-494.

(2) Gralla, R,J.; Itri, L.M.; Pisko, S.E.; Squillante, A.E.; Kelsen, D.P.; Braun Jr, D,W.; Bordin, L.A.; Braun, T.J.; Young, C.W. Antiemetic efficacy of high-dose metoclopramide: randomized trials with placebo and prochlorperazine in patients with chemotherapy-induced nausea and vomiting. *N. Engl. J. Med.* 1981; 305:905-909.

(3) Sims, J.L.; Sikorski, G.W.; Catino, D.M.; Berger, S.J.; Berger, N.A. Poly(adenosinediphosphoribose) polymerase inhibitors stimulate unscheduled deoxyribonucleic acid synthesis in normal human lymphocytes. *Biochemistry* 1982; 21: 1813-1821.

(4) Kjellen, E.; Wennerberg, J.; Pero, R. Metoclopramide enhances the effect of cisplatin on xenografted squamous cell carcinoma of the head and neck. *Br. J. Cancer* 1989; 59: 247-250.

(5) Lybak, S.; Kjellén, E.; Wennerberg, J.; Pero, R.W. Metoclopramide enhances the effect of ionizing radiation on xenografted squamous cell carcinoma of the head and neck. *Int. J. Rad. Onc. Biol. Phys.* 1990; 19:1419-1424.

(6) Cantoni, O.; Sestili, P.; Cattabeni, F.; Bellomo, G.; Pou, S.; Cohen, M.; Cerutti, P. Calcium chelator Quin 2 prevents hydrogen-peroxide-induced DNA breakage and cytotoxicity. *Eur. J. Biochem.* 1989;182:209-212.

Nuclear ADP-Ribosyltransferase as Target Antigen in Chronic Graft-Versus-Host Disease

Jozefa Wesierska-Gadek, Edward Penner, Peter Kier, Eva Hitchman and Georg Sauermann

Chronic graft-versus-host disease (c-GVHD), one of the major clinical complication of human allogeneic bone marrow transplantation has similarities with autoimmune diseases such as systemic lupus erythematosus or scleroderma (Ferrara and Deeg, 1991; Shulman and Sullivan, 1988). Screening by immunofluorescence microscopy suggested the presence of antinuclear autoantibodies in the sera of allogeneic bone marrow graft recipients (Shulman et al., 1980; Kier et al., 1990).

The present study based on immunoblotting experiments demonstrates that antibodies against nuclear ADP-ribosyl-transferase (ADPRT) occur in c-GVHD sera.

Results

Sera were obtained from 29 patients who underwent an allogeneic bone marrow transplantation. The indications for bone marrow transplantation and clinical features have previously been described (Kier et al., 1990). 16 of 29 recipients had extensive c-GVHD, 3 limited c-GVHD and 10 patients had no signs of c-GVHD. The majority of sera from patients with extensive c-GVHD stained nuclear and nucleolar structures in indirect immunofluorescence microscopy, as depicted in Fig.1

In order to identify the target antigens, immunoblotting experiments were performed. First, total nuclear proteins were electrophoretically separated, blotted unto nitro-cellulose sheets, and exposed to the patients' sera. After reaction with [^{125}I]protein A the formed immune complexes were detected by autoradiography (Wesierska-Gadek et al., 1990). Fig.2 shows the nuclear antigens recognized by c-GVHD sera. Interestingly, the sera detected a band at 110 to 120 kD, in the molecular weight range of nuclear ADP-ribosyltransferase. In the following experiments we tested whether this enzyme was indeed the target antigen. Since many of the sera stained nucleoli in immunofluorescence, the reaction of the sera with nucleolar proteins was also examined.

Fig.3 demonstrates the recognition of nuclear ADP-ribosyltransferase as target antigen by one of the c-GVHD sera. The serum reacted with a band at approximately 110-120 kD, independently whether total nuclear proteins, total nucleolar proteins, or purified nuclear ADP-ribosyl-transferase were used as subtrates. However, the serum did not recognize the nucleolar phosphoprotein C23 (nucleolin), a protein of comparable molecular mass.

In contrast, other c-GVHD sera detected nucleolin, but not nuclear ADP-ribosyltransferase, as illustrated in Fig.4.

The identity of nuclear ADP-ribosyltransferase as target antigen was additionally proven after 2-dimensional electrophoresis (Fig.5)

Apparently, two distinct proteins of about the same molecular weight, namely nuclear ADP-ribosyltransferase and nucleolin, can be recognized as target antigens by c-GVHD sera.

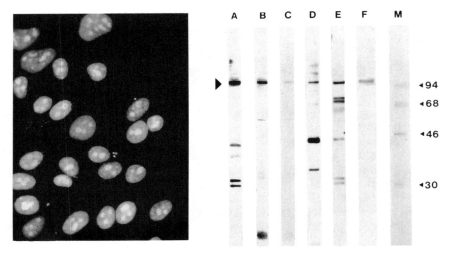

Fig.1. (left) c-GVHD serum giving homogenous nuclear plus nucleolar staining in immunofluorescence. HEp-2 cells were incubated with serum and fluorescein-conjugated anti-human IgG.

Fig.2. (right) Nuclear antigens recognized by c-GVHD sera.Immunoblots; total nuclear proteins as antigen source. 10% SDS-PAGE. Lanes A-F, c-GVHD sera; M, marker proteins.

Table 1 shows that antibodies against nuclear ADP-ribosyltransferase were only found in the sera of patients with extensive chronic graft-versus-host disease. Serial examination of sera revealed correlations between the occurence of antinuclear antibodies and the appearance of c-GVHD. Prior to and during the clinical manifestation of c-GVHD, antinuclear antibodies were observed in about 75 per cent of the cases. Approximately one half of the positive sera recognized nuclear ADP-ribosyltransferase.

Table 1. Antibodies to nuclear ADP-ribosyltransferase in chronic graft-versus-host disease

Patients	n	IF positive	Antibodies to nuclear ADPRT
No c-GVHD	10	0	0
Limited c-GVHD	3	0	0
Extensive c-GVHD	16	11	6
Normal controls	23	0	0

IF, immunofluorescence

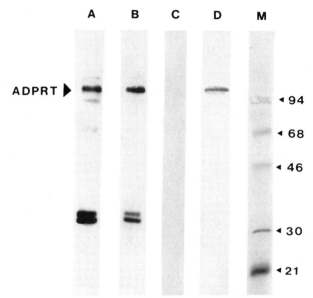

Fig.3 Nuclear ADP-ribosyltransferase recognized as target antigen by a c-GVHD serum. Immunoblot, 10% SDS-PAGE. Antigens: A, total nuclear proteins; B, total nucleolar proteins; C, purified nucleolar phosphoprotein C23 (nucleolin); D, nuclear ADP-ribosyltransferase purified on DNA cellulose, hydroxyapatite and Blue-Trisacryl-Sepharose columns; M, marker proteins.

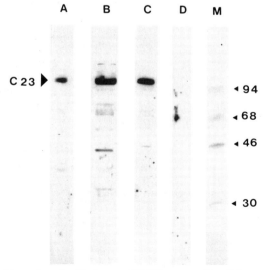

Fig.4. Recognition of nucleolin by a c-GVHD serum. Conditions as in Fig.3, except that another c-GVHD serum was used.

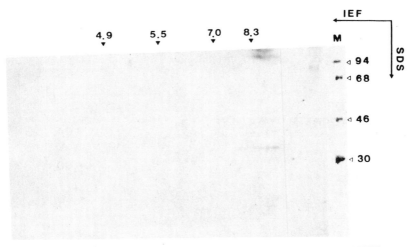

Fig.5. Recognition of purified nuclear ADP-ribosyltrans-ferase by a c-GVHD serum. 2-dimensional PAGE.

Discussion

The occurence of antibodies against nuclear ADP-ribosyl-transferase in sera of patients with rheumatologic autoimmune diseases, as systemic lupus erythematosus, Sjögren`s syndrome, rheumatoid arthritis or progressive systemic sclerosis, has been described (Negri et al., 1990; Yamanaka et al., 1987). Anti-ADP-ribosyltransferase anti-bodies have also been found in sera from patients with gastrointestinal autoimmune disease (Sauermann et al., 1991).

The present data demonstrate that autoantibodies to nuclear ADP-ribosyltransferase are also found in the sera of patients with extensive chronic graft-versus-host disease developing after allogeneic bone marrow transplantation. Chronic graft-versus-host disease occurs more than 100 days after bone marrow transplantation, with an incidence of up to 60 percent. In these patients severe disfunctions of the immune system are observed (Ferrara and Deeg, 1991; Shulman and Sullivan, 1988).

Interestingly, the c-GVHD sera examined in our study did not react with sleroderma-70 antigen, Sm, nuclear RNPs, Ro/SSB and La/SSB, antigens recognized in classic autoimmune diseases. On the other hand, the frequency of the presently described antibodies to nuclear ADP-ribosyl-transferase was relatively high.

References

1. Ferrara, J.; Deeg, H. Graft-versus-host disease. New Engl. J. Med. 324:667-674; 1991.

2. Shulman, H.; Sullivan, K. Graft-versus-host disease:allo- and autoimmunity after bone marrow transplantation. Concepts Immunopathol. 6:141-165; 1988.

3. Shulman, H.; Sullivan, K.; Weiden, P.; McDonald, G.; Striker, G.; Sale, G.; Hackman, R.; Tsoi, M.; Storb, R., Thomas, E. Chronic graft-versus-host syndrome in man: a long-term clinicopathologic study of 20 Seattle patients. Am.J.Med. 69:204-217; 1980.

4. Kier, P.; Penner, E.; Bakos, S.; Kalhs, P.; Lechner, K.; Volc-Platzer, B.; Wesierska-Gadek, J.; Sauermann, G.; Gadner, H.; Emminger-Schmidmeier, W.; Hinterberger, W. Autoantibodies in chronic GVHD: high prevalence of antinucleolar antibodies. Bone Marrow Transpl. 6:93-96; 1990.

5. Wesierska-Gadek, J.; Penner, E.; Hitchman, E.; Sauermann, G. Antibodies to nuclear lamin C in chronic hepatitis delta virus infection. Hepatology 12:1129-1133; 1990.

6. Negri, C.; Scovassi, A.; Cerino, A.; Negroni, M.; Borzi, R.; Meliconi, R.; Facchini, A.; Montecucco, C.;Astaldi-Ricotti, G. Autoantibodies to poly(ADP-ribose) polymerase in autoimmune diseases. Autoimmunity 6:203-209; 1990.

7. Yamanaka, H.; Willis; E.; Penning, C.; Peebles, C., Tan, E.; Carson, D. Human autoantibodies to poly(adeno-sine diphosphate-ribose) polymerase. J.Clin.Invest. 80:900-904; 1987.

8. Sauermann, G.; Penner, E.; Hitchman, E.; Wesierska-Gadek, J. Autoantibodies against nuclear ADP-ribosyl-transferase in autoimmune liver disease. Paul Mandel International Meeting on Poly(ADP-ribosyl)ation Reactions, Quebec, Canada, May 30-June 3, 1991, Abstr.24b.

INDUCTION OF POLY ADP-RIBOSYLATION IN HUMAN MALIGNANT CELLS

Neeta Singh, Nandini Rudra and Poonam Bansal
Dept.of Biochemistry, AIIMS, New Delhi, India

Introduction

Neoplastic cells have an imbalance in the control of their growth and differentiation pathways. Because signal transduction constitutes a critical component of differentiation-induction, the nature of the enzymes involved in this process is important and are being actively examined. Another important question is the nature of the gene changes that cause the cellular alterations and their mechanism. Although the precise mechanism of transformation and DNA repair in mammalian cells has not been elucidated, poly(ADP-ribose) synthesis has been noted for its close correlation to carcinogenesis and DNA repair. Poly ADPR transferase is an eukaryotic nuclear enzyme which catalyses the successive transfer of ADP-ribose moiety of NAD^+ to various nuclear proteins. A sudden increase in polymer can lead to rapid consumption of NAD^+ and ATP and therefore can prevent or delay cell proliferation.

Enhanced poly ADP-ribosylation in leukemia lymphocytes.

In this study we have tried to look at the involvement of poly ADP-ribosylation in carcinogenesis. We have studied the basal levels of poly ADPR transferase (poly ADPRT) and the correlation to NAD^+ levels in lymphocytes isolated from normal subjects, acute lymphocytic leukemia (ALL), acute myeloid leukemia (ADL) and chronic myeloid leukemia (CML). The peripheral blood samples from these subjects were taken prior to the commencement of chemotherapy. The lymphocytes were isolated by ficoll-hypaque gradient centrifugation (1). For the assay of poly ADPRT, the cells were permeabilized and the enzyme activity estimated as described earlier (2). The mean basal activity of poly ADPRT was 3 fold higher in lymphocytes of ALL cases, 3.5 fold higher for AML subjects and 2.5 fold higher for CML subjects (Table 1). Similar findings have been shown in leukemia granulocytes (3), colon cancer (4) and mouse fibroblasts transformed by S V 40 (5). Various DNA damaging agents increase polymer synthesis (2, 6). It seems plausible that these changes reflect alteration in chromatin structure or are in response to DNA change. Such alterations may play a role in state of predisposition or carcinogenesis. NAD^+ serves a dual role as a respiratory coenzyme and as a substrate for the post translational poly ADP-ribose modification of chromatin. Reports suggest that raising the intracellular NAD^+ concentration by 70% caused a five fold increase of chromatin associated poly ADPRT. NAD^+ was estimated, essentially as described by Jacobson et al (7). There was a drop of 53% in mean basal NAD^+ levels in the case of ALL patients, 63% for AML patients and 43% for CML patients (Table 2).

Table 1: Mean basal activity of poly ADPR transferase in lymphocytes

	incorporation of (^{3}H)NAD cpm x 10^3/5 x 10^6 cells	n	age group
Normal	22	14	15-50 years
AML	78	14	17-50 years
ALL	67	10	15-50 years
CML	54	8	25-45 years

It has been shown (2,8) that loss of NAD^+ is probably due to conversion of NAD^+ to poly ADP-ribose, rather than to inhibition of NAD^+ biosynthesis or to degradation by some other pathway.

Table 2: Mean basal levels of NAD^+ in lymphocytes

	pmoles of NAD^+/ 10^6 cells	n	age group
Normal	62.5	14	15-50 yrs
AML	24	14	17-50 yrs
ALL	29	10	15-50 yrs
CML	35.6	8	25-45 yrs

Kinetics of poly ADP-ribosylation in HL-60 cells on exposure to tumour promoters.

In another set of experiments the effect of tumour promoters was studied on poly ADPRT activity and NAD^+ levels in HL-60 cells. Tumour promoters do not bind to DNA but act by binding to membrane associated receptors and thus produce their initial effects at the epigenetic level. HL-60 cells were grown in RPMI medium containing antibiotics and 10% FCS. They were treated for varying length of time with PMA (25 ng/ml) and mezerein (125 ng/ml). Both the phorbol ester PMA (phorbol-12-myristate-13-acetate) and its related diterpene derivative mezerein, stimulated poly ADPRT activity but only by 1.6 fold at 90 min and 3 hrs respectively (Fig. 1). The enzyme activity dropped from 3 hrs onwards till 14 hrs but did not reach the initial basal levels (resuts not shown). The drop in enzyme activity could be due, partly, to partial inactivation of the enzyme undergoing automodification or to degradation of polymer by poly ADPR glycohydrolase. A drop of 18% at 3 hrs in

NAD+ levels was seen with both PMA and mezerein (Fig.2). This study provides insight into some of the molecular mechanisms of carcinogenesis and new strategies for cancer diagnostics and treatment.

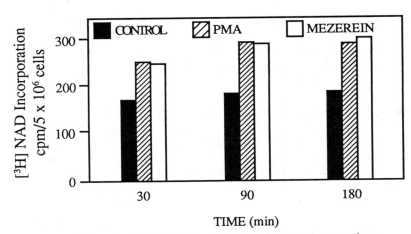

Fig. 1: Stimulation of poly (ADPRT) activity by PMA and mezerein on HL-60 cells.

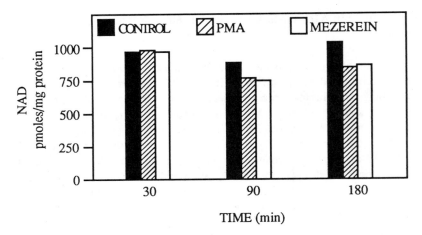

Fig. 2: Effects of NAD+ level after stimulation by PMA and mezerein on HL-60 cells

Acknowledgements

This work was supported by grants from All India Institute of Medical Sciences and the Indian Council of Medical Research.

References

1. Boyum, A. (1968) Isolation of mononuclear cells and granulocyte from human blood. Scand J Clin Lab Invest Suppl 21: 77-90.
2. Jacobson, M.K., Levi, V., Juarez Salinas, H., Barton, K.A., Jacobson, E.L. (1980) Effect of carcinogenic N alkyl-N-nitroso compounds on nicotinamide adenine dinucleotide metabolism. Cancer Res 40: 1797-1802.
3. Ikai, K., Ueda, K., Fukushima, M., Nakamura, T., Hayaishi, O. (1980) Poly(ADP-ribose)synthesis, a marker of granulocyte differentiation. Proc Natl Acad Sci USA 77: 682-3685.
4. Hirai, K., Ueda, K., Hayaishi, O. (1983) Aberration of poly(Adenosine diphosphate-ribose) metabolism in human colon adenomatous polyp and cancer. Cancer Res 43: 3441-3446.
5. Miwa, M., Oda, K., Segawa, K., Tanaka, M., Sugimura, T. (1977) Cell density dependent increase in chromatin associated ADP-ribosyl transferase activity in simian virus 40-transformed cells. Arch Biochem Biophys 181: 313-321.
6. Durkacz, B.W., Omidiji, O., Gray, D.A., Shall, S. (1980) (ADP-ribose) participates in DNA excision repair. Nature 283: 593-596.
7. Jacobson, E.L., Lange, R.A., Jacobson, M.K. (1979) Pyridine nucleotide synthesis in 3T3 cells. J Cell Physiol 99: 417-426.
8. Carson, D.A., Seto, S., Wasson, D.B. (1987) Pyridine nucleotide cycling and poly(ADP-ribose) synthesis in resting hyuman lymphocytes. J. Immunol 138: 1904-1907.

PROTEOLYTIC CLEAVAGE OF POLY(ADP-RIBOSE) POLYMERASE IN HUMAN LEUKEMIA CELLS TREATED WITH ETOPOSIDE AND OTHER CYTOTOXIC AGENTS

Scott H. Kaufmann, Serge Desnoyers[1], Brian Talbot[2], and Guy G. Poirier[1]

The Johns Hopkins Oncology Center, Baltimore, MD 21205

Poly(ADP-ribose) polymerase (pADPRp, EC 2.4.2.30) is a nuclear enzyme that catalyzes the transfer of ADP-ribose units from NAD to various protein acceptors (reviewed in 1,2). The activity of pADPRp in vitro is markedly stimulated by nicked DNA (3,4). Likewise, pADPRp activity increases when nicks and single-stranded DNA regions are generated in vivo during the repair of DNA damage caused by alkylating agents (5).

The major acceptors for poly(ADP-ribose) subunits in vivo appear to include histones, the polymerase itself, and topoisomerase I and II (6-8). Poly(ADP-ribosyl)ation of histone H1 appears to alter chromatin conformation (9). Poly(ADP-ribosyl)ation of the topoisomerases inhibits their activity (10,11). By increasing the accessibility of damaged chromatin to repair enzymes and by inhibiting DNA replication and transcription until repair has taken place, these changes conceivably contribute to the repair of damaged DNA. Consistent with this view, it has been observed that inhibiting pADPRp activity (NAD depletion or treatment with pADPRp inhibitors) increases the toxicity of certain alkylating agents (12,13).

On the other hand, it has also been suggested that pADPRp activity might contribute to the death of lethally damaged cells (14-17). Studies performed on lymphocytes or L1210 cells treated with high doses of N-methyl-N'-nitro-nitrosoguanidine (MNNG) have revealed that the appearance of DNA single-strand breaks is followed by activation of pADPRp, depletion of NAD and ATP, and finally loss of membrane integrity (16,18). Agents which introduce DNA double-strand breaks also activate pADPRp (19). In lymphocytes treated with cytotoxic concentrations of glucocorticoids, the formation of DNA double-strand breaks is likewise followed by depletion of NAD and ATP, and then by loss of membrane integrity (20).

It has generally been assumed in these studies that the depletion of NAD occurs as a consequence of activation of pADPRp by DNA strand breaks. In a previous study (21), however, it was observed that treatment of HL-60 human progranulocytic leukemia cells with the topoisomerase II-directed

[1]Laboratory of Poly(ADP-ribose) Metabolism, Molecular Endocrinology Dept., Laval University Medical Center, Ste-Foy, Quebec, G1V 4G2, Canada.
[2]Department of Biology, University of Sherbrooke, Sherbrooke, Quebec, J1K 2R1, Canada.

chemotherapeutic agent etoposide resulted in degradation of pADPRp to an 85 kDa proteolytic fragment. Our subsequent studies on the nature of this 85 kDa fragment and the conditions which lead to its formation are summarized below.

Characterization of etoposide-treated HL-60 cells

When HL-60 human progranulocytic leukemia cells were treated with etoposide, a series of morphological changes occurred. Phase contrast microscopy revealed the conversion of the normally round HL-60 cells (Fig. 1A) to dysmorphic cells (Fig. 1B) reminiscent of "apoptotic bodies" (22).

Electron microscopy (not shown) also revealed the morphologic changes of apoptosis (reviewed in 22). Two hours after the addition of etoposide, the cells displayed margination of their heterochromatin and loss of nucleoli.

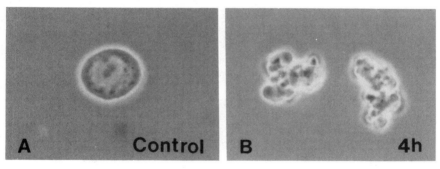

Figure 1. Morphological appearance of HL-60 cells before (A) or after (B) treatment with 68 µM etoposide for 4 h. Examination of control cells by phase contrast microscopy reveals round cells with round nuclei and prominent nucleoli. Etoposide-treated cells have the appearance of many interconnected membrane-enclosed vesicles. Staining of these cells (not shown) reveals that nuclear fragments are present in many of these vesicles. Electron microscopy (not shown) demonstrates the sequential changes described in other apoptotic cells (22). Greater than 95% of these cells exclude trypan blue 4 h after addition of etoposide.

This was followed by extensive vesiculation and finally by formation of multiple small membrane-bounded vesicles. It is important to stress that >90% of these cells excluded trypan blue six hours after the addition of etoposide.

As previously noted (21), these morphological changes were accompanied by endonucleolytic DNA degradation (Fig. 2A), a biochemical change which is characteristic of apoptotic cells (23). Within 90 minutes after addition of etoposide to HL-60 cells, a ladder of DNA fragments ranging from 180 bp upward in 200 bp intervals was observed (Fig. 2A, lane 3). These changes indicate the action of an endonuclease on the internucleosomal regions of DNA.

Proteolytic degradation of pADPRp in etoposide-treated HL-60 cells

To determine whether the degradation of DNA was accompanied by proteolysis, proteins from etoposide-treated HL-60 cells were subjected to SDS-polyacryl-amide gel electrophoresis. Proteolysis was not evident by Coomassie blue staining (Fig. 2B). Western blotting was performed with C-II-10, a monoclonal antibody which recognizes an epitope localized at the carboxyl terminus of the DNA binding domain in pADPRp (24). In untreated cells, this antibody reacted with a single 116 kDa polypeptide (Fig. 2C, lane 1).Concomitant with the DNA degradation,the signal for pADPRp at M_r 116,000 diminished and a prominent new signal at M_r 85,000 appeared (Fig. 2C, lanes 4-6).

The location of this apparent proteolytic cleavage was initially investigated by repeating the Western blotting using F-I-23, a monoclonal antibody which recognizes an epitope located in the amino terminal 29 kDa of pADPRp (24). F-I-23 recognized the intact 116 kDa polymerase (Fig. 2D, lanes 1-3) but not the 85 kDa fragment (Fig. 2D, lanes 4-6). This result suggests that the amino terminus of pADPRp has been removed to generate the 85 kDa fragment.

To confirm the location of the proteolytic cut, activity blotting (25) was performed. Previous results have shown that removal of as few as 45 amino acids from the carboxyl terminus of pADPRp inactivates the enzyme (25). Hence the 85 kDa polypeptide would be expected to be enzymatically inactive if significant proteolysis were occurring at the carboxyl terminus. In contrast, pADPRp would be expected to retain some enzymatic activity even if the entire amino terminal domain containing both zinc fingers were removed (25,26). To distinguish between these alternatives, polypeptides from etoposide-treated HL-60 cells were separated by SDS-PAGE, transferred to nitrocellulose, renatured, and incubated with [^{32}P]-labeled NAD in the presence of nicked DNA (Fig. 2E). In extracts from untreated cells, a band at M_r 116,000 was labeled (Fig. 2E, lane 1). This band represents automodified pADPRp (25). Preliminary experiments indicate that the 85 kDa pADPRp fragment in extracts from etoposide-treated cells also became labeled (Fig. 2E, lanes 3-6). Thus the carboxyl end of the polymerase must have remained largely or completely intact when the 85 kDa fragment was generated.

Proteolytic cleavage that removed 30 kDa from the amino terminus of the 116 kDa polymerase molecule would be expected to remove both zinc fingers (25). Accordingly, the 85 kDa fragment would be expected to demonstrate DNA-independent enzyme activity (25). Experiments to test this prediction are in progress.

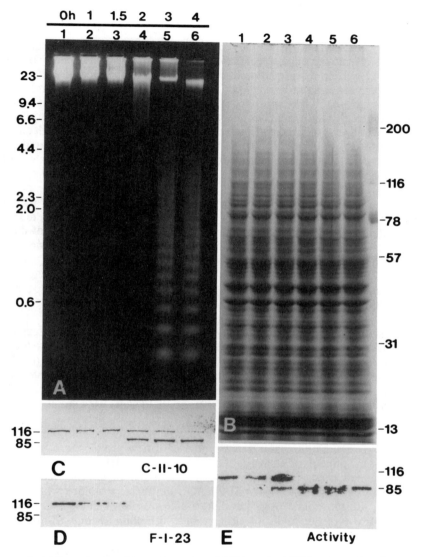

Figure 2. Effect of etoposide on integrity of DNA and pADPRp. HL-60 cells were treated with 68 μM etoposide for 0 (lane 1), 60 (lane 2), 90 (lane 3), 120 (lane 4), 180 (lane 5) or 240 min (lane 6). Replicate samples were lysed for analysis on agarose gels (panel A) or SDS-PAGE followed by staining with Commassie blue (panel B) or transfer to nitrocellulose (panels C-E). Polypeptides immobilized on nitrocellulose were blotted (ref. 21) with monoclonal antibodies (ref. 24) C-II-10 (panel C) or F-I-23 (panel D). Alternatively, the polypeptides immobilized on nitrocellulose were renatured and assayed for pADPRp activity (see ref. 25 for METHODS) in the presence of activated DNA (panel E).

An 85 kDa fragment of pADPRp generated after treatment with a variety of cytotoxic agents

To ascertain whether the proteolytic cleavage of pADPRP occurred only after treatment with etoposide, HL-60 cells were treated with a variety of agents including colcemid, cis-platinum, cytosine arabinoside, methotrexate, and 5-deaza-acyclo-tetrahydrofolate (an inhibitor of glycinamide ribonucleotide transformylase). The primary targets of these agents include diverse macromolecules in the nucleus and cytoplasm. As previously reported (21,27), all of these agents induced endonucleolytic DNA degradation, albeit with different time courses. In each case, the 85 kDa fragment of pADPRp appeared at the same time that a nucleosomal ladder of DNA fragments became detectable. (data not shown).

To rule out the possibility that the proteolytic cleavage of pADPRp was unique to HL-60 cells, a number of different leukemia cell lines were treated with etoposide. The 85 kDa fragment was rapidly generated in all of the cell lines tested including KG1a acute myelogenous leukemia cells and Molt 3 and Molt 4 T cell lymphocytic leukema cell lines (data not shown).

Induction of endonucleolytic DNA degradation and proteolytic cleavage of pADPRp by inhibitors of protein synthesis

In some experimental systems such as glucocorticoid-treated rat thymocytes, morphological and biochemical changes of apoptosis are delayed or prevented by treating cells with inhibitors of RNA or protein synthesis (reviewed in 28; see also 29). These results have given rise to the concept of "programmed cell death." In other experimental systems, however, it has not been possible to demonstrate an inhibitory effect of RNA or protein synthesis inhibitors (21, 30-32).

To determine whether the synthesis of new proteins was required in order for the proteolytic cleavage of pADPRp to occur, HL-60 cells were treated with cytotoxic concentrations of cycloheximide or puromycin and then subjected to agarose gel electrophoresis or SDS-PAGE and Western blotting with antibody C-II-10. Results obtained with cycloheximide are shown in Fig. 3. Treatment with 36 μM cycloheximide, a concentration which diminished [^{35}S]-methionine incorporation into protein by 90%, resulted in the appearance of a nucleosomal ladder of DNA fragments within 2 h (Fig. 3, lane 4). Even before the appearance of this DNA ladder, the 85 kDa proteolytic fragment of pADPRp became evident (Fig. 3, lane 3). Thus the endonucleolytic DNA degradation and the proteolytic cleavage of pADPRp do not appear to require ongoing protein synthesis.

Figure 3 Effect of cycloheximide on integrity of DNA and pADPRp. HL-60 cells were incubated with 36 μM cycloheximide for 1 (lane 3) or 2 h (lane 4). Untreated (lane 1) and etoposide-treated cells (lane 2) were included as controls. Cell lysates were subjected to agarose gel electrophoresis (upper panel) or Western blotting with C-II-10 antibody (lower panel). Degradation of DNA and pADPRp was evident 2 h after addition of cycloheximide to the culture (lane 4).

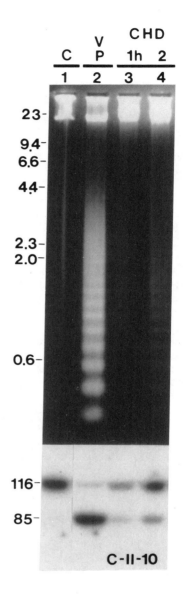

Biological significance of pADPRp proteolysis

In summary, we have shown that pADPRp is cleaved to an 85 kDa fragment after treatment of human leukemia cell lines with a number of cytotoxic agents (Figs. 2 and 3; see also refs. 21 and 27). This 85 kDa fragment has lost a major portion of the amino terminal DNA binding domain of pADPRp (Fig. 2D). As a consequence, the 85 kDa fragment displays pADPRp activity (Fig. 2E) that would be expected to be DNA-independent.

This apoptosis-associated cleavage of pADPRp must be distinguished from the recently described proteolytic degradation of pADPRp in untreated HeLa cells (33). In this latter system, proteolysis was reported to yield an enzymatically inactive 80 kDa fragment containing the DNA binding and automodification domains but lacking the catalytic domain.

The functional significance of the 85 kDa fragment described in the present study is currently unknown. Previous reports have indicated that cellular NAD pools are consumed in etoposide-treated HL-60 cells (34) and in other cells undergoing apoptosis (14-18,20). The relative contributions of the intact polymerase and the 85 kDa fragment to this NAD consumption are unclear. At least three possibilities exist. First, the intact polymerase might be activated by DNA strand breaks and might consume most of the cellular NAD prior to the proteolytic event described in the present paper. Alternatively, if the DNA strand breaks and the proteolytic cleavage of pADPRp occur simultaneously, the 85 kDa fragment might be generated before NAD levels decrease. If so, the 85 kDa fragment might itself be sufficiently active that it subsequently contributes to the lowering of NAD levels by indiscriminately ADP-ribosylating polypeptides thoughout the nucleus. Finally, it is conceivable that neither the intact polymerase nor the 85 kDa fragment consumes much NAD and that processes other than poly(ADP-ribosyl)ation also participate in the depletionof cellular NAD pools. Careful time course studies and judicious use of pADPRp inhibitors should help to distinguish between these possibilities.

It is important to stress that the proteolytic cleavage of pADPRp in chemotherapy-treated human leukemia cell lines occurs several hours prior to the loss of plasma membrane integrity (trypan blue uptake). Most cellular polypeptides remain intact during the course of chemotherapy-induced apoptosis (Fig. 2B). Nonetheless, previous studies have shown that the chemotherapy-induced endonucleolytic degradation of DNA is accompanied by the degradation of a number of other nonhistone nuclear polypeptides as well (21,27). Thus the proteolytic cleavage described in the present study is not unique to pADPRp.

The identity of the protease which cleaves pADPRp to the 85 kDa fragment is currently unknown. Studies with inhibitors of RNA synthesis (not shown) and protein synthesis (Fig. 3) suggest that the relevant protease is present in leukemia cell lines prior to drug treatment. Experiments to identify this protease are currently in progress.

It is unclear at present whether the proteolytic cleavage of pADPRp is limited to chemotherapy-induced cell death in leukemia cells or is more generally seen in cells undergoing apoptosis. Likewise, it is unclear whether the loss of the DNA binding domain of pADPRp alters its chromatin localization and/or the identity of its substrates in situ. These questions are also being actively investigated.

REFERENCES

1. Althaus, F.R., and Richter, C. (1987) ADP-ribosylation of proteins. Part I. Enzymology and biochemical significance. in Molecular Biology, Biochemistry and Biophysics, Vol. 37, Springer-Verlag, Berlin.
2. DeMurcia, G., Huletsky, A., and Poirier, G. G. (1988) Biochem. Cell Biol. 66: 626-635.
3. Benjamin, R. C., and Gill, D. M. (1980) J. Biol. Chem. 255: 10502-10508.
4. Ohgushi, H., Yoshihara, K, and Kamiya, T. (1980) J. Biol. Chem. 255: 6205-6211.
5. Singh, N., Poirier, G. G., and Cerutti, P. A. (1985) EMBO J. 4: 1491-1494.
6. Adamietz, P., and Rudolph, A. (1984) J. Biol. Chem. 259: 6841-6846.

7. Adamietz, P. (1987) Eur. J. Biochem. 169: 365-372.
8. Boulikas, T. (1988) EMBO J. 7: 57-67.
9. Poirier, G.G., de Murcia, G., Jongstra-Bilen, J., Niedergang, C., and Mandel, P. (1982) Proc. Nat. Acad. Sci. U.S.A. 79: 3423-3427.
10. Ferro, A.M., and Olivera, B.M. (1984) J. Biol. Chem. 259: 547-554, 1984.
11. Darby, M. K., Schmitt, B., Jongstra-Bilen, J., and Vosberg, H.-P. (1985) EMBO J. 4: 2129-2134.
12. Jacobson, E. L., Meadows, R., and Measel, J. (1985) Carcinogenesis 6: 711-714.
13. Durkacz, B.W., Irwin, J., and Shall, S. (1981) Eur. J. Biochem. 121: 65-69, 1981.
14. Wintersberger, U., and Wintersberger, E. (1985) FEBS. Let. 188: 189-191.
15. Berger, N. (1985) Radiation Res. 101: 4-15.
16. Carson, D.A., Seto, S., Wasson, D.B., and Carrera, C.J. (1986) Exp. Cell Res. 164: 273-281.
17. Gaal, J.C., Smith, K.R., and Pearson, C.K. (1987) Trends in Biochemical Sciences 12: 129-130.
18. Das, S.K., and Berger, N.A. (1986) Biochem. Biophys. Res. Commun. 137: 1153-1158.
19. Ikejima, M., Nogushi, S., Yamashita, R., Ogura, T., Sugimura, T., Gill, D.M., and Miwa, M. (1990) J. Biol. Chem. 265: 21907-21913.
20. Berger, N.A., Berger, S.J., Sudar, D.C., and Distelhort, C.W. (1987) J. Clin. Invest. 79: 1558-1563.21.Kaufmann, S.H. (1989) Cancer Res. 49: 5870-5878.
22. Wyllie, A.H., Kerr, J.F.R., and Currie, A.R. (1980) Int. Rev. Cytology 68: 251-306.
23. Wyllie, A.H. (1980). Nature 284: 555-556.
24. Lamarre, D., Talbot, B., de Murcia, G., Laplante, C., Leduc, Y., Mazen, A., and Poirier. G.G. (1988) Biochim. Biophys. Acta 950: 147-160.
25. Simonin, F., Menissier-de Murcia, J., Poch, O., Muller, S., Gradwohl, G., Molinete, M., Penning, C, Keith, G., and de Murcia, G. (1990) J. Biol. Chem. 265: 19249-19256, 1990.
26. Kameshita, I., Masuda, Z., Taniguschi, T., and Shizuta, Y. (1984) J. Biol. Chem. 259: 4770-4776, 1984.
27. Smith, G.K., Duch, D.S., Dev, I.K, Banks, S.D., and Kaufmann, S.H. (1991) Submitted.
28. Lockshin, R.A., and Zakeri, Z.F. (1990) J. Gerontology 45: B135-140.
29. Wyllie, A.H., Morris, R.G., Smith, A.L., and Dunlop, D. (1984) J. Pathol. 142: 67-77.
30. Vedekis, W.V., and Bradshaw, H.D., Jr. (1983) Mol. Cell. Endocrinol. 30: 215-227.
31. Wielckens, K., Delfs, T., Muth, A. Freese, V., and Kleeberg. H.J. (1987) J. Steroid Biochem. 27: 413-419, 1987.
32. Waring, P. (1990) J. Biol. Chem. 265: 14476-14480.

33. Kameshita, I., Mitsuuchi, Y., Matsuda, M., and Shizuta, Y. (1989) in ADP-ribose Transfer Reactions. Mechanisms and Biological Significance. M. K. Jacobson and E. L. Jacobson, eds. Springer-Verlag, New York, pp. 71-75.

34. Shimizu, T., Kubota, M., Tanizawa, A., Sano, H., Kasai, Y., Hashimoto, H., Akiyama, Y., and Mikawa, H. (1990) Biochem. Biophys. Res. Commun. 169: 1172-1177.

Supported in part by grants from the NIH, MRC, and FCAR.

ADP-Ribosylation of Topoisomerase II in Physiological Conditions

A.I. Scovassi, M. Negroni, C. Mariani, L. Clerici*, C. Negri and U. Bertazzoni

Istituto Genetica Biochimica Evoluzionistica CNR., 27100 Pavia; *Joint Research Center CEC, 21027 Ispra, Italy.

Introduction

ADP-ribosylation is known to play an important role in the modulation of chromatin structure, possibly by changing the affinity of histones to DNA and by interfering with the activity of enzymes which are responsible for condensation and decondensation processes (1,2). Several nuclear proteins have been found to be ADP-ribosylated in reconstituted systems (for a review, see 1). However, it is not yet clear which are the real acceptors in physiological conditions. In fact, beside poly(ADP-ribose)polymerase itself (3,4), only histones (5), topoisomerase I (6,7) and high and low mobility group proteins (8) have been recognized to be modified by poly(ADP-ribose) in intact eukaryotic systems.

In order to characterize non-histone proteins ADP-ribosylated *in vivo* in mammalian cells, we have followed different experimental procedures which allowed specific labeling of modified proteins and their immunological recognition. The experiments were essentially based on the isolation of ADP-ribosylated proteins from intact or permeabilized HeLa cells and from purified nuclei.

Evidence was obtained that ADP-ribosylation of DNA topoisomerase II occurs in intact mammalian cells.

Results

Different experimental approaches were used to study in mouse embryos and in human cells the non-histone nuclear proteins which are modified by ADP-ribosylation in physiological conditions.

Microinjection of ^{32}P-NAD in mouse embryos.

Mouse fertilized eggs were incubated 24 h at 37°C after microinjection with a solution containing ^{32}P-NAD. Cellular proteins were extracted from embryos by freezing and thawing and analyzed on SDS-PAGE. Autoradiography of the gel showed that proteins of 170, 115, 85 and 65 kDa were ADP-ribosylated (Fig.1, lane 1).

Labeling of ADP-ribose in HeLa cells.

ADP-ribose was labeled by incubating intact HeLa cells with ^{3}H-adenosine for 2 h at 37°C. To discard labeled nucleic acids, the cellular extracts were pretreated with DNase and RNase and proteins analyzed by SDS-PAGE. A typical autoradiogram of ADP-ribosylated proteins showed bands of 170, 115, 80 and 60 kDa (Fig.1, lane 2).

HeLa cell nuclei were incubated for 30 min at 25°C with a buffer containing ^{32}P-NAD and modified non-histone proteins were revealed as autoradiographic bands (Fig.1, lane 3): three major signals were visible, corresponding to peptides of 170, 115 and 85 kDa.

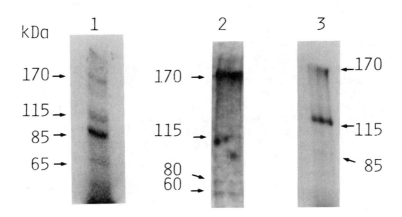

Fig.1. Autoradiographic analysis of proteins ADP-ribosylated in mammalian cells. Mouse embryos were microinjected with 500 pCi of ^{32}P-NAD (lane 1). HeLa intact cells (2×10^6) were incubated with 50 μCi of ^{3}H-adenosine (lane 2). ^{32}P-NAD (5 uCi) was incorporated in nuclei isolated from 2×10^7 HeLa cells (lane 3).

These results clearly indicate that the nuclear non-histone proteins ADP-ribosylated in mouse and human cells are specific and present the same pattern. It appears that the 115 kDa peptide results from the automodification of poly(ADP-ribose)poymerase (pADPRP) whereas the higher Mr peptide (170 kDa) exhibits the same size as topoisomerase II, which is known to be ADP-ribosylated *in vitro* (9).

Immunoprecipitation of topoisomerase II

In order to identify the 170 kDa protein as topoisomerase II, nuclei prepared from 3×10^8 cells were incubated for 30 min at 25° C with a buffer containing 50 µCi of ^{32}P-NAD. Topoisomerse II was immunoprecipitated with protein A-sepharose coated with polyclonal antibody against the enzyme. The eluted protein showed a single immunological band of 140 kDa by Western blot (Fig.2, lane 1). The difference in size from the 170 kDa species possibly arises from the acidic glycine treatment used during the procedure. When this immunoprecipitated protein was analyzed on SDS-PAGE, a single autoradiographic band of 140 kDa was obtained (Fig.2, lane 2).

1   2

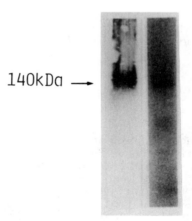

140kDa ⟶

Fig.2. Immunoprecipitation of topoisomerase II from nuclei incubated with ^{32}P-NAD. Immunoprecipitated protein was detected by Western blot (lane 1) and autoradiography (lane 2).

Boronate chromatography.

ADP-ribosylated proteins from intact HeLa cells were isolated by affinity chromatography on phenylboronate agarose gel (10) essentially as described by Krupitza and Cerutti (7). Nuclear proteins depleted of histones were loaded onto a boronate column. The retained fraction corresponding to ADP-ribosylated proteins was analyzed by Western blot by using antisera against topoisomerase II and pADPRP.

A single 170 kDa peptide was recognized by a monoclonal antibody produced in our laboratory (11) against p170 topoisomerase II (Fig.3, lane 1). When the nitrocellulose membrane was further incubated with a polyclonal antiserum to pADPRP, an additional band of 115 kDa was revealed (Fig.3, lane 2). No other peptides were recognized by these antibodies, thus indicating that the 85 and 65 kDa acceptor proteins observed in mouse and human cells should not correspond to proteolytic fragments of pADPRP nor of topoisomerase II.

In order to label the ADP-ribose moiety of proteins retained on boronate column, HeLa cells (3×10^8) were permeabilized by hypotonic shock (12) and incubated for 5 min at 25°C with 25 µCi of ^{32}P-NAD. SDS-PAGE analysis of proteins eluted from the column showed the presence of a major autoradiographic band of 170 kDa, corresponding to topoisomerase II, and of a signal of 120 kDa, corresponding to pADPRP (Fig.3, lane 3). Specific recognition of the 170 kDa band was obtained by Western blot with specific antibody against topoisomerase II (Fig.3, lane 4). The size of immunological band perfectly corresponded to that of the purified enzyme (Fig.3, lane 5).

Discussion

We have studied the non-histone nuclear proteins which are modified by ADP-ribosylation in mammalian cells in physiological conditions. Using different experimental approaches (microinjection of ^{32}P-NAD in mouse embryos, incubation of intact HeLa cells with ^{3}H-adenosine and of HeLa nuclei with ^{32}P-NAD) the same non-histone proteins of 170, 115, 85 and 65 kDa were found to be ADP-ribosylated.

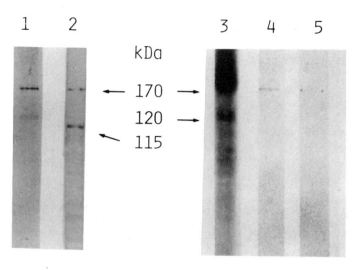

Fig.3. Characterization of proteins eluted from boronate
column. ADP-ribosylated proteins from intact HeLa cells were
separated by affinity chromatography and analyzed on Western
blot. Antibody to topoisomerase II (lanes 1,2) and to pADPRP
(lane 2) were used. Modified proteins from permeabilized HeLa
cells incubated with ^{32}P-NAD were eluted from boronate column
and analyzed by autoradiography (lane 3); lanes 4 and 5
correspond to immunological recognition of boronate eluted
fraction and HeLa purified topoisomerase II, respectively.

We have focused our attention on the 170 kDa
peptide, possibly corresponding to the p170 DNA
topoisomerase II of mammalian cells (13). As
suggested by previous reports, this enzyme is ADP-
ribosylated in reconstituted systems (9).
To obtain evidence for the *in vivo* ADP-
ribosylation of topoisomerase II, HeLa cell nuclei
were incubated with ^{32}P-NAD and the enzyme was
immunoprecipitated with polyclonal antibody and
found to be labeled. This observation was further
corroborated by the analysis of modified proteins
retained on boronate chromatography. The presence
of topoisomerase II among ADP-ribosylated proteins
was demonstrated both by western blot and
autoradiography.

This study has provided evidence that the
major non-histone nuclear proteins ADP-ribosylated
in physiological conditions present sizes of 170,
115, 85 and 65 kDa. The 115 kDa peptide was
identified as the autoribosylated form of pADPRP,
as reported by others (3,4), and the 170 kDa
modified protein was found to correspond to DNA
topoisomerase II. The nature of 85 and 65 kDa
peptides is still unclear since they cannot be
considered as proteolytic fragments of
topoisomerase and pADPRP.

Aknowledgements. We are greatly indebted to Prof. L.F. Liu for
providing rabbit anti-topoisomerase II antibody and to Dr. G.
De Murcia for polyclonal antibody against peptide FII of
pADPRP. We thank Dr. G. Ciarrocchi for purified HeLa cell
topoisomerase II and Dr. G.C.B. Astaldi Ricotti for suggestions
and discussions. This work was supported in part by contract
Bi7-034 from Radiation Protection Programme of C.E.C. C.N. is
recipient of a fellowship from Adriano Buzzati Traverso
Foundation.

References

1   Althaus, F.R., Richter, C. (1987) Molecular
     Biology Biochemistry and Biophysics, 37 M.
     Solioz (ed.), Springer-Verlag, Berlin,
     Heidelberg
2   Bertazzoni, U., Scovassi, A.I., Shall, S.
     (1989) Mutation Res 219: 303-307
3.   Ogata, N., Ueda, K., Kawaichi,
     M.,Hayaishi, (1981) J Biol Chem 256:
     4135-4137
4.   Adolph, K.W., Song, M.K.H. (1985)
     Biochemistry   24: 345-352
5.   Krupitza, G., Cerutti, P. (1989) Biochemistry
     28: 4054-4060
6.   Adamietz, P. (1985) ADP-ribosylation of
     proteins   F.R. Althaus, H. Hilz, S.
     Shall (eds.), Springer-Verlag, Berlin,
     Heidelberg, pp. 264-271
7.   Krupitza, G., Cerutti, P.(1989) Biochemistry
     28: 2034-2040
8.   Quesada, P., Farina, B., Jones, R. (1989)
     Biochim Biophys Acta 1007: 167-175
9.   Darby, M.K., Schmitt, B., Jongstra-Bilen, J.,
     Vosberg, H.P. (1985) EMBO J 4: 2129-2134

10. Adamietz, P., Hilz, H. (1984) Methods in Enzymology 106: 461-471
11. Negri, C., Chiesa, R., Cerino, A., Bestagno, M., Sala, C., Zini, N., Maraldi, N.M., Astaldi Ricotti G.C.B. Submitted for publication
12. Surowy, C.S., Berger, N.A. (1983) Biochim Biophys Acta 740: 8-18
13. Drake, F.H., Hofmann, G.A., Bartus, H.F., Mattern, M.R., Crooke, S.T., Mirabelli, C.K. (1989) Biochemistry 28: 8154-8160

Regulation of DNA polymerase β by poly(ADP-ribose) polymerase

Takashi Sugimura, Mitsuko Masutani, Tsutomu Ogura, Nobuko Takenouchi, Miyoko Ikejima, and Hiroyasu Esumi

The involvement of poly(ADP-ribose) polymerase in the DNA repair process has been suggested by various *in vitro* and *in vivo* studies (Althaus & Richter, 1987). However, the mechanism of the involvement of poly(ADP-ribose) polymerase in DNA repair is still unclear. Recently we isolated the 5'-flanking region of human poly(ADP-ribose) polymerase gene and found that it has a structural similarity to the 5'-flanking region of the genes of human and mouse DNA polymerase β which is one of the key enzymes of DNA repair (Ogura *et al.*, 1990a). We also found a striking similarity in the distribution patterns of the transcript and enzymatic activity of both enzymes in various mouse organs (Ogura *et al.*, 1990b). Recently we found that both poly(ADP-ribose) polymerase and DNA polymerase β activities dramatically increased during rat liver regeneration and after stimulation of human peripheral blood lymphocytes by phytohemagglutinin (to be published elsewhere). It is interesting to note that both enzymes utilize DNA nick for their enzymatic activities (Benjamin & Gill, 1980, Wang & Korn, 1980). All these characteristics of both enzymes suggest that there might be a similar regulatory mechanism of the expression of both genes and that both enzymes might functionally interact with each other. In this work, we report the interaction of poly(ADP-ribose) polymerase and DNA polymerase β *in vitro* and a possible interaction of *in vivo*.

Human poly(ADP-ribose) polymerase was purified from human placenta (Suzuki *et al.*, 1987) and purified recombinant rat DNA polymerase β (Date *et al.*, 1988) was a generous gift from Dr. A. Matsukage, Aichi Cancer Center. As shown in Fig. 1A, DNA polymerase β activity was remarkably inhibited by poly(ADP-ribose) polymerase. This inhibition was completely abolished by the addition of a 10-fold excess amount of template-primer (data not shown). These findings suggest that poly(ADP-ribose) polymerase inhibits DNA polymerase β activity by competing with template-primer. Similar inhibition of DNA polymerase α activity in reconstituted SV40 replication system was reported recently (Lee *et al.*, 1989).

We also used mutated poly(ADP-ribose) polymerase derivatives constructed previously (Ikejima *et al.*, 1990). As shown in Fig. 1B, mutated polymerase P88Δ, which has a deletion in the N-terminal portion including both zinc fingers and does not have ability to bind to the nick or the gap of DNA, did not inhibit the DNA polymerase β activity. On the other hand, DNA polymerase β activity was strongly inhibited by the addition of the wild type enzyme P113. These results also strongly indicate that poly(ADP-ribose) polymerase competes with DNA polymerase β in binding to the primer terminus. It is interesting to note that enzymatically inactive mutated polymerase P101 Δ, which has a chimeric zinc finger of zinc fingers 1 and 2 of poly(ADP-ribose) polymerase, also inhibited DNA polymerase β activity as strongly as wild type enzyme P113, indicating its strong binding to the DNA terminus (Fig. 1B). Furthermore we recently found that mutated polymerase at the second zinc finger, which is not activated by nicked DNA, inhibited nick-directed DNA synthesis of DNA polymerase β as strongly as the wild type poly(ADP-ribose) polymerase (data not shown). These results indicate that DNA binding itself is not sufficient to activate poly(ADP-ribose) polymerase.

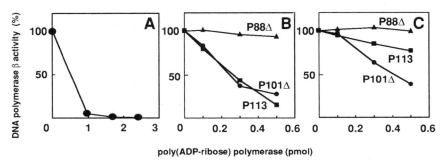

Fig.1 Inhibition of DNA polymerase β activity by poly(ADP-ribose) polymerase and its derivatives and their cancellation by NAD.
  Either 17-mer (Toyobo, dGTTTTCCCAGTCACGAC) annealed to M13 single stranded DNA at 1.6 µg/ml (A) or activated DNA at 1 µg/ml (B and C) was used as a template-primer. The reaction mixture (30 µl) contained 0.04 pmol of DNA polymerase β, 50 mM Tris-HCl (pH 7.5), 100 µM each of dATP, dGTP, TTP and 20 µM [3H]dCTP (640 dpm/pmol), 400 µg/ml bovine serum albumin, 7 mM MgCl2 and 1 mM dithiothreitol. Human placenta poly(ADP-ribose) polymerase was preincubated for 10 min at 25°C, first with template-primer in the above buffer, then DNA polymerase β was added and incubated further for 30 min at 37°C. The reaction was stopped by puting the tube on ice and the amount of incorporated [3H]dCTP was measured by DE paper. In B and C, poly(ADP-ribose) polymerase and its derivatives expressed in *E. coli* were used and the preincubation was not done. In C, the reaction was done in the presence of 200 µM NAD.

We examined the effect of NAD, which is a substrate for poly(ADP-ribose) polymerase, on the inhibition of DNA polymerase β activity. The inhibition of enzymatic activity of DNA polymerase β by wild type poly(ADP-ribose) polymerase was greatly released by the addition of 200 mM of NAD (Fig. 1C). However, the inhibition of the DNA polymerase β activity by mutated polymerase P101Δ, which has no poly(ADP-ribosyl)ation activity, was not released in the presence of NAD (Fig. 1B). These result strongly indicate that the cancellation of inhibition of DNA polymerase β by poly(ADP-ribose) polymerase by addition of NAD depends on poly(ADP-ribose) polymerase activity. Since auto-poly(ADP-ribosyl)ation of poly(ADP-ribose) polymerase results in a loss of affinity to DNA (Yoshihara *et al*. 1981), cancellation of inhibition of DNA polymerase β by poly(ADP-ribose) polymerase is considered to be due to the automodification of poly(ADP-ribose) polymerase.

The results in Fig. 1C also show that there was still significant inhibition in the presence of NAD, although most of the inhibition of DNA polymerase β by P113 was cancelled. The previous report indicated that poly(ADP-ribose)ation of DNA polymerase β resulted in a loss of DNA polymerase β activity (Yoshihara *et al.*, 1985). We examined the possibility that the residual inhibition of DNA polymerase β activity is due to poly(ADP-ribosyl)ation of DNA polymerase β in the present condition. DNA polymerase β was incubated with poly(ADP-ribose) polymerase in the presence of [^{32}P]NAD. After the incubation, DNA polymerase β was immunoprecipitated and fractionated by SDS polyacrylamide gel electrophoresis. A new band of about 44 kDa, which was associated with radioactivity, was observed in addition to a 40 kDa band corresponding to DNA polymerase β. It was estimated that 29% of DNA polymerase β was modified by comparing bands of 40 kDa and 44 kDa. The fraction of modification (29%) of DNA polymerase β coincided with the fraction of inhibition (30%) measured in the parallel experiment. Radioactive material was eluted and analyzed as described (Tanaka *et al.*, 1978) (Fig. 2). Mean size of oligomer and mean number of chain per DNA polymerase β were calculated as two and one, respectively. We conclude that the residual inhibition in the presence of NAD is due to poly(ADP-ribosyl)ation of DNA polymerase β as reported previously (Yoshihara *et al.*, 1985).

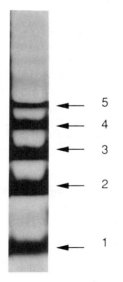

Fig. 2 Poly(ADP-ribosyl)ation of DNA polymerase β. DNA polymerase β was incubated with poly(ADP-ribose) polymerase under the conditions described in Fig. 1 in the presence of [^{32}P]NAD. DNA polymerase β was recovered by immunoprecipitation using polyclonal anti-DNA polymerase β antibody and was subjected to SDS polyacrylamide gel electrophoresis followed by autoradiography. Radioactive 44 kDa band, which was also positive for immunoblotting using anti-DNA polymerase β antibody, was recovered by cutting out from the gel. The gel was treated with 0.1 N NaOH and more than 50% of the radioactivity was solubilized. The radioactive material was subjected to polyacrylamide gel electrophoresis to determine the chain length (Tanaka *et al.*, 1978), and the bands were visualized by autoradiography.

From the above findings, it is clear that DNA polymerase β can be negatively regulated by poly(ADP-ribose) polymerase *in vitro* and that there are two modes of the inhibition, competition in binding to DNA terminus and poly(ADP-ribosyl)ation. These two modes are dependent on NAD concentration. Although biological significance of the inhibition of DNA polymerase β by poly(ADP-ribose) polymerase is not readily clear, it is possible that the inhibition of DNA polymerase β activity by poly(ADP-ribose) polymerase works *in vivo*. Since DNA polymerase β lacks 5' to 3' exonuclease activity (Wang & Korn, 1980), DNA chain elongation by DNA polymerase β from DNA nick could produce potential recombinogenic DNA fragments. Thus it is rational to imagine that poly(ADP-ribose) polymerase negatively regulated DNA polymerase β at the DNA nick.

In order to examine the interaction of DNA polymerase β and poly(ADP-ribose) polymerase *in vivo*, the effect of a specific inhibitor of poly(ADP-ribose) polymerase on DNA polymerase β activity was examined in HL-60 cells. When HL-60 cells were treated with 7.5 mM benzamide, DNA polymerase β activity was increased dramatically after 24 h. The increase of DNA polymerase β activity reached maximum at 72 h and remained constant up to 120 h as shown in Fig. 3.

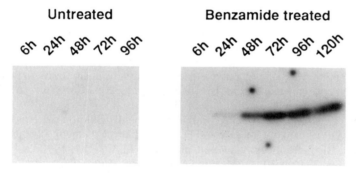

Fig. 3 Time-dependent increase in DNA polymerase β activity in HL-60 cells treated with benzamide.

DNA polymerase β activity was detected by activity gel method (Ogura *et al.*, 1990b). Fifty microgram protein of the crude cell extract from HL-60 cells, after treatment with or without 7.5 mM benzamide were separated on 10% SDS-polyacrylamide gel containing 150 μg/ml of sonicated salmon sperm DNA. After electrophoresis, the gel was soaked successively in a renaturation buffer (50 mM Tris-HCl, pH 8.0, 3 mM 2-mercaptoethanol), in 6 M guanidine•HCl in renaturation buffer and in the renaturation buffer, and incubated with reaction buffer consisting of 50 mM Tris-HCl, pH 7.5, 7 mM $MgCl_2$, 1 mM EDTA, 1 mM DTT, 12 μM each of dATP, dGTP, and TTP, and $[^{32}P]dCTP$ (3000 Ci/mmol) for 12 h at 37°C. The gel was washed with 5% trichloroacetic acid, dried, and visualized by autoradiography.

Northern blot analysis showed that the amount of DNA polymerase β mRNA did not increase during a 120 h treatment of benzamide indicating that DNA polymerase β activity increased by post-translational mechanism (data not shown). A preliminary experiment showed that the snake venom phosphodiesterase treatment of a crude extract from untreated HL-60 cells did not increase DNA polymerase β activity but the alkaline phosphatase treatment increased the activity dramatically. These results indicate that DNA polymerase β may be inactivated by phosphorylation, which in turn is regulated by poly(ADP-ribosyl)ation. Recentry Tokui *et.al.* reported that DNA polymerase β can be inactivated by phosphorylation with protein kinase C (Tokui *et al.*, 1991). This finding is well consistent with our data. There is also another possibility that ribose-phosphate moiety still attached to DNA polymerase β after digestion by snake venom phosphodiesterase inhibits DNA polymerase β activity and alkaline phosphatase treatment removes phosphate from ribose-phosphate resulting in a reactivatioin of DNA polymerase β. In any case, the present results indicate a possibility that DNA polymerase β activity is regulated by poly(ADP-ribose) polymerase *in vivo* by either direct or indirect mechanisms.

References

1. Althaus, F. R.; Richter, C., ed. ADP-Ribosylation of Proteins: Enzymology and Biological Significance. Berlin: Springer-Verlag; 1987:66-92.
2. Ogura, T.; Esumi, H. Characterization of a putative promoter region of the human poly(ADP-ribose) polymerase gene: structural similarity to that of the DNA polymerase β gene. Biochem. Biophys. Res. Commun. 167;701-710; 1990a.
3. Ogura, T.; Esumi, H. Striking similarity of the distribution patterns of the poly(ADP-ribose) polymerase and DNA polymerase β among various mouse organs. Biochem. Biophys. Res. Commun. 172;377-384; 1990b.
4. Benjamin, R. C.; Gill, D. M. Poly(ADP-ribose) synthesis *in vitro* programmed by damaged DNA. J. Biol. Chem. 255;10502-10508; 1980.
5. Wang, T. S-F.; Korn, D. Reactivity of KB cell deoxyribonucleic acid polymerases α and β with nicked and gapped deoxyribonucleic acid. Biochemistry 19;1782-1790; 1980.
6. Suzuki, H.; Miwa, M. Molecular cloning of cDNA for human poly(ADP-ribose) polymerase and expression of its gene during HL-60 cell differentiation. Biochem. Biophys. Res. Commun. 146;403-409; 1987.
7. Date, T.; Matsukage, A. Expression of active rat DNA polymerase β in *Escherichia coli*. Biochemistry 27;2983-2990; 1988.
8. Lee, S.-H.; Hurwitz, J. Studies on the DNA elongation inhibitor and its proliferating cell nuclear antigen-dependent control in simian virus 40 DNA replication *in vitro*. Proc. Natl. Acad. Sci. USA 86;4877-4881; 1989.
9. Ikejima, M.; Miwa, M. The zinc fingers of human poly(ADP-ribose) polymerase are differentially required for the recognition of DNA breaks and

nicks and the consequent enzyme activation. J. Biol. Chem. 265;21907-21913; 1990.

10. Yoshihara, K.; Kamiya, T. Mode of enzyme-bound poly(ADP-ribose)synthesis and histone modification by reconstituted poly(ADP-ribose) polymerase-DNA cellulose complex. J. Biol. Chem. 256;3471-3478; 1981.

11. Yoshihara, K.; Kamiya, T. Inhibition of DNA polymerase α, DNApolymerase β, terminal deoxynucleotidyltransferase, and DNA ligase II by poly(ADP-ribosyl)ation reaction *in vitro*. Biochem. Biophys. Res. Commun. 128;61-67; 1985.

12. Tanaka, M.; Sugimura, T. Demonstration of high molecular weight poly(adenosine diphosphate ribose). Nucleic Acids Res. 5;3183-3194; 1978.

13. Tokui, T.; Matsukage, A. Inactivation of DNA polymerase β by *in vitro* phosphorylation with protein kinase C. J. Biol. Chem. 266;10820-10824; 1991.

Extensive Purification of Nuclear Poly(ADP-ribose) Glycohydrolase

Sei-ichi Tanuma[1], Kazuhiko Uchida[2], Hisanori Suzuki[3], Hiroshi Nishina[1], Hideharu Maruta[1], Takashi Sugimura[4], and Masanao Miwa[2]

[1]Tokyo Institute of Technology, Kanagawa, Japan, [2] Institute of Basic Medical Sciences, University of Tsukuba, Ibaraki, Japan, [3] Instituo di Chimica Biologica, Universita di Verona, Itary, [4] National Cancer Center, Tokyo, Japan

Introduction

Catabolism of poly(ADP-tibose) attached to specific chromosomal proteins has been shown to occur during distinct nuclear processes such as DNA replication, repair and transcription (1). Thus, de-poly(ADP-ribosyl)ation is an important response of nuclei that reflect various cellular signals. Two different types of enzymes have been thought to be involved in de-poly(ADP-ribosyl)-ation of chromosomal proteins. One enzyme, poly(ADP-ribose) glycohydrolase catalyzes hydrolysis of glycosidic (1''-2') linkages of poly(ADP-ribose) to give mono(ADP-ribosyl)-protein and free ADP-ribose (2-9). A second type is ADP-ribosyl-protein lyase, which is capable of splitting mono(ADP-ribose)-protein linkages (10). The glycohydrolase has been purified from nuclei (4) and post-nuclear fractions (cytoplasm) (6-9) of several tissues and cultured cells. To distinguish the nuclear glycohydrolase from the cytoplasmic glycohydrolase, the nuclear enzyme is designated as poly(ADP-ribose) glycohydrolase I, and the cytoplasmic enzyme, poly(ADP-ribose) glycohydrolase II (5, 7, 9). The biological relationship between the two forms of glycohydrolase remains to be determined. As yet no procedure that makes available nuclear poly(ADP-ribose) glycohydrolase with sufficient purity and quantity to determine amino acid composition or sequence has been reported. Here, we report a reproducible and efficient method for extensive purification of the major poly(ADP-ribose) glycohydrolase present in mammalian cell nuclei and characterization of its properties.

Results and Discussion

A poly(ADP-ribose) glycohydrolase from human placental nuclei was purified approximately 55,000-fold to homogeneity (Table I). This enzyme was extracted from isolated nuclei only by sonication in 0.6 M ammonium sulfate, and then purified by sequentioal column chromatographies including HPLC. The purity of the glycohydrolase (step 8) was estimated to be more than 90% as judged by capillary electropherogram and silver staining of SDS-polyacrylamide gel. For analysis of the amino acid composition of poly(ADP-ribose) glycohydrolase, the pooled fractions from step 8 were further purified by reverse-phase HPLC on phenyl-5PWRP.

Table I. Purification of Nuclear Poly(ADP-ribose) Glycohydrolase from Human Placenta

Step	Protein (mg)	Activity (μmol/min)	Special activity (μmol/min/mg)	Purification (fold)	Yield (%)
1. Homogenate	31,800	33.3	0.0011	1	100
2. Nuclear lysate	1,820	31.2	0.0172	15.6	93.7
3. Nuclear extract	302	28.8	0.0954	86.7	86.5
4. Butyl Toyo-pearl 650	81.3	23.4	0.288	262	70.3
5. Red Sepharose	31.4	18.7	0.596	542	56.2
6. Single-stranded DNA agarose	7.44	15.8	2.12	1,930	47.5
7. Heparin-5PW	0.333	10.7	32.1	39,200	33.1
8. TSK-G2000SW	0.078	4.8	61.5	55,900	14.4
9. Phenyl-5PWRP	0.039	--	--	--	--

Table II. Comparison of Characteristics of Poly(ADP-ribose) Glycohydrolases

Source	Human	Guinea pig liver	Human erythrocyte	Pig thymus	Calf thymus
Localization	Nuclei	Nuclei	cytosol	cytosol	n.d.	n.d.
M_r by						
SDS polyacrylamide	71000	75300	59300	59300	61500 (67500, miner)	59300
Gel permiation	68000	72000	56000	56000	59000	56000
Acidic/basic amino acids	1,27	1.26	2.88	3.19	n.d.	n.d.
Mode of hydrolysis	exo	exo	exo	exo	exo	exo
Km for $(ADP\text{-}ribose)_n$	1.8	2.3	6.4	6.8	1.8	0.1-10
(μM) (n)	(15)	(15)	(15)	(15)	(20)	(10-30)
Vmax (μmol/min/mg)	67	36	15	18	19	50-100
Optimum pH	7.1	6.9	7.4	7.4	7.4	7.3
Inhibition by						
ADP-ribose	yes	yes	yes	yes	yes	yes
3',5'-cAMP	yes	yes	yes	yes	yes	yes
Reference		4	9	7	3	6

n.d., not determined

The purified glycohydrolase consists of a single polypeptide with Mr of 71,000 estimated by SDS-polyacrylamide gel. A native Mr of 68,000 was determined by gel permiation. These values were quite similar to those nuclear poly(ADP-ribose) glycohydrolase (I) purified from guinea pig liver (4) but clearly different from those of cytoplasmic poly(ADP-ribose) glycohydrolase (II) (3, 6-9) (Table II). The turn over number calculated from the specific activity and estimated protomeric Mr was about 70 mol ADP-ribose produced per sec per mol of enzyme. The pI value was 6.73 on isoelectric focusing. Amino acid analysis showed that enzyme had consistent numbers of acidic and basic amino acids and the ratio of acidic to basic amino acids of 1.27. In contrast, cytoplasmic glycohydrolase II from guinea pig liver (9) or human erythrocytes (7) contained lower content of basic amino acids and a relatively higher proportion of acidic amino acids.

The enzyme degraded poly(ADP-ribose) to ADP-ribose in exo-glycosidic mode. The Km value for poly(ADP-ribose) and the Vmax of the purified poly(ADP-ribose) glycohydrolase was 1.8 µM and 47 µmol per min per mg protein, respectively. Both glycohydrolase I and II showed variable activities that depended on the chain length of poly(ADP-ribose) (9). In the purified enzyme, the Km value for a longer plymer (n=27) was approximately one order of magnitude lower than that for a shorter polymer (n=6.7). The optimum pH for the enzyme activity was 7.1, while that for glycohydrolase II was around 7.4. The enzyme activity was inhibited by ADP-ribose and cAMP. These physical and catalytic properties are identical with those of nuclear glycohydrolase I but not with those of cytoplasmic glycohydrolase II (Table II).

The present results suggest that mammarian cells contain at least two distinct forms of poly(ADP-ribose) glycohydrolase exhibiting differences in properties and subcellular localization. The multiple forms may be related to diversity of function for the glycohydrolases. The nuclear glycohydrolase I is probably involved in chromatin associated events via de-poly(ADP-ribosyl)ation of histones and nonhistone chromosomal proteins. On the other hand, the cytoplasmic glycohydrolase II may play an important role on extranuclear de-poly(ADP-ribosyl)ation, such as mitochomdria and cytosolic free mRNA ribonucleoprotein (mRNP) particles. In fact, the purified glycohydrolase II from human erythrocytes had more activity on oligo(ADP-ribose) bound to mitochomdrial proteins and mRNP than on oligo(ADP-ribose) bound to histone H1 (7). The structural differences of glycohydrolase I and II should be clarified upon cloning and analysis of their genomes. The partial polypeptide sequences and cDNA cloning of glycohydrolase I and II are currently under investigation.

References

1. Ueda, K., Hayaishi, O (1985) Annu. Rev. Biochem. 54, 73-100
2. Miwa, M., Tanaka, M., Matsushima, T., Sugimura, T. (1974) J. Biol. Chem.249, 3475-3482
3. Tavassoli, M., Tavassoli, M.H., Shall, S. (1983) Eur. J. Biochem. 135,449-455
4. Tanuma, S., Kawashima, K., Endo, H. (1986) J. Biol. Chem. 261, 965-969
5. Tanuma, S., Kawashima, K., Endo, H. (1986) Biochem. Biophys. Res. Commun. 135, 979-986
6. Hatakeyama, S., Nemoto, Y., Ueda, K., Hayaishi, O. (1986) J. Biol. Chem.261, 14902-14911
7. Tanuma, S., Endo, H. (1990) Eur. J. Biochem. 191, 57-63
8. Thomassin, H., Jacobson, M., Guay, J., Verreault, A., Aboulela, N., Menard, L., Poirier, G.G. (1990) Nucleic Acid Res. 18, 4691-4694
9. Maruta, H., Inageda, K., Aoki, T., Nishina, H., Tanuma, S. (1991) Biochemistry in press
10. Oka, J., Ueda, K., Hayaishi, O., Kumura, H., Nakanishi, K. (1984) J. Biol. Chem. 259, 986-995

POLY(ADP-RIBOSE)POLYMERASE AND NMN ADENYLYL-TRANSFERASE INTERACTION: MOLECULAR AND IMMUNO-LOGICAL STUDIES

Raffaelli N., Emanuelli M., Lauro L., Magni G., Natalini P., Venanzi F.M. and Ruggieri S.

Despite the increasing interest on both the physiological and pathological role of ADP-ribosylation reactions, little appears to be known about the mechanism by which the intracellular level of NAD^+ is regulated with respect to its participation in the redox metabolism on the one hand and to its utilization in poly(ADP-ribosy)lation on the other. Conditions enhancing poly(ADP-ribosyl)ation can significantly lower NAD^+ concentration thereby influencing NAD^+ dependent metabolic processes in the whole cell, while the cellular redox status can be signaled to chromatin thereby influencing the level of ADP-ribosylation of nuclear proteins (1).

NAD^+ synthesis from NMN and ATP is catalyzed by the enzyme NMN adenylyltransferase (NMNAT), which, like poly(ADP-ribose) polymerase (ADPRP), is localized in the nucleus. It may be speculated that the common localization of the two enzymes could play a role in the cross-regulation of NAD^+ synthesis and degradation. Among possible mechanisms, regulation of NMNAT through its ADP-ribosylation, NAD^+ channelling from NMNAT to ADPRP, NMNAT-ADPRP association or interaction, could be hypothesized.

We had previously reported that incubation of homogeneous NMNAT from yeast with homogeneous ADPRP from bull testis in a reconstituted ADP-ribosylating system resulted in almost complete inhibition of ADPRP activity, while NMNAT activity was not affected (2). In order to test the significance of such observation in a homologous system, we have purified NMNAT from bull testis. This paper is concerned with the results obtained with the homologous reconstituted system, including enzymological, molecular and immunological studies on the NMNAT enzyme.

RESULTS AND DISCUSSION

Similar to the results obtained with the yeast NMNAT, when the enzyme from bull testis was incubated in the homologous ADP-ribosylating system, its activity was not affected, while ADPRP activity was markedly inhibited (Fig.1). As it can be seen from Fig.1A, the heat-inactivated NMNAT was not inhibitory, suggesting that, in contrast to the behavior of the yeast enzyme (2), the native conformation of the protein is required.

The inhibitory effect didn't seem to be dependent on the particular type of histones used in the reaction mixture, and it also occurred in the absence of histones.

The inhibition was not prevented by the protease inhibitor PMSF.

A

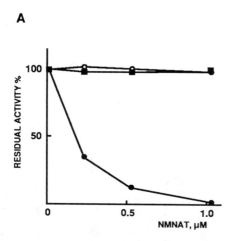

B

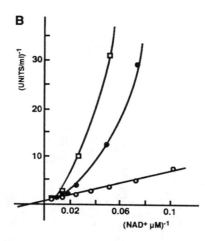

Fig 1A - Inhibition of ADPRP activity by NMNAT. The assay mixture contained 10mM MgCl$_2$, 1mM DTT, 0.02mM (adenine-2, 8-^{3}H)NAD$^+$ (80000 dpm/mol), 1μg histones (type II from SIGMA), 1 μg activated DNA (3) and 0.030 μg ADPRP in 100mM Tris buffer, pH 8.0. The reaction was carried out at 25°C for 15 minutes in the absence (O) and in the presence of increasing amounts of NMNAT: (●) native NMNAT, (■) heat inactivated (100°C, 10 min) NMNAT. The radioactivity incorporated into trichloroacetic acid-insoluble material was determined as described (2).

Fig. 1B - Effect of NAD$^+$ concentration on ADPRP activity in the absence (O) and in the presence of 0.1 μg (●) and 0.2 μg (□) NMNAT. The reaction mixture as in Fig. 1A, except for NAD$^+$ specific activity of 200000 dpm/mol, was incubated at 25°C for 2 minutes. One unit of enzymatic activity corresponds to the incorporation of 1 nmole of NAD$^+$ into acid insoluble material in 1 min at 25°C.

In order to investigate the type of inhibition, studies on the initial velocity of ADPRP reaction versus NAD$^+$ concentrations have been carried out either in the absence or in the presence of different concentrations of NMNAT. From the double reciprocal plot (Fig.1B) an apparent Km of 86 μM was obtained for NAD$^+$ in the absence of NMNAT; this value is in good agreement with the literature data for the same enzyme from various sources (4). In the presence of NMNAT, however, a deviation from linearity was observed and therefore a discrimination of the type of inhibition is not straightforward. Furthermore it can be seen that the inhibition by NMNAT is displayed only at NAD$^+$ concentrations lower than 130 μM, being absent at higher concentrations. This represents a peculiar feature of the homologous system and can constitute a novel kind of regulation of the pyridine nucleotide cycle since it can exert a modulatory effect on the rate of NAD$^+$ consumption (5). The present finding

%

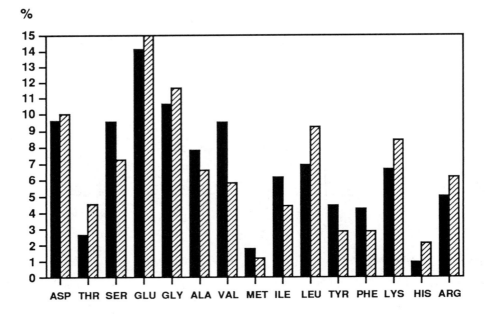

Fig. 2 - Comparison of the % amino acid composition of yeast ■ and bull testis ▨ NMNAT. For the amino acid analysis of the mammalian enzyme, duplicate samples of protein (1 µg) were hydrolyzed *in vacuo* for 45 and 90 minutes at 155°C in 6N HC1 (8). The analysis was performed on a Chromakon 550 (Kontron Instruments) amino acid analyzer and the amino acids were detected fluorometrically after post-column reaction with o-phtalaldehyde.

appears suggestive in view of the observations that ADPRP and NMNAT are in a 1:1 molar ratio within the chromatin (6) and that most ADPRP may be in a catalytically inactive state in situ (7).

The different behavior of the homologous system with respect to the heterologous one appears to reflect the different molecular, enzymological and immunological properties of the yeast and bull testis enzyme.

The molecular weight of the native enzyme from bull testis, estimated by gel-filtration, was 132 KDa; upon denaturing polyacrylamide gel electrophoresis the purified enzyme run as a single protein band of 33 KDa (unpublished results). The mol. wt. reported for the enzyme purified from yeast corresponds to four identical subunits of 50 KDa (9). Preliminary data show that both enzymes exhibit hydrophobic properties, with a marked tendency to aggregation, more pronounced for the mammalian enzyme.

Amino acids analysis studies on the bull testis NMNAT also provided evidence on the diversity of the two enzymes; a remarkable difference in the amino acid composition (Fig.2) has been, in fact, observed.

Consistent with the different characteristics of the two enzymes is the observation that antibodies raised in the rabbit against the bull testis NMNAT subunits do not cross- react with the yeast enzyme. The antibodies specifically recognize both denatured and native bovine NMNAT, as demonstrated by western blot experiments and immunoprecipitation of the activity.

Studies are in progress also with the aid of such antibodies for structural characterization of the putative inhibitory domain(s) of NMNAT.

REFERENCES

1. Loetscher P., Alvarez Gonzalez R. and Althaus F.R. (1987) Proc. Natl. Acad. Sci. USA, 84: 1286-1289.
2. Ruggieri S., Gregori L., Natalini P., Vita A., Emanuelli M., Raffaelli N. and Magni G. (1990) Biochemistry, 29: 2501-2506.
3. Loeb L.A. (1969) J. Biol. Chem., 244: 1672-1681.
4. Mandel P., Okazaki H. and Niedergang C. (1977) FEBS Letters, 84: 331-336.
5. Rechsteiner M., Hillyard D. and Olivera B.M. (1976) Nature, 259: 695-696.
6. Uhr M.L. and Smulson M. (1982) Eur. J. Biochem., 128:435-443.
7. Yamanaka H., Penning C.A., Willis E.H., Wasson D.B. and Carson D.A. (1988) J. Biol. Chem., 263: 3879-3883.
8. Hare P.E. (1977) Methods Enzymol., 47: 3-18.
9. Natalini P., Ruggieri S., Raffaelli N. and Magni G.,(1986) Biochemistry, 25: 3725-3729.

Covalent modification of poly (ADP-Ribose) polymerase by reactive benzamides.

Manoochehr Tavassoli and Sydney Shall.

Cell and Molecular Biology Laboratory, School of Biological Sciences, University of Sussex, Brighton, East Sussex, BN1 9QG, ENGLAND.

Introduction

In a continuing effort to understand both the molecular enzymology and the cell physiology of poly (ADP-Ribose) polymerase, we have been developing a new class of enzyme inhibitors. We describe in this chapter some new enzyme inhibitors which are competitive inhibitors of the enzyme, but are also constructed so that they can react covalently with the enzyme protein to form covalent products. Such compounds should be of use in identifying the residues at or near the active site of the enzyme.

Results

We have synthesized a number of new compounds which are designed to be both competitive inhibitors as well as irreversible inhibitors of the polymerase. These compounds are designed as competitive inhibitors with chemically reactive groups. Table III shows a list of such compounds, together with their chemical formulae. All these new compounds are based on the benzamide nucleus. They were shown to be competitive inhibitors of the polymerase by incubation with permeabilized mouse leukaemia cells (L1210) over very short periods of time (5 minutes); the inhibitors were used at a concentration of 5 μM, with an NAD concentration of 100 μM (Table I). The results show that all these compounds were indeed competitive inhibitors. We therefore determined the K_i values of some of these compounds; and these results are shown in Table II. We found that these new compounds had lower K_i values than did 3-aminobenzamide. Consequently, we have a basis for supposing that these compounds might show some selectivity in their ability to alkylate cell proteins.

We then incubated intact mouse leukaemia cells (L1210) with these compounds for 24 hours, except for 3-(methyl nitroso ureido) benzamide which was incubated for only 4 hours because of its chemical instability. At the end of the incubation period the cells were washed three times with phosphate-buffered saline and the enzyme activity of the polymerase was measured in permeabilised cells. There are two control compounds included in this experiment; firstly, we used 3-formylaminobenzamide which we know is a moderately good inhibitor of the polymerase. The results (Table III) show that the washing procedures successfully removes a large excess of this reversible inhibitor and that there is now inhibition remaining from the irreversible inhibitor. Moreover, inhibition of the polymerase for 24 hours by a reversible, competitive inhibitor does not alter the level of enzyme activity in the cell. Secondly, we included 200 uM of ethyl-3-acryloyl- aminobenzoate which is a chemically reactive compound, but is

Table I.   Effect of reactive benzamides on poly (ADP-Ribose) polymerase
            enzyme activity.

Compound Used	Inhibition. (%)
3-Chloroacetylaminobenzamide	53
3-Bromoacetylaminobemzamide	47
3-(3-Chloropropionyl)aminobenzamide	57
3-(2-Chloropropionyl)aminobenzamide	53
3-(3-Bromopropionylaminobenzamide	58
3-(4-Chlorobutanyl)aminobenzamide	38
3-acryloylaminobenzamide	47
3-Crotonylaminobenzaide	20
3-(N-methyl-N-nitroso)ureidobenzamide	60

not an enzyme inhibitor; and would not be expected to bind specifically to the
enzyme. Since the chemically reactive compounds that are not enzyme
inhibitors, do not appear to inhibit the enzyme, we infer that the enzyme
inhibition is specific and is probably a consequence of the binding of the
inhibitor to the enzyme; if so, then it is possible that the covalent adduct is

Table II. K_i values for selected irreversible and competitive inhibitors of
             poly(ADP-Ribose) polymerase. Only the aminobenzamide
             substituent is listed

Compound	K_i (μM)
3-(3-Chloropropionyl)aminobenzamide	1.88 ± 0.14
3-(2-Chloropropionyl)aminobenzamide	2.01 ± 0.16
3-(3-Bromopropionyl)aminobenzamide	1.73 ± 0.15
3-Acryloylaminobenzamide	3.14 ± 0.29
3-Aminobenzamide	11.3 ± 1.15

associated with an amino acid residue near to the enzyme active site. Thus we
can conclude that the inhibition that we observed was due to specific binding of
the relevant compound to the polymerase protein with subsequent covalent
interaction with the enzyme protein. Among the compounds shown in Tables I
and III, 3-acryloylaminobenzamide yielded the most inhibition of the
polymerase; after exposure of L1210 cells to 100 μM for 24 hours, there
remained only 16% residual enzyme activity. This treatment does not affect the
cell viability even during 10 days of exposure (Table IV), although there seems
to be a small inhibition in the growth rate. 3-(3-bromopropionyl)
aminobenzamide at 200 μM for 24 hours gave almost as much enzyme
inhibition. However, this is likely to be a more reactive compound than the
acryloyl derivative and therefore, perhaps less specific. The 3-(3-chloro-
propionyl) aminobenzamide is predictably less reactive; this might be associated
with greater specificity. The 3-(N'-methyl-N-nitroso)ureidobenzamide is a very

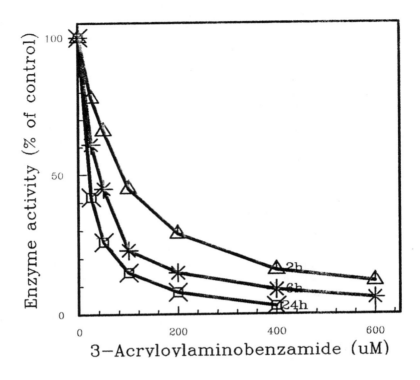

Figure 1 Irreversible inhibition of poly(ADP-Ribose) polymerase by 3-acryloyaminobenzamide, effects of concentration of inhibitor. Duration of incubation: Triangles; 2 hours:Asterisks; 6 hours: Crossed squares; 24 hours.

unstable compound, but even after only 4 hours incubation with 200 μM of the compound for 4 hours, there is already 40% inhibition of the enzyme activity. Finally, it is noted that 3-(4-chlorobutanyl) aminobenzamide is non-inhibitory even at 500 μM. Also 3-(2-chloropropionyl)aminobenzamide and 3-crotonylaminobenzamide are inactive as irreversible inhibitors, possibly due to the steric hindrance of the methyl group.

Kinetics of Enzyme Inactivation.

To study the kinetics of the irreversible inhibition of poly (ADP-Ribose) polymerase by 3-acryloylaminobenzamide, L1210 cells were treated with increasing concentrations of the inhibitor for 2, 6 or 24 hours, and then the inhibitor was washed out with three washes of 10 ml. of phosphate-buffered saline, and then the enzyme activity was measured in permeabilised cells. As a control the cells were also incubated with 3-aminobenzamide, to demonstrate that the inhibitor was totally washed out before the assay, thus ensuring that we were only estimating covalent inhibition (Figure 1). Examination of the data in Figure 1 clearly shows that the degree of inhibition is dependent both on the concentration of the inhibitor as well as on the duration of the exposure; the latter feature is characteristic of an irreversible chemical reaction. In addition, of course, we showed that the washing procedure did remove all of the non-

Table III   Irreversible inhibitors of poly(ADP-Ribose) polymerase

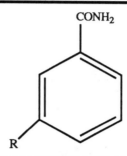

Addition	Concentration (μM)	R	Inhibition (%)
3-acryloylaminobenzamide	100	-NH-CO-CH=CH$_2$	84
3-Crotonylaminobenzamide	100	-NH-CO-CH=CH-CH$_3$	3
3-(3-Bromopropionyl-aminobenzamide	200	-NH-CO-CH$_2$-CH$_2$Br	79
3-(3-Chloropropionyl)-aminobenzamide	200	-NH-CO-CH$_2$-CH$_2$Cl	41
" " "	500	" " " " "	64
3-(2-Chloropropionyl)-aminobenzamide	200	-NH-CO-CHCl-CH$_3$	1
3-(4-Chlorobutanyl)-aminobenzamide	500	-NH-CO-(CH$_2$)$_2$-CH$_2$Cl	1
3-(N'-methyl-N'-nitroso)-ureidobenzamide	200	-NH-CO-NH(NO)-CH$_3$	40
Ethyl-3-acryloylaminobz*	200	-NH-CO-CH=CH$_2$	0
3-formylaminobenzamide**	1000	-NH$_2$	2

* non inhibitory analogue
** competitive inhibitor of poly(ADP-Ribose) polymerase

covalent inhibitor, 3-aminobenzamide. By 6 hours most of the reaction had occurred at most concentrations. The inhibition of the enzyme activity was both dose- and time-dependent (Figure 1).

Table IV Effect of the irreversible polymerase inhibitor, 3-acryloyl-aminobenzamide on growth and viability of mouse L1210 cells.

Concentration of 3-acryloylamino-benzamide. (μM)	Incubation duration			
	24 h		48 h	
	cell number x10^{-5}	viability (%)	cell number x10^{-5}	viability (%)
0	3.98	>98	12.00	>98
100	3.55	>98	7.80	>98
200	2.93	96	4.50	90
400	2.02	84	2.26	<1

Table V Effect of the irreversible polymerase inhibitor, 3-acryloyl-aminobenzamide on the polymerase enzyme activity and on the cloning efficiency of mouse L1210 cells.

3-acryloylamino-benzamide. (μM)	Cloning Efficiency. (%)	Residual Enzyme Activity. (%)
0	100	100
100	100	15
200	22	8
400	<1X10^{-5}	3

In order to show that indeed these inhibitors do covalently modify the enzyme protein, we synthesized radioactive 3-acryloylaminobenzamide. We then reacted this material (with a specific radio-activity of 290 mCi/mmole, 46 μCi/ml) with either nearly purified polymerase (at 0.1 mg/ml) or with partially purified polymerase (at 1 mg/ml) at 37 °C for 4 hours. After the 4 hours reaction, the proteins were precipitated with trichloroacetic acid, and the pellet was analyzed by polyacrylamide gel electrophoresis (Figure 2). The stained protein gel shows for the purified enzyme, a weak band at 116 kDa, and the partially purified preparation shows numerous bands; this preparation would seem to have suffered proteolysis, and it retained no enzyme activity. The autoradiogram shows a very strong band at the position of the purified enzyme molecule, and a number of much weaker bands, which may also be proteolytic products or impurities. The partially purified enzyme shows a moderate band of radioactivity at about 85 kDa; and several of the bands in the two preparations have similar molecular weights.

Effect of these irreversible polymerase inhibitors on DNA repair.

One of the irreversible polymerase inhibitors which we have synthesized is interesting because it has a dual functionality. Ethyl-3-(N-methyl-N-nitroso) ureido benzoate is a chemically reactive alkylating agent but is not an inhibitor of the polymerase; this compound would be expected to damage DNA,

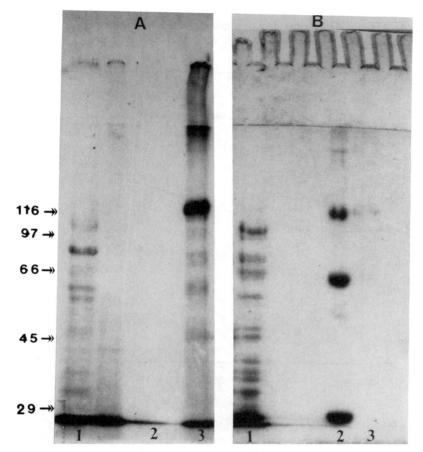

Figure 2 Covalent modification of poly(ADP-Ribose) polymerase by (^{3}H) 3-acryloyaminobenzamide.
A: Autoradiograph; B: Protein staining. Lane 1: Partially purified enzyme; lane 2: Molecular weight markers; Lane 3: purified enzyme.

but does not inhibit repair. In contrast, the chemical analogue, 3-(N-methyl-N-nitroso) ureido benzamide is both an alkylating agent as well as an inhibitor of the polymerase; this latter compound would be expected to damage DNA to about the same degree as the ethyl ester, but in addition would be expected to interfere with the DNA repair processes. Mouse L1210 cells were treated with 200 μM of either of these two alkylating agents. At the indicated times, the progress of DNA damage and repair was followed with the nucleoid sedimentation technique (Figure 3). The non-inhibitory ethyl ester clearly alkylated the DNA and activated the DNA repair response in these cells. The benzamide derivative also damaged the DNA, and to about the same extent, but in this case there is clear evidence that the progress of DNA repair has been retarded. Thus, it is possible to demonstrate that irreversible polymerase inhibitors also retard DNA repair, as do the reversible inhibitors.

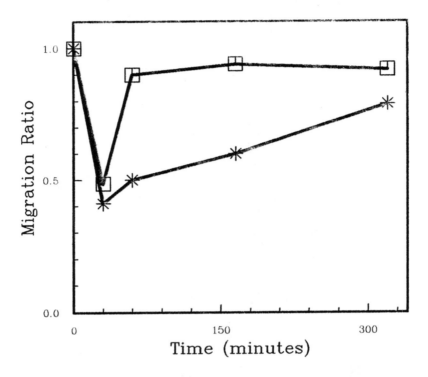

Figure 3   Effect of 3-(N-methyl-N-nitroso)ureido-benzamide on DNA damage and repair in the mouse lymphoblatoid cell line L1210. Symbols: Squares: non-inhibitory alkylating compound; Asterisks: enzyme inhibitory alkylating compound

Discussion.

In this brief report we describe the properties of several new compounds which appear to be irreversible, covalent inhibitors of the poly (ADP-Ribose) polymerase. The evidence that they form covalent products is that the inhibition is time-dependent, it remains after the inhibitors have been washed out, and that radioactive products are seen of the correct molecular weight, after reaction with purified enzyme. Clearly, it will be necessary to establish with which amino acid residue in the enzyme these inhibitors are reacting in order to provide formal proof of their mode of action.

It is also interesting that it is possible to synthesise compounds that are simultaneously DNA damaging agents and inhibitors of DNA repair, such as 3-(N-methyl-N-nitroso) ureido benzamide.

We thank the Wellcome Trust and the Royal Society for support, as well as the British Cancer Research Campaign, and the Medical Research Council.

ALTERATIONS IN REPAIR OF DNA DAMAGE
IN POLY(ADP-RIBOSE) POLYMERASE DEFICIENT CELL LINES

Nathan A. Berger, Satadal Chatterjee, Ming-Fang Cheng
Shirley J. Petzold and Sosamma J. Berger

Department of Medicine and Ireland Cancer Center,
University Hospitals of Cleveland and
Case Western Reserve University, Cleveland, Ohio 44106

ADPRT 54 and ADPRT 351, are Chinese hamster, V79, derived mutant cell lines, selected for their marked deficiency in poly(ADP-ribose) polymerase activity (1). These cells were employed to analyze the effect of poly(ADP-ribose) polymerase deficiency on responsiveness to a series of DNA damaging agents including monofunctional and bifunctional alkylating agents, topoisomerase I and II inhibitors and X- and UV-irradiation (1-4). In contrast to previous studies, the use of these poly(ADP-ribose) polymerase deficient, mutant cell lines and comparison to events in normal, parental V79 cells allows for studies of the role of poly(ADP-ribose) polymerase without the use of enzyme inhibitors which could have multiple nonspecific effects.

Enzymatic Characteristics of Poly(ADP-Ribose) Polymerase Deficient V79 Cell Lines

Poly(ADP-ribose) polymerase activity measured in DNase treated, permeabilized cells of the mutant cell lines ADPRT 54 and ADPRT 351 is 5 to 10% of that measured in parental V79 cells (1). Under the conditions employed, DNase treatment provides sufficient DNA strand breakage to stimulate maximum activity of endogenous poly(ADP-ribose) polymerase. These mutant cells show a decrease in the initial rate of poly(ADP-ribose) synthesis as well as a decrease in the total amount of polymer synthesized (1). The mutant cells synthesize less than 5% of the poly(ADP-ribose) synthesized by parental cells. Treatment of parental cells with DNA damaging agents can result in depletion of NAD, the substrate for poly(ADP-ribose) synthesis. However, depletion of NAD in response to DNA damaging agents is retarded in the mutant cells and auto poly(ADP-ribosylation) measured by autoradiography is markedly decreased in quantity and polymer size (1). Immunofluorescent staining for poly(ADP-ribose) polymerase using monoclonal antibodies shows intense nuclear staining in the parental cell line whereas the mutant cells show marked deficiency of enzyme protein. The activities of NAD glycohydrolase, poly(ADP-ribose) glycohydrolase and phosphodiesterase are normal to decreased in the mutants relative to the parental cell lines (1).

Cellular Characteristics of Poly(ADP-Ribose) Deficient V79 Cell Lines

When the cellular characteristics of the mutant cell lines are compared to the parental V79 cells, the mutants are found to have their DNA content doubled and their protein content doubled. They also have a prolonged doubling time relative to the parental cell line. For example, doubling time of ADPRT 54 and ADPRT 351 are 68 and 50 hours respectively compared to a 10 hour doubling time for the V79 cells. Mutant cells also show a marked increase in spontaneous sister chromatid exchanges (1).

Response to DNA Damage

Table 1 outlines the general response of the poly(ADP-ribose) polymerase deficient mutant cell lines to a variety of DNA damaging agents relative to the response shown by the parental V79 cell lines (2-4). The mutant cell lines are hypersensitive to all mono and bifunctional alkylating agents tested. They are also hypersensitive to X-irradiation and X-ray mimetic agents such as bleomycin. They are hypersensitive to UV-irradiation and to UV-mimetic agents such as 4-nitroquinolinoxide. The mutant cell lines are hypersensitive to the topoisomerase I active agent camptothecin. In contrast, they are markedly resistant to the topoisomerase II active agents VP-16, adriamycin and m-AMSA.

Table 1. Sensitivity to DNA Damaging Agents in ADPRT 54, ADPRT 351, Poly(ADPR) Polymerase Deficient, Cell Lines Relative to Parental V79 Line

Alkylating Agents (Mono & Bifunctional)	Hypersensitive
X-Irradiation	Hypersensitive
X-Ray-Mimetic	Hypersensitive
UV-Irradiation	Hypersensitive
UV-Mimetic	Hypersensitive
Topo I Active Agents	Hypersensitive
Topo II Active Agents	Resistant

Figure 1 compares survival curves of the mutant and parental V79 cell lines following treatment with the alkylating agent methylmethane sulfonate. These curves are notable for the marked increase in sensitivity shown by the mutant cell lines. It is also noteworthy that the mutant cell lines show loss of the shoulder regions in the survival curves. A similar loss of shoulder regions in survival curves was also noted following treatment with DMS and EMS as well as with MNU, ENU and MNNG (4). In addition, similar results were obtained with clinically useful alkylating agents such as phenylalanine mustard and BCNU (3). The shoulder regions in the survival curves following UV irradiation and 4-nitroquinolinoxide were also abolished in the mutant cell lines. The normally occurring shoulder regions in survival curves presumably represent a dose range where repair of sublethal DNA damage is occurring. The hypersensitivity shown by the mutant cells provides a strong indication that poly(ADP-ribose) polymerase is required in the normal process of DNA repair. Furthermore, the disappearance of the shoulder regions suggests that poly(ADP-ribose) polymerase is required for repair of sublethal DNA damage inflicted at low concentrations of alkylating agents and other DNA damaging treatments. This observation is of particular importance since many systems used to measure poly(ADP-ribose) polymerase activity and its response to DNA damage have required the use of high levels of DNA damage to produce sufficient alterations for biochemical measurements. The disappearance of the shoulder regions demonstrated by the mutant cell lines in these experiments indicates a role for poly(ADP-ribose) polymerase even at low levels of DNA damage.

Figure 2 illustrates the increased resistance shown by the mutant cells to topoisomerase II inhibitors. In this figure, cells were exposed continuously to the indicated concentration of VP-16. Since cells are most sensitive to the cytotoxic effects of VP-16 during S phase, these experiments were conducted with continuous exposure to the drug to avoid any possible complication due to the difference in doubling times. However, similar results were obtained with a one hour exposure as well (2).

Table 2 summarizes the relative sensitivity of the V79 cells and the poly(ADP-ribose) deficient mutants to a series of cytotoxic agents. Relative sensitivity for each agent was determined by dividing the IC50 for V79 cells by the IC50 for each of the mutant cell lines (2-4). All values greater than 1 represent hypersensitivity whereas values lower than 1 indicate resistance by the mutant cells relative to the parental V79 cells.

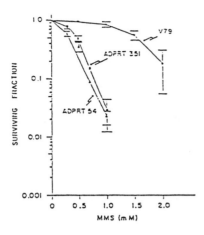

Figure 1

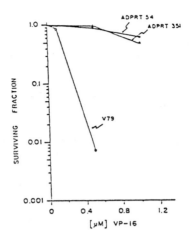

Figure 2

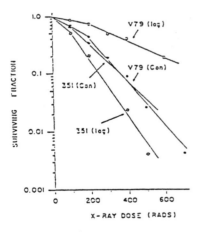

Figure 3

Figure 1. Cytotoxicity of MMS.

Figure 2. Cytotoxicity of VP-16.

Figure 3. Cytotoxicity of X-ray.

Table 2. Relative Sensitivities of V79 Cells and
Poly(ADP-Ribose) Polymerase Deficient Mutants
to Various Cytotoxic Agents

Agent	Relative Sensitivity		
	V79	ADPRT 54	ADPRT 351
DMS	1.0	3.8	3.8
MMS	1.0	4.7	3.1
EMS	1.0	6.4	6.4
MNU	1.0	9.0	3.3
ENU	1.0	8.6	5.2
MNNG	1.0	3.8	5.2
4NQO	1.0	2.5	2.5
Melphalan	1.0	16.5	9.5
BCNU	1.0	3.7	3.4
Mitomycin	1.0	5.8	3.9
Bleomycin	1.0	3.8	3.8
Adriamycin	1.0	0.3	0.3
m-AMSA	1.0	0.2	0.2
VP-16	1.0	0.1	0.1
Camptothecin	1.0	3.7	2.5
X-ray	1.0	1.6	1.8
UV	1.0	6.2	9.3

In further experiments we used these mutant cell lines to evaluate the role of poly(ADP-ribose) polymerase in the repair of potentially lethal DNA damage. Confluent cells were subject to X-irradiation and then held at confluence for subsequent time intervals before being subcultured to analyze survival. The V79 cells showed a progressive increase in survival over a 6 hour holding period. Similar increases in survival occurred with the ADPRT 54 and ADPRT 351 cell lines. These results suggest that poly(ADP-ribose) does not make a significant contribution to potentially lethal damage repair in these cell lines at confluence.

In further experiments we compared the X-ray survival curves of logarithmically growing and confluent cells from the parental and mutant lines. Figure 3 shows, that at confluence, the mutant ADPRT 351 cell line showed a similar survival curve to that shown by the parental V79 cell line. However, when the cells were examined in logarithmic growth the ADPRT 351 cell line was hypersensitive relative to the V79 cell line. This occurs because the logarithmically growing V79 cell lines are more resistant to X-irradiation than they are at confluence. In contrast, the ADPRT 351 cell line is hypersensitive to X-irradiation in the logarithmically growing state relative to the confluent state. Similar results were obtained with the ADPRT 54 cell line.

These studies provide important insights as to the requirement for poly(ADP-ribose) polymerase in the response to DNA damage and the DNA repair process. The use of mutant cell lines allows these processes to be analyzed in the absence of enzyme inhibitors which have the potential for many nonspecific effects. In summary, these results show that the poly(ADP-ribose) polymerase deficient cell lines are hypersensitive, even at low doses, to all agents capable of inducing frank DNA strand breaks. These results suggest a requirement for involvement of poly(ADP-ribose) polymerase in the repair of or response to all DNA strand breaks. The hypersensitivity of the poly(ADP-ribose) polymerase deficient cell lines to topoisomerase I active agents suggests that poly(ADP-ribose) polymerase may be involved in topoisomerase I activity. The resistance of the ADPR deficient cell lines to topoisomerase II active agents may represent a compensatory mechanism in these cell lines for altered topoisomerase I activity. The similarity of potentially lethal damage repair curves and survival curves in the confluent V79 and ADPRT mutant cell lines suggests that poly(ADP-ribose) polymerase has negligible involvement in repair of radiation damage in confluent cells. In contrast, the marked hypersensitivity to X-irradiation shown by log phase mutant cells relative to log phase V79 cells suggests an important role for poly(ADP-ribose) polymerase in recovery from DNA damage during cell proliferation.

References

1. Chatterjee, S.; Hirschler, N.V.; Petzold, S.J.; Berger, S.J.; Berger, N.A. Mutant cells defective in poly(ADP-ribose) synthesis due to stable alterations in enzyme activity or substrate availability. Exp. Cell Res. 194:1-15; 1989.

2. Chatterjee, S.; Trivedi, D.; Petzold, S.J.; Berger, N.A. Mechanism of epipodophyllotoxin-induced cell death in poly(adenosine diphosphate-ribose) synthesis-deficient V79 Chinese hamster cell lines. Cancer Res. 50:2713-2718; 1990.

3. Chatterjee, S.; Cheng, M.F.; Berger, N.A. Hypersensitivity to clinically useful alkylating agents and radiation in poly(ADP-ribose) polymerase-deficient cell lines. Cancer Commun. 2:401-407; 1990.

4. Chatterjee, S.; Cheng, M.F.; Berger, S.J.; Berger, N.A. Alkylating agent hypersensitivity in poly(adenosine diphosphate-ribose) polymerase deficient cell lines. Cancer Commun. 3:71-75; 1991.

Part 3
Enzymology

Amino Acid Specific Modification of Poly(ADP-ribose)polymerase with Monomers and Polymers of ADP-ribose

Ma. Guadalupe Martinez-Cadena, Mario Pedraza-Reyes, and Rafael Alvarez-Gonzalez

Department of Microbiology and Immunology, Texas College of Osteopathic Medicine, The University of North Texas, Fort Worth, Texas, 76107-2690, U.S.A.

The primary protein target for poly(ADP-ribose) modification in mammalian chromatin is poly(ADP-ribose)polymerase (PADPRP) itself (1,2). However, our current understanding of this reaction at the biochemical level is very limited. Here, we have utilized 2' and 3'-deoxyNAD analogs as substrates for the amino acid-specific covalent modification of PADPRP with monomers and polymers of ADP-ribose. Specific mono(ADP-ribosyl)ation of PADPRP at arginine residues was achieved by incubating pure polymerase with mono(ADP-ribosyl)transferase A (3) from turkey erythrocytes and 2'-deoxyNAD as an ADP-ribosylation substrate (4,5). In contrast, the auto[poly(ADP-ribosyl)ation] of PADPRP was performed with 3'-deoxyNAD. Utilization of this NAD analog is advantageous because it does not alter the physicochemical properties of the polymerase upon modification (6,7).

Results

Arginine-specific mono(ADP-ribosyl)ation of poly(ADP-ribose)polymerase: Incubation of PADPRP with [^{32}P] radiolabeled 2'-deoxyNAD and arginine-specific mono(ADP-ribosyl)transferase from turkey erythrocytes results in the covalent modification of the polymerase (Fig. 1). However, incubation in the absence of the transferase showed no detectable labeling of PADPRP (4,5). These observations agree with previous results indicating that 2'-deoxyNAD is a good substrate for mono but not poly(ADP-ribosyl)ation (5).

We next subjected the PADPRP-mono(2'-deoxyADP-ribose) adducts to partial proteolysis in order to identify the peptides containing arginine-bound monomers. Partial proteolysis of mono(2'-deoxyADP-ribosyl)ated-PADPRP into functional domains was accomplished by enzymatic digestion with either papain or α-chymotrypsin (8-10). Shizuta and co-workers showed that proteolytic treatment of PADPRP with papain results in the formation of two peptide fragments of 74 and 46 kD. In contrast, proteolysis with α–chymotrypsin results in the formation of peptide fragments of 66 and 54 kD. The 46 and 66 kD peptides arise from the amino-terminus and possess the DNA-binding domain. However, the 74 and 54 kD fragments arise from the carboxy-terminus and contain the NAD-binding domain.

Both the DNA-binding and the NAD-binding domain are mono(2'-deoxyADP-ribosyl)ated at arginine residues following incubation with 2'-deoxyNAD and mono(ADP-ribosyl)transferase A (Fig. 1). Interestingly, the same peptide fragments were modified when PADPRP was subjected to partial proteolysis prior to mono(2'-deoxyADP-ribosyl)ation (data not shown). Moreover, mono(2'-deoxyADP-ribosyl)ated-PADPRP was not able to polymerize ADP-ribose when incubated with $[^{32}P]$ NAD in the presence of DNA (data not shown). It remains to be shown whether this arginine-specific mono(ADP-ribosyl)ation of PADPRP affects DNA-binding, NAD-binding and catalysis, or both. It is of interest to note that Shimoyama and co-workers have isolated and characterized an arginine-specific mono(ADP-ribosyl)transferase activity in hen liver nuclei (11). Their observations suggest that PADPRP activity might be modulated *in vivo* by a mono(ADP-ribosyl)ation cycle. Experiments are currently in progress to address this question.

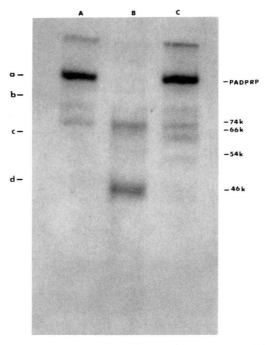

Fig.1. Arginine-specific mono(2'-deoxyADP-ribosyl)ation of PADPRP and its proteolytic peptide mapping. Lane A of the autoradiograph shows the electrophoretic migration of mono(2'-deoxyADP-ribosyl)ated-PADPRP following incubation of PADPRP with an arginine specific mono(ADP-ribosyl)transferase and $[^{32}P]$ 2'-deoxyNAD. Peptide fragments generated from PADPRP-(2'-deoxyADP-ribose) adducts following incubation with papain (lane B) and with α-chymotrypsin (lane C) (8-10). a, b, c, and d show the migration of the high molecular weight narkers (see Fig. 2).

Automodification of poly(ADP-ribose)polymerase with 3'-deoxyNAD as a substrate: Pure PADPRP was subjected to auto-poly(ADP-ribosyl)ation with 3'-deoxyNAD as a substrate in the presence of DNA. We have previously shown that this reaction does not significantly change the electrophoretic mobility of the polymerase (6,7). Therefore, high substrate concentrations can be used to saturate the enzymatic automodification sites. In addition, the smaller size of the polymers synthesized with this NAD analog (6) does not result in the steric hindrance of PADPRP-oligo(3'-deoxyADP-ribose) adducts to partial proteolysis (*vide infra*).

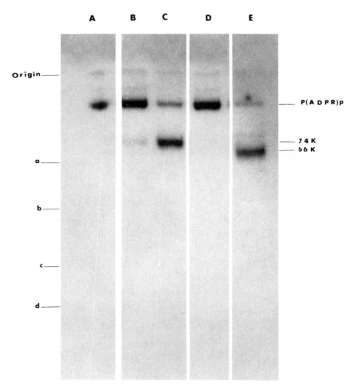

Fig. 2. Electrophoretic analysis of native and proteolytically degraded PADPRP following automodification with [^{32}P] 3'-deoxyNAD. The relative migration of the high molecular weight markers is indicated to the left of the autoradiograph where a, b, c, and d correspond to β-galactosidase, phosphorylase b, bovine serum albumin, and ovalbumin, respectively. Lane A shows the position of automodified PADPRP. Lanes B and C show oligo(3'-deoxyADP-ribosyl)ated-PADPRP adducts following incubation for 8 min at 25 °C in the absence (B) or presence (C) of 40 µg/ml of papain. Lanes D and E show PADPRP and the peptide fragments generated following incubation for 30 min at 25 °C with (D) or without (E) 40 µg/ml of α-chymotrypsin.

Incubation of pure PADPRP with 200 μM [^{32}P] 3'-deoxyNAD and nicked DNA results in enzyme automodification as judged by the radiolabeled band of identical electrophoretic mobility to PADPRP on the autoradiograph (Fig. 2, Lane A). Proteolytic treatment of automodified enzyme adducts with papain revealed that only the carboxy-terminus (74 kD) was oligo(3'-deoxyADP-ribosyl)ated. We also found that only the 66 kD amino-terminus was oligo(3'-dADP-ribosyl)ated (Fig. 2, lane E) following treatment with α–chymotrypsin. Our results agree well with those reported by Shizuta and co-workers (8-10). They also identified the same peptide fragment(s) as the automodification site(s) when using low concentrations of NAD as a substrate.

In addition, we subjected PADPRP-oligo(3'-deoxyADP-ribose) conjugates to a time-dependent incubation with 0.1 N NaOH at 37 °C. About 70% of the chemical bonds were cleaved under these conditions (data not shown). Therefore, a considerable fraction of the conjugates appears to be of the monoester type involving either glu and/or asp residues. These results also agree with the data obtained by Kawaichi et al.(12) who reported the same sensitivity of PADPRP-poly(ADP-ribose) adducts synthesized with 100 μM NAD to strong alkali, e.g., pH=13. In summary,our results suggest that both mono and poly(ADP-ribosyl)ation of PADPRP may modulate the versatile functions of this enzyme in eukaryotic chromatin structure and function.

Acknowledgments. This work was supported by grants BRSG 2S07RR05789-06 and BRSG 2S07RR05879-07 from NIH, the Texas Enhancement Research Program, and the Samuel Roberts Noble Foundation, Inc., to RAG. MGMC and MPR were partially supported by doctoral and postdoctoral fellowships from CONACyT.

References

1. Yoshihara, K., Hashida, T., Yoshihara, H., Tanaka, Y., and Oghushi, H. (1977) Biochem Biophys Res Commun 78: 1281-1288
2. Ogata, N., Ueda, K., Kawaichi, M., and Hayaishi, O. (1981) J Biol Chem 256: 4135-4137
3. Moss, J., Stanley, S. J., and Watkins, P. A. (1980) J Biol Chem 255: 5838-5840
4. Alvarez-Gonzalez, R. (1988) J Chromatogr 444: 89-95
5. Alvarez-Gonzalez, R., Moss, J., Niedergang, C., and Althaus, F. R. (1988) Biochem 27: 5378-5383
6. Alvarez-Gonzalez, R. (1988) J Biol Chem 263: 17690-17696
7. Pedraza-Reyes, M., and Alvarez-Gonzalez, R. (1990) FEBS Letters 277: 88-92
8. Nishikimi, M., Ogasawara, K., Kameshita, I., Taniguchi, T., and Shizuta, Y. (1982) J Biol Chem 257: 6102-6107
9. Kameshita, I., Matzuda, Z., Taniguchi, T., and Shizuta, Y. (1984) J Biol Chem 259: 4770-4776

10. Kameshita, I., Matsuda, M., Nishikimi, M., Ushiro, H., and Shizuta, Y. (1986) 261: 3863-3868
11. Tanigawa, Y., Tsuchiya, M., Imai, Y., and Shimoyama, M. (1984) J Biol Chem 259: 2022-2029
12. Kawaichi, M., Ueda, K., and Hayaishi, O. (1981) J Biol Chem 256: 9483-9489

Enzymatic properties of poly(ADP-ribose) polymerase and poly(ADP-ribose) glycohydrolase on chromatin.

Hélène Thomassin, Jean Lagueux, Luc Ménard, Jim Kirkland, Christophe Hengartner, Paul Cook* and Guy G. Poirier.

Laboratoire du métabolisme du poly(ADP-ribose), Endocrinologie Moléculaire, Centre de Recherche du CHUL et Université Laval, Québec, G1V 4G2, Canada
*Dept. of Microbiology, Texas College of Osteopathic Medicine, University of North Texas, Forth Worth, Texas 76107, USA

Introduction

The catabolism of NAD$^+$ in cells is carried out by poly(ADP-ribose) polymerase (PARP) which converts NAD$^+$ into poly(ADP-ribose) in response to DNA damage. All the evidence that has accumulated up to now indicates that the degradation of poly(ADP-ribose) *in vivo* is carried out by poly(ADP-ribose) glycohydrolase (the glycohydrolase) and not by a phosphodiesterase. Moreover, we have recently found that inhibition of the glycohydrolase during heat shock in C3H 10T1/2 fibroblasts causes a 5-fold accumulation of polymer levels without any effect on the levels of NAD$^+$ (1). The half-life of poly(ADP-ribose) *in vivo* after a carcinogenic treatment is very short, being close to 1 minute (2). Nevertheless, the decrease in NAD$^+$ levels, which reflects the activity of the PARP, is not proportional to the accumulation of polymer suggesting that high turnover rates occur in intact cells. These high rates of synthesis and degradation of the polymer support the hypothesis of a close physical and temporal relationship between the PARP and the glycohydrolase. This hypothesis is further supported by the physical attachment of the PARP to nuclear structures (chromatin and matrix) and the presence of the glycohydrolase in the nucleus.

All of the *in vitro* studied carried out to date to understand the combined effect of PARP and glycohydrolase on poly(ADP-ribose) levels have used them sequentially. We have reconstituted a turnover system in which both enzymes could be active at the same time on chromatin , in order to recreate a situation similar to the one that exists *in vivo*. We report in this paper that the glycohydrolase affects the apparent molecular weight and activity of poly(ADP-ribose) on the PARP in the presence of nucleosomes. Furthermore, a mathematical analysis of PARP glycohydrolase interaction was performed in various turnover systems.

RESULTS AND DISCUSSION

Previous studies carried out *in vitro* to characterize the effect of poly(ADP-ribosylation) on native chromatin were limited to the action of the PARP alone (3) or to the sequential action of PARP and poly(ADP-ribose) glycohydrolase (4, 5). However, synthesis and degradation of poly(ADP-ribose) are closely related and take place simultaneously *in vivo*. In this study, poly(ADP-ribosylation) of native chromatin was considered in reconstituted *in vitro* turnover systems where highly purified PARP and pure poly(ADP-ribose) glycohydrolase were added at the same time. Native chromatin was poly(ADP-ribosylated) with a ratio of 1 unit of PARP to 1 A_{260} unit of polynucleosomes (20-40N) in absence or in presence of glycohydrolase. Different systems have been studied that will be referred to as P/G:1/0 for non-turnover system or P/G:1/X depending on the PARP/glycohydrolase units ratio used.

A turnover system has been previously reconstituted using partially purified PARP and glycohydrolase (6). A mathematical model was developed to describe the kinetics of polymer accumulation in such a system. Poly(ADP-ribose) turnover system can be considered as a coupled enzymes system in which the poly(ADP-ribose) product of the PARP catalyzed-reaction, serves as a substrate for the glycohydrolase. Equations 1 and 2 summarize such a mechanism:

$$E_1 + S \; \rightleftharpoons \; E_1S \; \longrightarrow \; E_1 + B + C$$

$$E_2 + B \; \rightleftharpoons \; E_2B \; \longrightarrow \; E_2 + P$$

Where E_1 is the PARP, E_2 is the glycohydrolase, S is NAD$^+$, C nicotinamide, B polymer and P is ADP-ribose. On the basis of this theory, equations were derived, similar to that obtained by Yang and Schulz (7) to predict the time course of coupled-enzyme reactions. Poly(ADP-ribose) accumulation was calculated from this equation.

The mathematical model was tested to predict the polymer accumulation on the PARP during poly(ADP-ribosylation) of polynucleosomes under the reconstituted turnover systems defined above. Calculated time-courses for polymer accumulation were compared to the experimental results. Values for simulation of the systems were deduced from experimental results and from the specific activities of pure enzymes. Ménard et al. (6) have shown that the presence of poly(ADP-ribose) acceptor protein had no effect on nicotinamide accumulation. Additional poly(ADP-ribosylation) sites did not affect the rate of hydrolysis of NAD$^+$ and the apparent Km for NAD$^+$ remained unchanged. However, poly(ADP-ribose) accumulation increased when polymer acceptor proteins were present.

Our results confirm previous observations by Ménard et al. (6), i.e. the amount of poly(ADP-ribose) formed under turnover conditions was greatly influenced by the glycohydrolase. In all of the systems considered a plateau (steady state) was reached. Increasing amounts of glycohydrolase reduced with the plateau level and the time at which the steady state was reached. Experimental values under the P/G: 1/10 systems closely corresponded to the theoretical values. A discrepancy between experimental and theoretical values appeared after 2 minutes of reaction under the 1/1 turnover system. This discrepancy could be attributed to inactivation of the PARP by extensive automodification. Studies have shown that high levels of poly(ADP-ribose) on the PARP caused the detachment of the enzyme from its activating DNA.

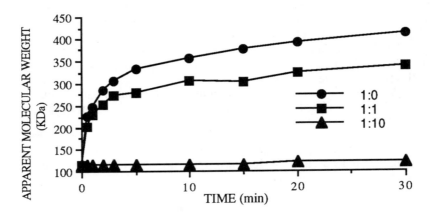

Fig. 1. Increase of PARP apparent molecular weight due to poly(ADP-ribose) accumulation under different turnover system. Ratios of PARP/glycohydrolase in units 1:0, 1:1 and 1:10. Electrophoresis was performed on a acrylamide-agarose gel pH 6.0 (6). PARP and glycohydrolase were purified to homogeneity as described in de Murcia et al. (4) and Thomassin et al. (8).

The mathematical model that has been used does not take into account the inactivation of PARP by extensive automodification. However correction of the Kcat corrected this problem.

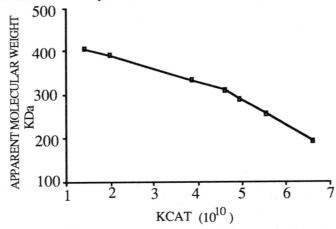

Fig. 2. Change in the Kcat of PARP in function of change of it molecular weight. The curve fit was used in the mathematical model to predict the poly (ADP-ribose) accumulation on the enzyme itself

This discrepancy was corrected by calculating the extent of automodification and the differences in Kcat PARP. Indeed we found a drop in Kcat as the apparent molecular weight of PARP increase (Fig.2). Furthermore upon high turnover no change in molecular weight of PARP increase could be observed (Fig. 1). Finally a perfect correlation could be made when the corrected Kcat was used in the equation (Fig. 3).

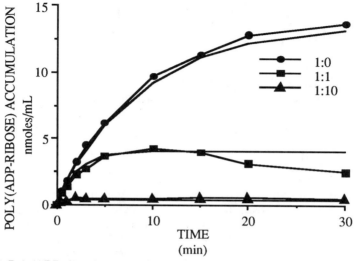

Fig. 3. Poly(ADP-ribose) accumulation under the different turnover systems. The theoretical fitting of PARP activity take in account the change in Kcat.

ACKNOWLEDGEMENTS

We want to thank Elaine Leclerc for her excellent secretarial assistance. This work was supported by The Natural Sciences and Engineering Research Council of Canada, The Medical Research Council of Canada and the programme pour la formation de chercheurs et d'aide à la recherche (Québec).

REFERENCES:

1. Jonsson, G.G., Ménard, L., Jacobson, E.L., Poirier, G.G. and Jacobson, M.K. (1988) Cancer Res. 48: 4240-4243.
2. Jacobson, E.L., Antol, K.M., Juarez Salinas, H., and Jacobson, M.K. (1983) J. Biol. Chem. 258: 103-107.
3. Poirier, G.G., de Murcia, G., Jongstra Bilen, J., Niedergang, C., and Mandel, P. (1982) Proc. Natl. Acad. Sci. USA. 79: 34223-3427.
4. de Murcia, G., Huletsky, A., Lamarre, D., Gaudreau, A., Pouyet, J., Daune, M., and Poirier, G.G. (1986) J. Biol. Chem. 261:7011-7017.
5. Gaudreau, A., Ménard, L., de Murcia, G. and Poirier G.G. (1986) Biochem. Cell. Biol. 64:146:153.
6. Ménard, L., Thibault, L., and Poirier, G.G. (1990) Biochim. Biophys. Acta. 1049:45-58.
7. Yang, S.-Y. and Schulz, H. (1987) Biochem. 26:5579-5584.
8. Thomassin, H., Jacobson, M.K., Guay, J., Verreault, A., Aboul-Ela, N., Ménard, L., and Poirier, G.G. (1990) Nucleic Acids Res. 18:4691-4694.

A New "Transition-State" Inhibitor Specific for Poly(ADP-ribose) Glycohydrolase

James T. Slama, Anne M. Simmons, M. E. Hassan,
Nasreen Aboul-Ela, and Myron K. Jacobson

Introduction

The degradation of poly(ADP-ribose) *in vivo* by hydrolysis of the glycosidic
(1′′-2′) ribosyl-ribose linkage is catalyzed by poly(ADP-ribose)
glycohydrolase (1,2). Specific inhibitors for this glycosidase have not been
available (3), despite suggestions that such inhibitors would have useful
pharmacological   properties   (4).   Adenosine   diphosphate
dihydroxypyrrolidine (ADP-DHP)(Figure 1) was shown to be a potent and
selective inhibitor for the bovine thymus poly(ADP-ribose) glycohydrolase.
Glycohydrolase activity is inhibited 50 % at 0.3 μM ADP-DHP. ADP-DHP
is further shown to inhibit the *Bungarus fasciatus* venom NAD
glycohydrolase competitively, with an inhibitor dissociation constant (K_i)
four-fold lower than the product dissociation constant (K_d) of adenosine
diphosphate ribose (ADP-ribose). ADP-DHP was designed to mimic the
structure of the oxo-carbonium ion intermediate that has been supported
experimentally in the catalyzed hydrolysis of NAD to ADP-ribose and
nicotinamide by vertebrate NAD glycohydrolases. Similar oxo-carbonium
ions are expected as intermediates in the hydrolysis of poly(ADP-ribose)
catalyzed by poly(ADP-ribose) glycohydrolase. ADP-DHP is therefore the
first of a series of "transition-state" inhibitors which will be produced for
poly(ADP-ribose) glycohydrolase.

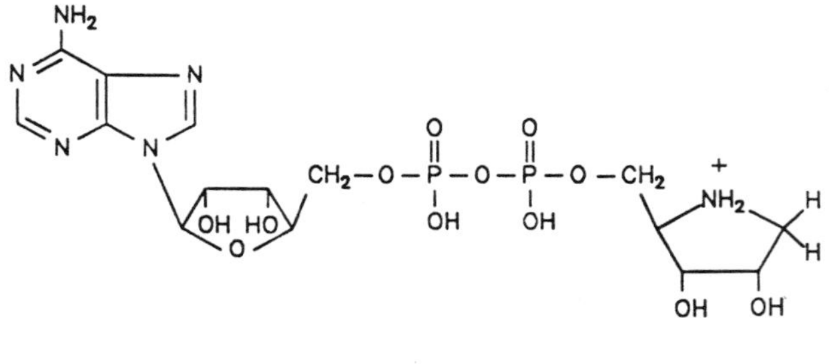

Figure 1. Structure of Adenosine diphosphate dihydroxypyrrolidine (ADP-DHP).

Results

Synthesis

The synthesis of the target inhibitor is depicted in Figure 2. The starting material is a protected 2-hydroxymethyl-3,4-pyrrolidine diol. The parent pyrrolidine is available from stereoselective synthesis starting with either 4-hydroxy-L-proline (5,6) or L-glutamic acid (7) or from multi-step synthesis starting from carbohydrates: L-lyxose (8) or gulonolactone (9). 1-Carbobenzyloxy-2-hydroxymenthy-3,4-pyrrolidine diol acetonide was phosphorylated using phosphoryl chloride (POCl3) in tetrahydrofuran, and the monophosphate purified by anion exchange chromatography. Completion of the synthesis requires coupling of the anion of the protected phosphoester with an activated adenosine 5′-monophosphate to form the unsymmetrical anhydride. AMP activated as the adenosine 5′-phosphomorpholidate (10) coupled with the pyrrolidine phosphate ester in high yield forming a protected derivative of ADP-DHP. The acetionide protecting group was removed on treatment with dilute aqueous acid, and the resulting N-carbobenzyloxy(ADP-DHP) purified by anion exchange chromatography and characterized spectroscopically. The carbobenzyloxy protecting group was removed easily on catalytic hydrogenation with a 10 % palladium on carbon catalyst, producing ADP-DHP in quantitative yield. The inhibitor, ADP-DHP, was purified by anion exchange chromatography, and its identity confirmed by the characteristic ^{1}H-NMR and fast atom bombardment mass spectrum.

Enzymology

ADP-DHP was tested as an inhibitor of the *Bungarus fasciatus* venom NAD glycohydrolase (11) and the porcine brain NAD glycohydrolase (12) by adding it to the assay mixtures and measuring the resulting rate of NAD hydrolysis. The result (Figure 3) was that ADP-DHP inhibited the activity of the venom NADase significantly at concentrations of 0.1 mM. Surprisingly, a closely related enzyme, the porcine brain NAD glycohydrolase was unaffected by concentrations of ADP-DHP up to 1 mM. The mechanism of inhibition of *B. fasciatus* venom NADase by the inhibitor was determined by Lineweaver-Burk analysis of the effect of a fixed inhibitor concentration on

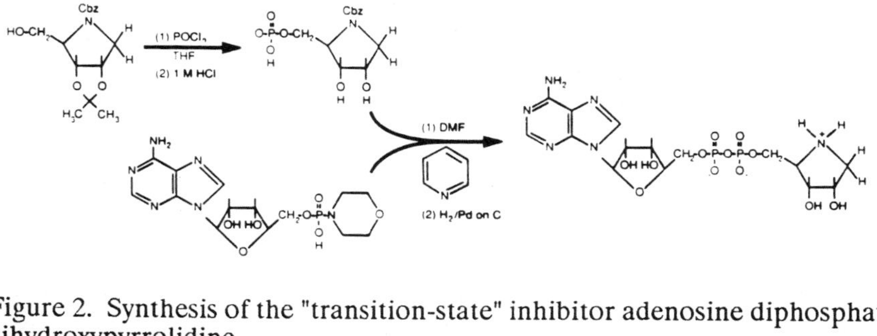

Figure 2. Synthesis of the "transition-state" inhibitor adenosine diphosphate dihydroxypyrrolidine.

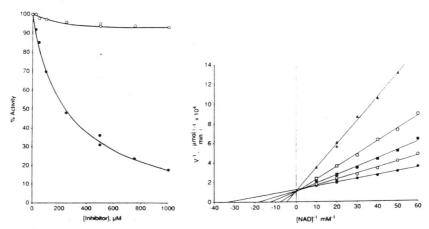

Figure 3. (left) Inhibition of the NAD glycohydrolase activity of the *B. fasciatus* venom NADase (●) and the porcine brain NADase (○) by addition of increasing amounts of the "transition-state" inhibitor ADP-DHP. The assay was conducted in 33 mM phosphate, pH 7.5 at 37 °C in a total volume of 0.3 mL. The concentration of NAD was 50 μM and contained 50,000 dpm [*carbonyl*-^{14}C]NAD. The amount of [carbonyl-^{14}C]nicotinamide released was measured radiometrically as described in our previous work (15).

Figure 4. (right) Competitive inhibition of the *B. fasciatus* NAD glycohydrolase by addition of ADP-DHP. The assay was conducted as described under Figure 3. The concentration of NAD was varied from 16.7 to 100 μM. Inhibitor concentrations were: (●) no inhibitor; (○) 48.7 μM; (■) 103.4 μM; (□) 200.8 μM; (▲) 499 μM.

the initial rates of enzymatic hydrolysis (Figure 4). ADP-DHP was found to be a linear competitive inhibitor of the venom NADase with an inhibitor dissociation constant of 94 ±4 μM.

ADP-DHP was evaluated as an inhibitor of the bovine thymus poly(ADP-ribose) glycohydrolase (13). Varying amounts of inhibitor were added to the enzyme assay (13) in which the substrate [^{32}P]poly(ADP-ribose) was present at a concentration of 10 μM in monomer residues. The result (Table 1) was that the glycohydrolase was inhibited significantly at concentrations of the inhibitor below 1 μM.

The specificity as well as the potency of an inhibitor is also of importance. ADP-DHP was tested as an inhibitor for poly- and mono(ADP-ribose) synthetases. ADP-DHP did not inhibit the activity of the bovine thymus poly(ADP-ribose) synthetase significantly at concentrations up to 1 mM (Table 1). Further, concentrations of inhibitor up to 1 mM did not affect the NAD glycohydrolase activity of the model mono(ADP-ribosyl) transferase, diphtheria toxin A subunit (data not shown). Inhibitory activity is therefore specific for the hydrolytic enzymes, with the highest potency shown against the poly(ADP-ribose) glycohydrolase.

Table 1. Enzyme Inhibition by ADP-DHP and ADP-ribose.

Enzyme and condition	Activity[a] (units per 60 μL assay)	% Control
Poly(ADP-ribose) glycohydrolase[b]		
added ADP-DHP		
0. μM (control)	17.2×10^{-3}	100
150. μM	0.52×10^{-3}	3
15. μM	1.03×10^{-3}	6
1.5 μM	2.58×10^{-3}	15
0.15 μM	10.49×10^{-3}	61
added ADP-ribose		
100. μM	10.8×10^{-3}	63
10. μM	17.9×10^{-3}	104
1. μM	17.0×10^{-3}	99
Poly(ADP-ribose) synthetase[c]	(units per 90 μL assay)	
added ADP-DHP		
0. μM (control)	0.44	100
1000. μM	0.40	90
100. μM	0.47	106

(a) One unit of enzyme is the amount which liberates one nmole of ADP-ribose per min at 37 °C.
(b) Poly(ADP-ribose) glycohydrolase was assayed in the presence of 10 μM poly(ADP-ribose) as described in (13).
(c) Poly(ADP-ribose) polymerase was assayed as described in (13) using 100 μM NAD; one unit of enzyme incorporates one nmole of ADP-ribose into polymer per min.

Discussion

Adenosine diphosphate dihydroxypyrrolidine (ADP-DHP) was designed to mimic an oxo-carbonium ion produced as an intermediate in the enzymatic hydrolysis of NAD or in the enzymatic hydrolysis of poly(ADP-ribose). Protonated ADP-DHP resembles the oxo-carbonium ion in its general shape and with respect to its charge. It, however, differs from the hybridization expected on the oxo-carbonium ion C-1 carbon. In accord with the expectation that a transition state analogue will behave as an enzyme inhibitor, the dissociation constant for ADP-DHP (94 μM) with B. fasciatus venom NADase was found to be significantly lower than the product dissociation constant (360 μM)(14) for ADP-ribose.

Poly(ADP-ribose) glycohydrolase represents an important biological and pharmacological target. Hydrolysis of nuclear poly(ADP-ribose) is necessary to complete DNA repair and other processes dependant on poly(ADP-ribosyl)ation. No specific inhibitors for this enzyme are yet known although a variety of studies (4) suggest that such compounds would be informative. Although poly(ADP-ribose) glycohydrolase is still poorly characterized and its mechanism is unknown, it is probable that hydrolysis involves formation of an oxo-carboniun ion like intermediate. The "transition-state inhibitor" was therefore tested as an inhibitor of this glycohydrolase. ADP-DHP was found to inhibit hydrolysis with an IC_{50} of 0.3 μM, a concentration 1000 times lower than the IC_{50} for the product of hydrolysis, ADP-ribose. The low dissociation is indicative of the tight binding expected from a true transition state analogue. The efficacy of inhibition in the micromolar range suggests that ADP-DHP will be useful for inhibiting poly(ADP-ribose) glycohydrolase in cultured cells in vitro, in developing specific photoaffinity labels, and in developing specific affinity absorbents. ADP-DHP will also serve as a paradigm for the development of inhibitors which resemble both the transition state charge and geometry, and which incorporate features that enable them to more readily cross the plasma membrane and be taken up by living cells.

Acknowledgements. Support was provided by National Institutes of Health Grant GM-32821 to J.T.S. and by National Institutes of Health Grant CA-43894 to M.K.J.

References

1. Althaus, F.R., Richter, C (1987) ADP-ribosylation of Proteins. Enzymology and Significance, Springer-Verlag, Berlin, Heidelberg. 237 p. Molecular Biology, Biochemistry and Biophysics, Vol 37.
2. Hatakeyama, K., Nemoto, Y., Ueda, K., Hayaishi, O. (1986) J Biol Chem 261: 14902-14911.
3. Tavassoli, M., Tavassoli, M.H., Shall, S. (1985) Biochem Biophys Acta 827: 228-234.
4. Jurez-Salinas, G., Duran-Torres, G., Jacobson, M.K., (1984) Biochem Biophys Res Comm 122: 1381-1388.
5. Hassan, M.E., Slama, J.T. (1986) American Chemical Society, Southwest Regional Meeting, Houston, TX; Nov 19-23: Paper No 276, Abstracts.
6. Guillerm, G., Varkados, M., Auvin, S., Le Goffic, F. (1987) Tetrahedron Lett 28: 535-538.
7. Ikota, N., Hanaki, A. (1987) Chem Pharm Bull 35: 2140-2143
8. Reist, E.J., Gueffroy, D.E., Blackford, R.W., Goodman, L., (1966) J Org Chem 31: 4025-4030.
9. Fleet, G.W., Son, J.C., (1988) Tetrahedron 44: 2637-2647.
10. Moffatt, J.G., (1964) Can J Chem 42: 599-604.
11. Yost, D.A., Anderson, B.M., (1981) J Biol Chem 256: 3647-3653.
12. Windmueller, H.G,. Kaplan, N.O., (1962) Biochem Biophys Acta 56: 388-391.
13. Tomassin, H., Jacobson, M.K., Guay, J., Verreault, A., Aboul-ela, N., Menard, L., Poirier, G.G. (1990) Nucleic Acids Research 18: 4691-4694.
14. Yost, D.A., Anderson, B.M., (1982) J Biol Chem 257: 767-772.
15. Slama, J.T., Simmons, A.M. (1988) Biochemistry 27, 183-193.

Histones Affect Polymer Patterns Produced by Poly(ADP-ribose)polymerase

Hanspeter Naegeli and Felix R. Althaus

University of Zürich-Tierspital, Institute of Pharmacology & Biochemistry, Winterthurerstrasse 260, CH-8057 Zürich, Switzerland

The post-translational poly ADP-ribosylation of proteins involves a complex pattern of ADP-ribose polymers. Several variables in these polymers, such as polymer numbers, polymer sizes, and branching frequencies may be important for differentially regulating the function of a given protein. Little is known about the molecular factors regulating these parameters (for review see 1-3).

In the present study, we have reconstituted an *in vitro* system to identify the molecular factors regulating the polymer size patterns produced by poly(ADP-ribose)polymerase (4, 5).

Histones Cause Distinct Adaptations in the Reaction Products of Poly(ADP-ribose)Polymerase

Fig. 1 shows that histones may regulate both the quantities (1A) and the sizes (1B) of ADP-ribose polymers produced by purified poly(ADP-ribose)polymerase. These effects reached a plateau when polymer synthesis was maximally stimulated (5). Surprisingly, poly(ADP-ribose)polymerase produced a specific size pattern for each histone species tested (1B). The patterns showed minimal interexperimental variation, i.e. $\leq$ 5% variation for the relative frequency of a particular size class of a given histone species.

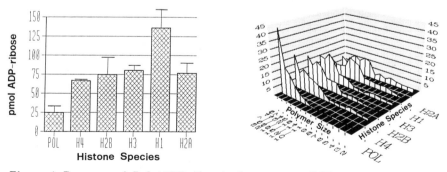

Figure 1: Reponse of Poly(ADP-ribose)polymerase to Different Histones. 1A, polymeric ADP-ribose production (pmols) in the automodification reaction ("POL") or in the presence of maximally effective amounts of histones H4 (27 pmols), H2B (22 pmols), H3 (20 pmols), H1 (13 pmols), or H2A (22 pmols). *1B,* polymer size distributions as a function of histone species. "POL" denotes the polymer size distribution obtained in the automodification reaction. The amount of ADP-ribose in each size class is expressed as a percentage of total ADP-ribose incorporation. Polymer size classes: 2, dimer, 3, trimer, etc. (for methods see 5).

Specificity of Histone-Regulated Polymer Size Adaptation
We have previously shown that polymer addition to acceptor proteins involves a processive reaction also in the presence of histones (4). Thus, the polymer size distributions shown in Fig. 1B are independent of incubation time, and the stimulation of polymer synthesis (Fig. 1A) is due to elevated polymer numbers and not increased polymer sizes (4).

Apart from histones, a number of factors are known to modulate poly(ADP-ribose)polymerase activity (for review see 1-3). We therefore tested whether any of these factors would also cause adjustments in ADP-ribose polymer sizes. Varying the substrate concentration (0.01 to 1.0 mM NAD^+), the DNA fragment size (146 bp to approx. 20 kb), or the number of nicks in DNA affected the polymer numbers but not the polymer size distributions (data not shown). The polyamine spermine (6), and Mg^{2+} (7,8) also stimulated polymerase activity. However, while we could reproduce these stimulatory effects in our *in vitro* system, neither spermine (0 to 2 mM) nor $MgCl_2$ (0 to 10 mM) changed the size distributions of polymers (data not shown). These experiments suggest that the polymer size modulating activity is not related to the general stimulation of poly(ADP-ribose)polymerase activity by histones. In addition, other basic proteins (e.g. lysozyme, cytochrome c), acidic proteins (bovine serum albumin, DNase I), and DNA-binding proteins (DNase I, nuclease S1) were inactive in this *in vitro* test system (data not shown). These results suggested that the modulation of ADP-ribose polymer sizes is due to a specific adaptation of poly(ADP-ribose)polymerase.

Size Adaptations of Polymers Bound to Automodified Poly(ADP-ribose)Polymerase
We next examined the possibility that the relative increase of small polymer molecules was due to heterologous modification of histones, and not due to an altered size pattern of ADP-ribose polymers bound to the automodified enzyme itself. For this purpose, we used histones immobilized on agarose beads, which could be rapidly separated from automodified polymerase (for details see 5). The results in Fig. 2 clearly demonstrate that immobilized histones induce adaptations in the polymer pattern associated with automodified polymerase. These and other results (5) established that histones regulate polymer synthesis both in the auto- and heteromodification reaction of poly(ADP-ribose)polymerase.

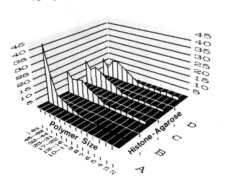

Figure 2: Effects of Immobilized Histones on the Polymer Pattern Associated with Automodified Polymerase. The amount of ADP-ribose in each polymer size class is expressed as in Fig. 1. A, polymerase incubation in the presence of underivatized agarose. B-C, polymers produced in the presence of increasing amounts of immobilized histones. The ratios of agarose to histone-agarose were 10:0 (in A), 9:1 (B), 8:2 (C), and 5:5 (D); for methods see 5.

323

Allocation of the Polymer Number and Polymer size Modulating Activities to Different Polypeptide Domains of Histone H1
Histone H1 had the most dramatic impact on the numbers and a major effect on the size distribution of the polymerase reaction products. Therefore, histone H1 was selected to obtain futher information on the molecular properties of this polymer-regulating activity. Three different proteolytic fragments were prepared (9, 10) as shown in Fig. 3, and purified by gel filtration (11).

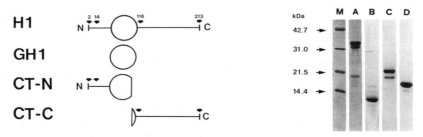

Figure 3: Proteolytic Fragments of Histone H1. A, schematic illustration of the structural domains of histone H1 and its proteolytic fragments. The histone H1 fragments (GH1, CT-C and CT-N) were prepared as described (9-11). The arrows indicate potential sites of ADP-ribosylation. B, sodium dodecyl sulfate/gel electrophoretic analysis of histone H1 fragments. Lane A: crude perchloric acid extract of rat liver nuclei used as starting material for fragment preparation; Lane B: GH1. Lane C: CT-C. Lane D: CT-N. The molecular weight markers (Lane M) were: hen egg white ovalbumin (42.7 kDa), bovine carbonic anhydrase (31.0 kDa), soybean trypsin inhibitor (21.5 kDa), and hen egg white lysozyme (14.4 kDa).

Like histone H1, the GH1 fragment induced the polymerase to produce larger polymer numbers, but did not exhibit polymer size regulating activity. This activity was found when the carboxy and amino-terminal fragments were tested (Fig. 4). Both the CT-C and CT-N fragments increased polymer numbers (5), and in contrast to GH1, caused distinct changes in the polymer size distributions (Fig. 4) . A very dramatic shift to the smaller molecular size classes was observed with the CT-N fragment. These data indicate that the two regulatory activities of histone H1 can be attributed to different polypeptide domains. Preliminary results show that domains rich in lysine or arginine, may cause very strong shifts in the polymer size distributions whithout serving as polymer acceptors, while those rich in histidine, glutamate, or aspartate were completely ineffective in this *in vitro* system.

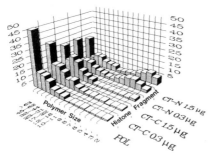

Figure 4: Effects of CT-C and CT-N on the Size Distribution of ADP-ribose Polymers. POL, automodification reaction; CT-C, CT-N, in the presence of the indicated amounts of CT-C or CT-N fragments, respectively. The relative amount of ADP-ribose in each size class is expressed relative to total quantity of poly(ADP-ribose) formed.

In conclusion, we have observed that the nuclear enzyme poly(ADP-ribose)polymerase responds to the presence of histones by producing histone-specific size patterns of ADP-ribose polymers. This observation may be conceptualized in several ways. First, this may be regarded as a unique case of enzyme regulation involving structural modifications in the reaction products. Second, apart from their role in chromatin organization, histones may act as specific regulators of protein poly ADP-ribosylation in the nucleus. Third, specific polymer size patterns may be an important component of the molecular signal priming poly ADP-ribosylated chromatin proteins for an active role in DNA excision repair.

The relevance of polymer adaptations to protein poly ADP-ribosylation *in vivo* is not known. However, the pattern of ADP-ribose polymers produced in isolated nuclei (5) is almost identical to the polymer pattern found in nucleosomal core particles (4), and to the composite pattern synthesized by poly(ADP-ribose)polymerase *in vitro* in the presence of all four core histones and histone H1. These results suggest that histones are the predominant regulators of ADP-ribose polymer numbers and sizes. However, it remains to be seen whether the enzyme poly(ADP-ribose)glycohydrolase, the major polymer catabolizing enzyme in mammalian cells, may also affect these parameters *in vivo*.

Acknowledgements
This work was supported by Grant 31-31.20391 (to F.R.A.) from the Swiss National Foundation for Scientific Research, financial support from the Krebsliga des Kantons Zürich, and the Jubiläumsstiftung of the University of Zürich.

References
1. Ueda, K. (1986) In *Pyridine Nucleotide Coenzymes* (Dolphin, D., Poulson, R. & Avramovic, O., eds.) Part B, pp. 549-597, John Wiley & Sons, New York
2. Althaus, F.R. & Richter, C. (1987) *ADP-ribosylation of Proteins: Enzymology and Biological Significance*, Springer-Verlag, Berlin
3. Jacobson, M.K. & Jacobson, E.L. (1989) *ADP-ribose Transfer Reactions: Mechanisms and Biological Significance*, Springer-Verlag, Berlin
4. Naegeli, H., Loetscher, P. & Althaus, F.R. (1989) *J. Biol. Chem. 264*, 14382-14385
5. Naegeli, H. & Althaus, F.R. (1991) *J. Biol. Chem. 266*, in press
6. Kawamura, M., Tanigawa, Y., Kitamura, A., Miyake, Y. & Shimoyama, M. (1981) *Biochim. Biophys. Acta 652*, 121-128
7. Tanaka, Y., Hashida, T., Yoshihara, H. & Yoshihara, K. (1979) *J. Biol. Chem. 254*, 12433-12438
8. Ferro, A.M. & Olivera, M.O. (1982) *J. Biol. Chem. 257*, 7808-7813
9. Hartman, P.G., Chapman, G.E., Moss, T. & Bradbury, E.M. (1977) *Eur. J. Biochem. 77*, 45-51
10. Bradbury, E.M., Chapman, G.E., Danby, S.E., Hartman, P.G. & Riches, P.L. (1975) *Eur. J. Biochem. 57*, 521-528
11. Thoma, F., Lose, R. & Koller, T. (1983) *J. Mol. Biol. 167*, 619-640

Proteolytic Degradation of Poly(ADP-ribose)polymerase by Contaminating Proteases in Commercial Preparations of DNAse I.

Rafael Alvarez-Gonzalez, Guy G. Poirier,[&] and Marcos Martinez

Department of Microbiology and Immunology, Texas College of Osteopathic Medicine, The University of North Texas, Fort Worth, Texas, 76107-2690, U.S.A.

The stimulation of poly(ADP-ribose)polymerase (PADPRP) activity by the addition of agents that induce DNA damage and formation of DNA strand-breaks *in vitro* and *in vivo* is well established (1). Not surprisingly, DNAse I is commonly added enzyme to stimulate PADPRP activity in biological samples such as cell ghosts (2) and nuclei preparations (3). A high level of DNA-strand breaks helps to accurately determine the amount of PADPRP activity in these samples.

Recently, it has been reported that both PADPRP activity (4-6) and ADP-ribose polymers (6,7) co-purify with the nuclear matrix. We observed that 15-30% of PADPRP co-isolated with the nuclear matrix from adult rat liver (6). However, other investigators reported that only 2-5% of PADPRP was associated with this fraction (4,5). Nuclear Matrices are typically isolated by first enzymatically digesting DNA with either exogenous (4,5) or endogenous (6) endonucleases. Subsequently, nucleotides are extracted with low salt buffer and chromatin proteins (histones) with high salt buffer (4-6). A typical enzyme chosen to digest nuclear DNA is DNAse I from Sigma (4,5). Here, we compare the spectrum of poly(ADP-ribosyl)ated-proteins in rat liver nuclei in the absence or presence of commercial preparations of DNAse I obtained from either Sigma (type IV) or Worthington (type DPFF).

Results and Discussion

Effect of increasing amounts of DNAse I type IV from Sigma on the poly(ADP-ribosyl)ation of rat liver chromatin proteins. Fig. 1 shows the autoradiograph of [^{32}P] poly(ADP-ribosyl)ated-proteins in adult rat liver following incubation of rat liver chromatin with 1 μM NAD at 25 $^{\circ}$C for 5 min and subsequent low pH lithium dodecyl sulphate polyacrylamide gel electrophoresis (LiDS-PAGE). Lane (a) shows the total amount of [^{32}P] poly(ADP-ribosyl)ated-PADPRP in the absence of DNAse I. However, increasing amounts of DNAse I in the incubation reaction mixture resulted in the disappearance of labeled PADPRP (lanes b through g). Concomitantly with the lowering of modified PADPRP, we observed an increase in the radiolabel co-migrating with bovine serum albumin of 67 kD. Therefore, our results suggest that proteolysis of PADPRP takes place in a DNAse I dose dependent-manner. Interestingly, when we determined the total level of PADPRP activity in duplicate samples as the total amount of [^{32}P] radiolabeled NAD incorporated into 20%

[&] Department of Biochemistry, Faculty of Medicine, Universite Laval Quebec, Quebec, Canada.

trichloroacetic acid (TCA) insoluble material, we observed the typical increase in PADPRP activity (Fig. 2). This was not too surprising since a portion of the radiolabeled PADPRP proteolytic fragments would still be expected to be acid insoluble.

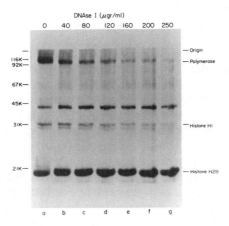

Fig.1. Effects of increasing amounts of DNAse I on the extent of poly(ADP-ribosyl)ation of rat liver chromatin proteins. The relative electrophoretic migration of the low molecular weight markers is indicated to the left of the autoradiograph. Lanes a, b, c, d, e, f, and g show the extent of poly(ADP-ribosyl)ation of PADPRP, histone H1, histone H2b and a 42 kD protein following incubation of adult rat liver chromatin with 1 µM [^{32}P] NAD in the presence of 0, 40, 80, 120, 160, 200, and 250 µg/ml of DNAse I type IV from Sigma, respectively.

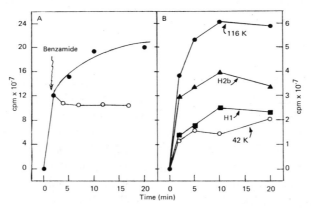

Fig. 2. Time-dependent poly(ADP-ribosyl)ation of chromatin proteins from adult rat liver following incubation with 1 µM [^{32}P] NAD and 100 µg/ml of DNAse I type IV from Sigma at 25 °C. Panel (A) shows the incorporation of radiolabeled substrate in total chromatin and panel (B) shows the incorporation of radiolabeled substrate onto individual acceptor proteins as determined by LiDS-PAGE and scintillation counting of individual band slices of the same gel.

<u>Effects of two commercial preparations of DNAse I on the automodification reaction catalyzed by pure DNA-dependent PADPRP from calf thymus.</u> In order to confirm that DNAse I type IV from Sigma was contaminated with protease activities that preferentially degraded PADPRP, we performed experiments with a purified preparation of DNA-dependent PADPRP from calf thymus. In these experiments, we compared the effects of commercial preparations of DNAse I type IV from Sigma and type DPFF from Worthington. Fig. 3b shows the autoradiograph of automodified PADPRP incubated in the presence of increasing amounts of the Sigma type IV DNAse I preparation. In this experiment, $[^{32}P]$ radiolabeled proteolytic peptides of low molecular weight were observed even at 40 µg/ ml of DNAse I. Fig. 3b also shows that increasing amounts of this endonuclease resulted in higher levels of proteolysis. It should be noted that the same results were observed with other Sigma preparations of DNAse I (not shown). In contrast, the same concentrations of DNAse I type DPFF from Worthington did not change the electrophoretic migration of automodified enzyme as shown on the autoradiograph of Fig. 3a. Thus, it can be concluded that DNAse I type DPFF from Worthington is not contaminated with proteases.

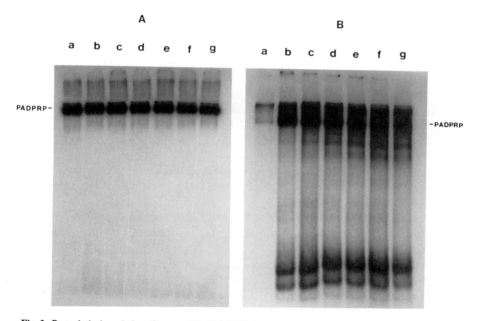

Fig. 3. Proteolytic degradation of automodified PADPRP by proteases contaminating commercial preparations of DNAse I. Panel (A) shows the effect of 0, 40, 80, 120, 160, 200 and 250 µg/ml of DNAse I type DPFF from Worthington (lanes a through g, respectively). Panel (B), on the other hand, shows the effect of 0, 40, 80, 120, 160, 200 and 250 µg/ml of DNAse I type IV from Sigma (lanes a through g, respectively).

In summary, our results show that DNAse I preparations from Sigma are usually contaminated with proteases. Unfortunately, several investigators utilize this enzyme source to degrade nuclear DNA during the isolation of nuclear matrices from cells and tissues (4,5). Therefore, under these conditions, preferential proteolysis of automodified PADPRP may lead to an underestimation of the actual levels of PADPRP activity associated with the nuclear matrix (4,5). In our nuclear matrix purification protocol (6,8), we digest the DNA of rat liver nuclei with endogenous endonucleases by incubating isolated nuclei at 37 $^{\circ}$C for 45 min (9,10). This treatment does not lead to the proteolytic degradation of nuclear matrix associated PADPRP. Therefore, the data presented here strongly suggest that the level of PADPRP activity associated with the nuclear matrix fraction oscillates between 15-30% as we have previously described (6). Our results also agree with those of Moy and Tew (11) who found 15% of the total PADPRP activity co-purifying with the nuclear matrix following digestion of chromatin DNA with 5 μg/ml of DNAse I. According to our data, the amount of Sigma DNAse I utilized by this group, does not contain large amounts of contaminating proteases (see Fig. 1).

Acknowledgments. This project was supported by a grant from the Samuel Roberts Noble Foundation, Inc. to Rafael Alvarez-Gonzalez.

References

1. Althaus, F. R., and Richter, C. (1987) ADP-ribosylation of Proteins: Enzymology and Biological Significance. Molecular Biology, Biochemistry, and Biophysics (Springer Verlag Heidelberg New York Tokyo), Vol. 37
2. Benjamin, R. C., and Gill, D. M. (1980) J Biol Chem 255: 10493-10501
3. Cesarone, C. F., Scarabelli, L., Giannoni, P., and Oruneso, M. (1990) Mutation Res 245: 157-163
4. Wesierska-Gadek and Sauermann, G. (1985) Eur J Biochem 153, 421-428
5. Adolph, K. W., and Song, M.-K. H. (1983) Biochem Biophys Res Commun 115: 938-945
6. Alvarez-Gonzalez, R. and Ringer, D. P. (1988) FEBS Letters 236: 362-366
7. Cardenas-Corona, Ma.-E., Jacobson, E. L., and Jacobson, M. K. (1987) J Biol Chem 262, 5857-5865
8. Pedraza-Reyes, M. and Alvarez-Gonzalez, R. (1990) FEBS Letters 277: 88-92
9. Berezney, R., and Buchholtz, L. A. (1981) Exp Cell Res 132: 1-13
10. Panzeter, P. L., Etheredge, J. L., Kizer, D. E., and Ringer, D. P. (1987) Biochem Biophys Res Commun 149: 27-37
11. Moy, B. C. and Tew, K. D. (1985) Chem Biol Interactions 54: 209-222

Inactivation of the Polymerase but not the DNA Binding Function of ADPRT by Destabilization of one of its Zn^{2+} Coordination Centers by 6-Nitroso-1,2-Benzopyrone

Kalman G. Buki, Pal I. Bauer, Jerome Mendeleyev, Alaeddin Hakam and Ernest Kun

Laboratory for Environmental Toxicology and Chemistry and the Octamer Research Foundation, Romberg Tiburon Centers, San Francisco State University, Tiburon, CA 94920, USA, and the Department of Chemistry and Biochemistry, San Francisco State University, San Francisco, CA, 94132, USA.

Introduction

In the course of studies on the metabolism of inhibitors of poly (ADP- ribose) transferase (ADPRT), we observed that the inhibitor 6-amino-1,2-benzopyrone (6-ABP) (1) when metabolized by rat liver microsomal preparations is oxidized to the corresponding 6-nitroso derivative (6-NOBP) as the main metabolite (2). Since ADPRT is located in the nuclear matrix (3,4) and most of the cytochrome P450 resides in the endoplasmic reticulum which is contiguous with the nuclear membrane (5), generation of 6-NOBP near ADPRT is likely. This suggested that the exact cellular significance of this oxidation path be investigated with respect to ADPRT.

Results and Discussion

Inhibitory effect of 6-NOBP on ADPRT was tested in a standard ADPRT assay (1) determining V_{init} (2 min). It was found that inhibition by 6-NOBP is somewhat weaker than by the parent compound 6-ABP (K_i for 6-NOBP is 40 μM vs 28 μM for 6-ABP), but this result nevertheless shows that 6-NOBP binds to ADPRT.

ADPRT contains two coordinate Zn^{2+} complexes in two zinc finger polypeptide motifs (6-8). Site-directed mutagenesis of human ADPRT (8) has shown that the two zinc fingers are not identical. Disruption of zinc finger FII dramatically reduced DNA binding, while that of FI resulted in only slight alteration of DNA binding (8).

Incubation of ADPRT with increasing concentrations of 6-NOBP for 2 hrs at 22°C resulted in a sharp decrease of the enzyme's polymerase activity coincidental with dissociation of $^{65}Zn^{2+}$ from

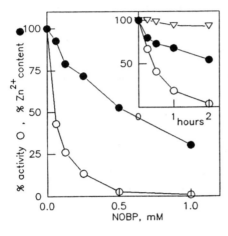

Figure 1. Effect of 6-NOBP on enzymatic and zinc binding activity of ADPRT. Ordinate: polymerase activity (open circles) and $^{65}Zn^{2+}$ content (closed circles) of ADPRT following incubation for 2 hrs in MES buffer (pH 6) with varying concentrations of 6-NOBP (abscissa), compared with ADPRT incubated without 6-NOBP. Inset: time course in the presence of 0.5 mM 6-NOBP at 22°C. Top curve (open triangles) shows protection of enzymatic activity by 0.5 mM $ZnCl_2$ or $CdCl_2$. Closed cirles show $^{65}Zn^{2+}$ ejection; lower curve (open circles) shows enzymatic activity as % of controls.

the $^{65}Zn^{2+}$-loaded enzyme (Fig. 1). If 2-mercaptoethanol is included in the incubation, it protects -SH groups of ADPRT against oxidation by the nitroso compound and favors selective attack on the zinc finger sites of ADPRT, consistent with the observed binding of 6-NOBP to ADPRT. Coinciding with an almost complete loss of polymerase activity, nearly half of the zinc content of ADPRT was ejected by the action of 6-NOBP, and this effect was not altered by DNA. Protection against ADPRT inactivation by 6-NOBP was afforded by either 0.5 mM $ZnCl_2$ or $CdCl_2$ (top curve, inset of Fig. 1), but no protection was conferred by Fe^{2+}, Co^{2+}, or Mg^{2+} (not shown). The Zn^{2+} ejection caused by 6-NOBP, evidently from one zinc finger of ADPRT, was reversible, but reactivation within the same time frame as the Zn^{2+} ejection required incubation of the enzyme with a combination of added Zn^{2+}, dithiothreitol and DNA (results not shown).

From the putative Zn^{2+} coordinate complex in ADPRT (6-8) a simple mechanism we suggest for zinc release is the nitroso group in 6-NOBP oxidizes two of the cysteine sulfur(thiolate) ligands of Zn^{2+} to the disulfide with concomitant reduction of nitroso group to hydroxylamino, for which there is chemical precedent (9).

Disulfides are known to have diminished capacity to coordinate metals compared to thiolates (10,11), and configurational constraints may also be important. A phenomenon somewhat analogous to the one reported here has been observed with E. coli RNA polymerase, where p-hydroxymercuriphenylsulfonate differ-

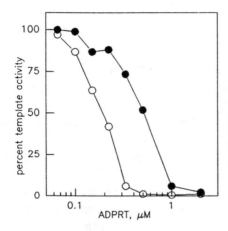

Figure 2. Inhibition of DNA template activity by native ADPRT (open circles) and 6-NOBP-treated ADPRT (closed circles). The ordinate shows template activity, assayed as[^{32}P]dCMP incorporation (30 min) by Klenow fragment of DNApolymerase I on M13 DNA annealed to template primer #1212 (14).Parallels agreed within 10%.

entially removes one of the two zinc ions from this enzyme (12). Inequality with respect to the intimate environments of the metal centers in the two zinc fingers of the mammalian glucocorticoid receptor has been observed in NMR relaxation times which differ markedly (13). We tested p-hydroxymercuribenzoate on ADPRT and found that treatment with this reagent removes both zinc ions indiscriminately and denatures ADPRT, and thus 6-NOBP exhibits distinct advantages for experimental studies.

Since site-directed mutagenesis, disrupting zinc fingerFII, dramatically reduced DNA binding of ADPRT (4), we also determined the effects of the ejection of Zn^{2+} from ADPRT on DNA binding. A quantitative assessment of the consequences of the binding of ADPRT to DNA was obtained by determining the inhibition of template activity with M13 DNA as template (14) as illustrated in Fig. 2. Untreated ADPRT inhibited template activity as a function of ADPRT concentration and at 0.5 μM ADPRT nearly complete inhibition occurred (I_{50} is 0.17 μM). Removal of one Zn^{2+} ion per molecule of ADPRT reduced the inhibitory effect of ADPRT(I_{50} is 0.5 μM) but did not abolish it. Based on the similarity to results with site-directed mutation of FI which only slightly modified DNA binding (8), we conclude that in all probability 6-NOBP preferentially destabilizes Zn^{2+} from FI and that FI is essential in the coenzymic polymerase activation of ADPRT by DNA, since DNA-dependent polymerase activity of ADPRT was lost coincidental with the loss of half of the total Zn^{2+}. In agreement with the apparent inequality of the two zinc fingers in ADPRT, preliminary kinetic tests showed a biphasic exchange rate of $^{65}Zn^{2+}$ with two distinct $t_{1/2}$ values (approximately 2.5 min and 7-14 hrs). It is noteworthy that the concentration of ADPRT that produces significant inhibition of template activity (Fig. 2) was well within the physiological concentration of this protein in cell nuclei, which is 1μM (15), therefore the in vitro system (Fig.2) does appear to have cell biological relevance. We also tested the effect of isolated polypeptide components of ADPRT (16) in the M13 DNA template system and found that only the 29 kDa terminal polypeptide, containing both zinc fingers (6-8), was inhibitory (results not shown). These observations support the proposed DNA template binding site of ADPRT to be located at the 29 kDa polypeptide.

Acknowledgments

This work was supported by grants from the US Air Force Office of Scientific Research (AFOSR-89-0221) and the Aaron Diamond Foundation (New York University School of Medicine) and by Octamer, Inc. (Tiburon, CA, USA).

References

1. Hakam, A., McLick, J., Buki, K. G. and Kun, E. (1987) FEBS Lett. 212: 73-78
2. Buki, K.G., Bauer, P.I., Mendeleyev, J., Hakam, A. and Kun, E. (1991) submitted for publication
3. Kirsten, E., Minaga, T. and Kun, E. (1982) FEBS Lett. 139: 117-120
4. Kaufmann, S. H., Brunet, G., Talbot, B., Lamarr, D., Dumas, C.,Shaper, J. H. and Poirier, G. (1991) Exp. Cell Res. 192: 524-535
5. Baron,J., Kawabata, T. T., Redick, J. A., Knapp, S. A. and Guengrich, F.P. (1983) in Rydstrom, J., Montelias, J., and Bengtsson, M. (eds.) Extrahepatic Drug Metabolism and Chemical Carcinogenesis. Elsevier, New York, p. 73
6. Mazen,A., Menessier-de Murcia, J., Molinet, M., Simonin, F. Gradwohl, G., Poirier, G.and de Murcia,G.(1989) Nucleic Acids Res. 17: 4689-4697
7. Menessier-de Murcia, J., Molinet, M., Gradwohl, G., Simonin, F. and de Murcia, G. (1989) J. Mol. Biol. 210: 229-233

8. Gradwohl, G., Menessier-de Murcia, J., Molinet, M., Simonin, F.,Koken, M., Heijmakers, H. J. and de Murcia, G. (1990) Proc. Natl. Acad. Sci. USA 87: 2990-2994

9. Dolle, B., Topner, W. and Neuman, H. G. (1980) Xenobiotica 10: 527-536

10. Martin, R. B. (1979) In Sigel, H. (ed.) Metal Ions in Biological Systems. Dekker, New York, Vol. 9, pp. 15-16

11. Rabenstein, D. L., Guevremont, R. and Evans, C. A. (1979) ibid., pp. 132-134

12. Giedroc,D. P., Keating, K. M., Martin, C. T., Williams, K. R. and Coleman, J. E. (1986) J. Inorg. Biochem. 28: 155-169

13. Pan,T., Freedman, L.P.and Coleman, J. E.(1990) Biochemistry 29: 9218-9225

14. Nobori,T.,Yamanaka, H. and Carson, D. A. (1989) Biochem. Biophys. Res. Commun. 163: 1113-1118

15. Bauer, P.I., Buki, K. G. and Kun, E. (1990) Biochem. J. 270: 17-26

16. Buki, K. G. and Kun, E. (1988) Biochemistry 27: 5990-5995

Conversion of Poly(ADP-ribose) Polymerase Activity to NAD-Glycohydrolase During Retinoic Acid Induced Differentiation of HL60 Cells.

Eva Kirsten, Pal I. Bauer, and Ernest Kun

Laboratory for Environmental Toxicology and Chemistry, Romberg Tiburon Centers, and Department of Chemistry and Biochemistry, San Francisco State University, San Francisco, CA 94132.

Introduction

Cells of the promyelocytic leukemia line HL60 can be induced to differentiate to mature granulocytes by polar-planar compounds (e.g.,DMSO)(1), or by retinoic acid (RA)(2). Involvement of poly ADP-ribose polymerase (ADPRT) in this process had been suggested by work from different laboratories; notably, it was observed (3) that during DMSO induced differentiation the cellular endogenous $(ADPR)_n$ polymer content increased sharply in the final stages of differentiation while simultaneously and in apparent contradiction ADPRT activity (with or without added DNAase I) declined and remained very low in the differentiated state. We now studied the changes of the ADPRT system that occur upon RA induced differentiation of HL60 cells (4).

Results and Discussion

ADPRT activities, as tested in permeabilized (5) cells in absence or presence of DNAase I, declined precipituously during treatment of HL60 cells with $1\mu M$ RA; after 4 days of treatment only approximately 3% of the activities of untreated control cells remained detectable. ADPRT activity in the presence of DNAase I has in many cases been found to parallel cellular ADPRT enzyme content. In contrast, our results of immunological analysis (6,7) of ADPRT protein in HL60 cells indicated no change upon treatment with RA : 1.6 ($\pm$0.5) x 10^5 and 1.5 ($\pm$ 0.5) x 10^5 molecules per cell in untreated HL60 cells and in cells treated with RA for 4 days, respectively. The activity of poly ADP-ribose

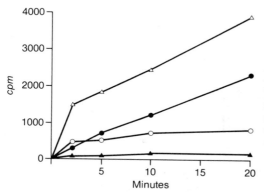

Figure 1. NAD-glycohydrolase and $(ADPR)_n$ polymer forming activities in nuclei from untreated and RA treated HL60 cells (see text.) Reprinted with permission (4)

glycohydrolase (8), tested in nuclei isolated from HL60 cells (12), was also unchanged : 24.0 ±2.9 and 23.7 ±2.7 pmol ADPR released per μg DNA (8 min) prior to and after 4 days of RA treatment, repectively. On the other hand, analysis of cellular contents of endogenously poly-ADP-ribosylated protein, isolated from "citric acid nuclei" (9) by affinity chromatography on phenylboronate columns(10,11), revealed an approximately 3.5-fold increase after RA treatment: from 0.65 ± 0.2 to 2.25 ± 0.4 μg ADP-ribosylated protein per 10^7 cells. These proteins cross-reacted with anti-ADPRT IgG in dot-blot assays (7).

Figure 1 shows the results of TLC analysis of reaction products when HL60 cell nuclei (12) were incubated with radiolabelled NAD. The rate of liberation of free ADPR was increased in nuclei from RA treated cells (open triangles) while $(ADPR)_n$ polymer formation was strongly depressed (filled triangles). For untreated HL60 cells polymerase activity is shown as filled circles and NAD glycohydrolase activity as open circles. In order to test whether this apparent increase in nuclear NAD glycohydrolase by RA treatment could have been the result of contaminating cell wall NADases we also tested the hydrolysis of etheno NAD (13,14) and found similarly negligible degrees of hydrolysis in nuclei from untreated and retinoic acid treated cells, while whole cells did show an increase of etheno NAD hydrolysis after RA treatment,as expected from earlier observations (15) (results not shown). Hence it appears that nuclear NAD glycohydrolase activity of RA treated

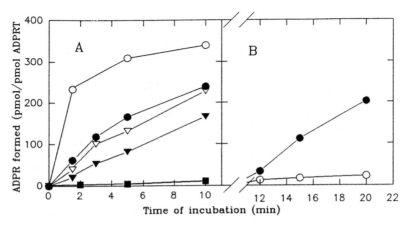

Figure 2. Polymerase and NAD- Glycohydrolase Activities of Purified ADPRT prior to (A) and after (B) Auto-ADP- ribosylation (for explanations see text). Reprinted with permission (4).

HL60 cells is most likely ascribable to an NAD glycohydrolase function of ADPRT. In principle, free ADPR generated during the assay could also arise from turnover of endogenous polymer by poly-ADP-ribose glycohydrolase; an increase in the turnover rate could arise from the emergence of longer $(ADPR)_n$ chains which turn over faster than shorter oligomers, as has been demonstrated *in vitro* by reconstitution models (16). However, this mechanism is unlikely to play a major role in the case of RA induced differentiation, since the very low polymer generating activity in RA treated HL60 cells appears to impose a "bottleneck" limit onto this turnover cycle.

Figure 2 shows the poly-ADP-ribose polymerase activities (open symbols) and the NAD glycohydrolase activities (filled symbols) of isolated purified (17) ADPRT in the native state (A) and after auto-ADP-ribosylation by pre-incubation with NAD (B), as monitored by TLC analysis of reaction products. All circles represent activities in the presence of histone (1.9 μM H3) and octameric deoxyribonucleotide (750 nM)(18). Part A also shows the activities in the presence of octameric deoxyribonucleotide only (triangles) and activities in the absence of both histone and octamer (squares, bottom curves). As figure 2B shows, auto-ADP-ribosylated ADPRT functioned solely as an NAD glycohydrolase. The changes in ADPRT and NAD-glycohydrolase activities described here appear to be specific for the granulocyte-oriented differentiation pathway: when HL60 cells were induced to differentiate to macrophage-like cells by phorbol ester (cf.19), or to monocytes by sodium butyrate (20), we could not observe similar phenomena (unpublished experiments).

On the basis of our present results we propose that during RA induced differentiation of HL60 cells to granulocytes endogenous ADPRT becomes functionally an NAD glycohydrolase following *in situ* auto-ADP-ribosylation.

Acknowledgments. This work was supported by a grant from the Air Force Office of Scientific Research (AFOSR-89-0231).

References

1. Collins,S.J.,Ruscetti,F.W.,Gallagher,R.E.,and Gallo,R.C.(1978)   Proc. Natl. Acad.Sci.(USA)75, 2458-2462.
2. Breitman,T.R.,Selonik,S.E.,andCollins,S.J.(1980)Proc.Nat.Acad. Sci. (USA) 77, 2936-2940.
3. Kanai,M., Miwa,M., Kondo,T., Tanaka,Y., andSugimura,T.(1982) Biochem. Biophys.Res.Commun.105, 404-411.
4. Kirsten,E.,Bauer,P.I.,and Kun,E.(1991) Exper.Cell Res.194,1-8.
5. Surowy,C.S.and Berger,N.(1983) Biochim.Biophys.Acta 740,8-18.
6. Yamanaka,H.,Penning,C.A.,Willis,E.H.,Wasson,D.B.,and Carson,D.A. (1988) J.Biol.Chem. 263, 3879-3883.
7. Ludwig,A.,Behnke,B.,Holtlund,J.,and Hilz,H. (1988) J.Biol.Chem. 263, 6993-6999.
8. Tanuma,S., Kawashima,K.,and Endo,H.(1986)J.Biol.Chem.261,965- 969.
9. Adamietz,A.and Rudolph A.(1984) J.Biol.Chem.259, 6841-6846.
10.Jacobson,M.K.,Payne,D.M.,Alvarez-Gonzales,R.,Juarez-Salinas,H., Sims,J.L., and Jacobson,E.L. (1984) in Methods in Enzymology (Wold,F.and Moldave,K.,Eds.), Vol.106,p.491, Academic Press, New York.
11.Jackowski,G.and Kun,E.(1983) J.Biol.Chem.158, 12587-12593.
12.Ausubel,F.M.,Brent,R.,Kingston,R.E.,Moore,D.,Smith,J.A.,Seidman, G. and Struhl,K.U.(Eds.)(1985) Current Protocols in Molecular Biology, Section 4,10,2, Wiley Interscience, New York.
13.Muller,H.M.,Muller,C.D.,and Schuber,F.(1983)Biochem.J.212,459-   464.
14.Masmoudi,A.and Mandel,P.(1987) Biochemistry 26,1965-1969.
15.Hemmi,H. and Breitman,T.R. (1982) Biochem.Biophys.Res.Commun.109, 669-674.
16. Menard,L.,Thibault,L.,and Poirier,G.G.(1990) Biochim.Biophys.Acta 1049, 45-58.
17.Buki,K.G., Kirsten,E.,and Kun,E.(1987) Analyt.Biochem. 167,160-166.
18.Hakam,A.,McLick,J.,Buki,K.G.,and Kun,E.(1987) FEBS Lett. 212,73-78.
19.Rovera,G.,Santoli,D.,and Damsky,C. (1979)Proc.Nat.Acad.Sci.(USA)76,27792783
20.Boyd,A.W.and Metcalf,D.(1984) Leukemia Res.8,27-43.

Improved Assays of Poly(ADP-Ribose) Metabolizing Enzymes

Takahiro Kido, Jun'ichi Inagawa, Marek Banasik, Isao Saito., and Kunihiro Ueda

Introduction

During the study of tissue distribution of poly(ADP-ribose) metabolizing enzymes, *i.e.*, poly(ADP-ribose) synthetase, poly(ADP-ribose) glycohydrolase, and ADP-ribosyl protein lyase (1), we found that their substrates and products, that are derivatives of ADP-ribose, disappeared rapidly from the reaction mixture containing tissue homogenates or extracts. As judged by the fact that 5'-AMP, a known inhibitor of phosphodiesterase (2), diminished remarkably the disappearance or degradation of ADP-ribosyl compounds, a phosphodiesterase activity seemed to be responsible for the phenomenon. In this chapter, we report the results of our studies on this degradative enzyme activity and the discovery of potent inhibitors against it.

Methods

ADP-ribosyl protein lyase and phosphodiesterase activities were assayed by measuring conversion of acid-insoluble ^{14}C of [*Ade*-^{14}C]ADP-ribosylhistone H2B to acid-soluble ADP-3"-deoxypentos-2"-ulose (ADP-X) (1, 3) or 5'-AMP, respectively (Fig. 1). The two products were differentiated from each other or else, based on the behavior of elution from an AG 1 column, as described below.

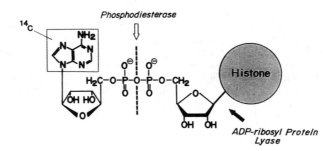

Fig. 1. Assays of ADP-ribosyl protein lyase and phosphodiesterase activities.

The standard mixture common to both assays contained 100 mM potassium phosphate buffer (pH 6.0), 1 mM dithiothreitol, 6.4 μM [^{14}C]ADP-ribosyl H2B (4.7 cpm/pmol), and enzyme in 100 μl. After incubation for 10 min at 37°C, the reaction was stopped by addition of 20% (final) CCl$_3$COOH, and the mixture, neutralized with Tris-HCl buffer and diluted with water, was applied to an AG 1-formate column (0.8 X 3.0 cm). The column was rinsed with water, and then eluted sequentially with 0.05 M, 0.6 M, and (four times) 6 M HCOOH. Aliquots of the eluates were examined for ^{14}C by the liquid scintillation method.

Results and Discussion

Incubation of [*Ade*-^{14}C]ADP-ribosyl histone H2B with a homogenate or extract (low-speed supernatant) of rat liver made some amount of ^{14}C acid-soluble, suggesting a release of the labelled moiety from histone. Analysis with AG 1 column chromatography (Fig. 2) and paper chromatography (Fig. 3) revealed that the released material comprized several distinct products. Among them, adenosine and 5'-AMP were tentatively identified in the 0.05 M and 0.6 M HCOOH eluates, respectively. The main fraction of ^{14}C, eluting at 0.05 M HCOOH, separated into two peaks upon paper chromatography. Neither of the peaks coincided with marker ADP-ribose, ADP-X, ADP, 5'-AMP, 5'-IMP, adenosine, inosine, or adenine. Although their structures remain to be identified, the chromatographic behavior suggested their having not more than one phosphate. Because both ADP-ribose, a product of non-enzymatic hydrolysis, and ADP-X, a product of ADP-ribosyl protein lyase action, were recovered in the 6 M HCOOH eluate (Fig. 2), the ^{14}C eluting at lower concentrations of HCOOH was indicative of a phosphodiesterase action on the ADP-ribosyl residue. An apparent decrease, rather than an increase, in ^{14}C eluting at 6 M HCOOH in the presence of tissue extract supported this notion.

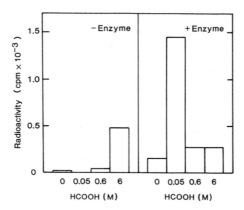

Fig. 2. AG 1 column chromatography of degradation products of ADP-ribosyl histone H2B by rat liver extract.

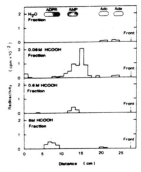

Fig. 3. Paper chromatography of AG 1 column fractions of degradation products of ADP-ribosyl histone H2B by rat liver extract.

Furthermore, we found that, in the assay of poly(ADP-ribose) synthetase or poly(ADP-ribose) glycohydrolase, NAD, poly(ADP-ribose), and ADP-ribose were rapidly degraded to 5'-AMP (and adenosine) in relatively large amount, and that a phosphodiesterase activity (expressed as mol/min/mg protein) acting on

Table 1. Effects of EDTA and/or AMP and IMP on hydrolysis of ADP-ribose or ADP-ribosyl histone H2B by rat liver extract.

Addition(s)			Hydrolysis (%) of	
			ADP-ribose	ADP-ribosyl H2B
(−)			100	100
EDTA	1	mM	53	-
	3	mM	39	-
	5	mM	29	-
AMP	1	mM	18	-
	3	mM	9	-
	5	mM	8	-
EDTA	1	mM	1	2
+ AMP	1	mM		
IMP	1	mM	6	-
	3	mM	4	-
	5	mM	1	< 1

ADP-ribose, bis(*p*-nitrophenyl)phosphate or *p*-nitrophenyl-dTMP proved to be about ten times higher than the ADP-ribosyl protein lyase activity in rat liver extract. Under such conditions that both substrate and product were degraded by contaminating enzyme(s), assays could not be accurate nor reproducible. We, therefore, performed a survey of inhibitors of phosphodiesterase of animal origin, using [^{14}C]ADP-ribose as the substrate. Starting from a reported inhibitor, *i.e.*, 5'-AMP (2), we came to a descovery of two (sets of) potent inhibitors, that is, EDTA + 5'-AMP and 5'-IMP (Table 1). EDTA or 5'-AMP, as used separately, was not strong enough, even at 5 mM, to abolish the phosphodiesterase activity in rat liver extract, but their combination, at 1 mM each, inhibited 99% of ADP-ribose or 98% of ADP-ribosyl H2B hydrolysis as estimated by ^{14}C eluting at 0.05 M and 0.6 M HCOOH from the AG 1 column. Almost identical or stronger inhibition was effected by 5 mM IMP. Although EDTA (1 mM) + AMP (1 mM) and IMP (5 mM) were comparable in inhibition of phosphodiesterase, the latter seemed to be more useful for application to other enzyme assays; our preliminary experiment indicated that the former combination brought about 50% inhibition of ADP-ribosyl protein lyase activity and more of poly(ADP-ribose) synthetase activity, whereas the latter exhibited essentially no effect on the lyase activity and only about 15% inhibition of the synthetase activity.

In the presence of 5 mM IMP, the reaction of ADP-ribosyl histone H2B cleavage proceeded linearly for, at least, several minutes in the presence of rat

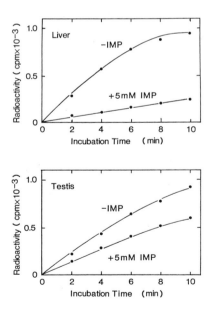

Fig. 4. Time course of ADP-ribosyl protein lyase reaction in the presence and absence of IMP.

liver extract (Fig 4). Analysis of reaction products by column and paper chromatographies showed that the main product behaved closely, but not identically, with marker ADP-ribose, a feature characteristic of ADP-X. A minor product, from nonenzymatic hydrolysis, was identified as ADP-ribose (*cf.* Fig 2). The difference between the values with and without IMP may indicate the products by the action of phosphodiesterase.

Based on these observations, we chose 5 mM IMP as an inhibitor of phosphodiesterase, and, using this inhibitor, examined tissue distribution of poly-(ADP-ribose) metabolizing enzyme activities. Preliminary results indicated that, among five major organs examined (thymus, liver, kidney, pancreas, and testis), the thymus had the highest, and the pancreas the lowest activity of ADP-ribosyl protein lyase. By contrast, the pancreas and the kidney had the highest, while the liver the lowest activity of phosphodiesterase. The highest activity of poly-(ADP-ribose) synthetase was found in the pancreas, followed by the thymus and the testis.

The new assay systems containing IMP have been successfully applied also to extracts of various cultured cells.

References

1. Oka, J., Ueda, K., Hayaishi, O., Komura, H., Nakanishi, K. (1984) J. Biol. Chem. 259: 986-995
2. Futai, M., Mizuno, D. (1967) J. Biol. Chem. 242: 5301-5307
3. Komura, H., Iwashita, T., Naoki,H., Nakanishi, K., Oka, J., Ueda, K., Hayaishi, O. (1983) J. Am. Chem. Soc. 105: 5164-5165

Specific Inhibitors of Poly(ADP-Ribose) Synthetase

Marek Banasik, Hajime Komura, and Kunihiro Ueda

Inhibitors of poly(ADP-ribose) synthetase are useful tools for the research on biological roles of poly(ADP-ribose). Many compounds have been shown to inhibit the activity of poly(ADP-ribose) synthetase in *in vitro* systems, and several of them have been used *in vivo* (1-12). Most of these inhibitors, however, including 3-aminobenzamide, nicotinamide, and thymidine, accompany significant side effects on the cell (1, 13, 14), and the results obtained with them remain, more or less, inconclusive. Therefore, researchers in this field have awaited for more specific and potent inhibitors.

In this chapter, we report *in vitro* effects of a large variety of compounds on the activity of poly(ADP-ribose) synthetase purified from bovine thymus. Furthermore, by comparing the effects on poly(ADP-ribose) synthetase and monomer/arginine-specific ADP-ribosyltransferase [mono(ADP-ribosyl)transferase] (EC 2. 4. 2. 31) from hen heterophils (15, 16), we show that many of the inhibitors newly found are highly specific for poly(ADP-ribose) synthetase. A preliminary report of this study was published elsewhere (10).

Tables 1-4 summarize results of our large-scale survey of inhibitors of poly(ADP-ribose) synthetase. Compounds are classified according to chemical structures. Table 1 (and part of Table 4) contains derivatives of known inhibitors, that is, benzamide, acetophenone, nicotinamide, and pyrimidine. Other Tables contain new types of inhibitors, although the compounds listed in Tables 2 and 3, having a carbonyl group built in a polyaromatic skeleton as an N-substituted carbamoyl group or a C- extended acetyl group, may be considered as analogues of benzamide or acetophenone, respectively. Table 4 contains other compounds that are not structurally related to the aforementioned inhibitors.

All of newly found potent inhibitors, except for benzamides, were polyaromatic heterocyclics. Among them, the derivatives of 1,8-naphthalimide, 6(5*H*)-phenanthridinone, and isoquinoline were the most potent inhibitors so far found, followed by those of quinazoline, carsalam, phthalazine, and chromone. A common structural feature for all of these potent inhibitors is a carbonyl group built in a second ring system that is conjugated with a six-membered aromatic ring. The carbonyl oxygen atom seems to serve as an electron donor, and the carbon atom as an electron acceptor in the interaction with the enzyme molecule (10).

Table 1. Effects of benzamide, acetophenone, and nicotinamide derivatives and analogues on poly(ADP-ribose) synthetase activity[a]

Compound	IC_{50} (µM)	Inhibition (%) at	
		1 mM	5 mM
1. BENZAMIDE DERIVATIVES			
Benzamide	22	92	98
2-Acetamidobenzamide	1000	50	71
3-Acetamidobenzamide	12	97	100
4-Acetamidobenzamide*	- - #	6	26
2-Aminobenzamide (anthranilamide)	650	56	74
3-Aminobenzamide	33	88	97
4-Aminobenzamide	1800	37	66
2-Bromobenzamide	2900	32	57
3-Bromobenzamide*	55	87	97
4-Bromobenzamide*	2200	36	61
2-Chlorobenzamide	1000	50	73
3-Chlorobenzamide	22	92	98
4-Chlorobenzamide	300	67	86
3-(N,N-Dimethylamino)benzamide	120	81	95
4-(N,N-Dimethylamino)benzamide*	- -	4	- -
2-Fluorobenzamide*	120	85	97
3-Fluorobenzamide	20	92	98
4-Fluorobenzamide*	200	71	88
2-Hydroxybenzamide (salicylamide)	82	78	90
3-Hydroxybenzamide	9.1	95	99
4-Hydroxybenzamide	280	74	90
2-Methoxybenzamide	20	89	96
3-Methoxybenzamide	17	92	97
4-Methoxybenzamide	1100	47	72
2-Methylbenzamide	1500	43	67
3-Methylbenzamide	19	95	99
4-Methylbenzamide	1800	37	67
2-Nitrobenzamide	- -	15	37
3-Nitrobenzamide	160	74	90
4-Nitrobenzamide*	- -	2	16
Phthalamide (1,2-benzenedicarboxamide)	1000	50	69
m-Phthalamide (isophthalamide)*	50	89	98
p-Phthalamide (terephthalamide)*	- -	3	- -
Phthalamidic acid (2-carboxybenzamide)	- -	7	31
3-Acetamidosalicylamide*	2000	34	64
5-Acetamidosalicylamide	45	84	96
5-Aminosalicylamide	100	79	92
5-Chlorosalicylamide*	190	77	86
N-Phenylsalicylamide*	- -	–1+	> 9$
2,4-Dichlorobenzamide*	- -	11	22
2,6-Dichlorobenzamide	- -	8	28
2,6-Difluorobenzamide	180	73	90
2,4-Dimethoxybenzamide*	- -	21	46
3,5-Dimethoxybenzamide*	1200	46	70

3,5-Dinitrobenzamide	2500	36	64
N,N-Dimethylbenzamide	- -	- 4	- 5
3-Nitrosalicylamide*	1600	41	73
3,5-Dibromosalicylamide*	560	62	93
cf. Acetamide	- -	0	1
cf. 3-Aminobenzamidine dihydrochloride	- -	3	23
cf. Benzaldehyde*	- -	3	21
cf. Cyclohexanecarboxamide	620	59	77
cf. Phenylacetamide	- -	4	0
cf. p-Phthalic acid (terephthalic acid)*	- -	3	- 88
cf. Salicylaldoxime	- -	6	16
cf. Thiobenzamide	620	57	77
2. ACETOPHENONE DERIVATIVES			
Acetophenone*	2300	35	64
o-Acetamidoacetophenone*	- -	2	6
m-Acetamidoacetophenone	930	65	81
p-Acetamidoacetophenone*	- -	2	13
o-Aminoacetophenone*	- -	0	- 1
m-Aminoacetophenone	1900	38	69
p-Aminoacetophenone*	- -	2	4
o-Hydroxyacetophenone*	- -	20	42
m-Hydroxyacetophenone	600	65	81
p-Hydroxyacetophenone	- -	16	37
3. NICOTINAMIDE DERIVATIVES AND ANALOGUES			
Nicotinamide	210	72	87
α-Picolinamide	250	68	85
Isonicotinamide	990	51	74
6-Aminonicotinamide	1100	48	72
1-Methylnicotinamide chloride	3800	28	55
5-Methylnicotinamide	350	66	82
N '-Methylnicotinamide	- -	1	14
cf. Nipecotamide	- -	1	1
cf. Thionicotinamide	1800	43	59
Picolinic acid	- -	6	34
Nicotinic acid	- -	3	9
Isonicotinic acid	- -	- 2	9
Trigonelline hydrochloride	- -	11	3 0

a Enzyme activity was assayed as described previously (11,12).

* 2% (final) Me_2SO.

Not done or not measurable.

+ Stimulation.

$ Minimum value estimated under the condition of limited solubility.

In general, the presence of a carbonyl, carbamoyl, or acetyl group conjugated with an aromatic ring is important to, but not indispensable for, the inhibitory action. For example, phthalazine, quinazoline, and norharman lack any of these groups but have a C=N double bond in the skeleton in analogous positions. In addition, an aromatic ring is contributive to, but not indispensable for, the inhibition, as exemplified by cyclohexanecarboxamide and trans-1-decalone. More details of the structure-activity relationship will be published elsewhere.

Table 2. Effects of various benzamide equivalents and their derivatives or analogues on poly(ADP-ribose) synthetase activity

Compound	IC$_{50}$ (μM)	Inhibition (%) at	
		1 mM	5 mM
1. QUINAZOLINE DERIVATIVES			
Quinazoline (1,3-benzodiazine)	2000	38	63
4-Hydroxyquinazoline	9.5	92	98
2-Mercapto-4(3*H*)-quinazolinone*	44	94	- -
2-Methyl-4(3*H*)-quinazolinone*	5.6	99	100
2,4(1*H*,3*H*)-Quinazolinedione (benzoyleneurea)*	8.1	98	> 98
2-Trichloromethyl-4(3*H*)-quinazolinone*	2200	32	> 74
6,7-Dimethoxyquinazoline-2,4-dione*	- -	22	> 34
cf. Quinoxaline (1,4-benzodiazine)*	- -	1	8
2. CARSALAM DERIVATIVES			
Carsalam (carbonylsalicylamide)*	460	66	88
Chlorthenoxazin*	8.5	98	> 98
6-Acetamidocarsalam*	- -	6	> 35
8-Acetamidocarsalam*	1400	43	77
8-Nitrocarsalam	- -	15*	22**
cf. Isatoic anhydride*	< 3900	4	> 60
3. PHTHALAZINE DERIVATIVES			
Phthalazine (2,3-benzodiazine)	150	78	91
1(2*H*)-Phthalazinone*	12	98	100
1,4-Dihydrazinophthalazine hydrochloride**	- -	2	- 7
Phthalhydrazide*	30	95	99
3-Aminophthalhydrazide (luminol)*	23	96	100
4-Aminophthalhydrazide (isoluminol)*	290	70	93
3-Nitrophthalhydrazide*	72	90	99
4-Nitrophthalhydrazide*	510	63	91
cf. Maleic hydrazide	- -	3	13
cf. Phthalimide potassium salt	- -	-1	11
cf. Succinic hydrazide*	- -	0	-1
4. ISOQUINOLINE DERIVATIVES AND ANALOGUES			
Isoquinoline (2-benzazine)*	- -	18	47
1-Hydroxyisoquinoline (isocarbostyril)	7.0	98	100
1,5-Dihydroxyisoquinoline	0.39	100	100
6(5*H*)-Phenanthridinone*	0.30	> 99	> 99
2-Nitro-6(5*H*)-phenanthridinone*	0.35	> 93	- -
cf. Phenanthrenequinone*	- -	> 30	> 43
1,8-Naphthalimide*	1.4	98	> 99
4-Amino-1,8-naphthalimide*	0.18	> 99	> 99
N-(2-Chloroethyl)-1,8-naphthalimide*	< 1800	> 37	> 72
N-Hydroxynaphthalimide	450	> 96	> 99
cf. 1,8-Naphthalic anhydride*	- -	> 38	- -
5. OTHERS			
10,11-Dihydrodibenz[*b*,*f*][1,4]-oxazepin-11-one*	< 2300	34	> 62

** 10% (final) Me$_2$SO.

Table 3. Effects of various acetophenone equivalents and their derivatives or analogues on poly(ADP-ribose) synthetase activity

Compound	IC_{50} (μM)	Inhibition (%) at 1 mM	Inhibition (%) at 5 mM
1. QUINOLINE DERIVATIVES			
4-Hydroxyquinoline*	80	86	97
8-Hydroxyquinoline*	- -	11	> 42
2-Methylquinoline (quinaldine)*	- -	3	−1
3-Quinolinecarbonitrile*	- -	9	> 37
3-Quinolinecarboxamide*	- -	20	44
2-Quinolinecarboxylic acid (quinaldic acid)*	- -	29	37
3-Quinolinecarboxylic acid*	- -	26	35
4-Hydroxy-2-methylquinoline*	74	87	98
4-Hydroxy-3-quinolinecarboxamide*	- -	39	> 47
Kynurenic acid*	670	61	83
Xanthurenic acid*	190	77	99
2. CHROMONE DERIVATIVES			
Chromone-2-carboxylic acid	560	62	85
Flavone (2-phenylchromone)*	22	88	- -
2-Methylchromone	45	87	96
2-Methyl-3-acetylchromone*	- -	0	16
Chrysin (5,7-dihydroxyflavone)*	- -	> 11	> 42
Apigenin (4',5,7-trihydroxyflavone)*	< 1500	> 42	> 68
Apiin*	- -	34	> 40
β-D-Glucopyranosylapigenin	- -	36*	49**
cf. 4-Chromanone*	720	55	79
3. COUMARIN DERIVATIVES			
Coumarin (1,2-benzopyrone)*	2800	33	60
6-Aminocoumarin*	850	53	82
4-Hydroxycoumarin*	570	62	89
Novobiocin	2200	16	90
4. NAPHTHALENE DERIVATIVES			
1,3-Dihydroxynaphthalene*	1300	43	79
1,4-Naphthoquinone (1,4-naphthalenedione)*	250	87	> 97
Vitamin K_5*	< 1300	> 31	> 92
2-Hydroxy-1,4-naphthoquinone (lawsone)*	330	74	97
5-Hydroxy-1,4-naphthoquinone (juglone)*	250	> 93	- -
2-Methyl-1,4-naphthoquinone (vitamin K_3)*	420	> 70	> 84
2-Amino-3-chloro-1,4-naphtoquinone*	820	> 53	> 80
2,3-Dichloro-1,4-naphthoquinone*	260	> 62	- -
5-Hydroxy-2-methyl-1,4-naphthoquinone*	700	> 62	> 90
Vitamin K_1**	520	> 72	- -
cf. 1,4-Benzoquinone	400	78	> 82
cf. trans-1-Decalone*	4300	22	52
cf. α-Tetralone*	310	67	87
cf. β-Tetralone*	- -	16	35
cf. 1-Indanone*	810	53	75

Table 4. Effects of various compounds on poly(ADP-ribose) synthetase activity

Compound	IC$_{50}$ (μM)	Inhibition (%) at	
		1 mM	5 mM
1. PYRIDINE DERIVATIVES			
3-Acetylpyridine	- -	16	35
4-Hydroxypyridine	2300	34	64
Pyridoxal 5-phosphate	4250	13	56
2. PYRIMIDINE DERIVATIVES			
Pyrimidine*	- -	0	2
4(*3H*)-Pyrimidone (4-hydroxypyrimidine)	- -	6	18
2-Thiouracil*	- -	5	30
Uracil	- -	3	33
5-Aminouracil*	- -	8	> 42
Barbituric acid	- -	3	24
5-Bromouracil	160	74	88
5-Chlorouracil	270	71	88
5-Fluorouracil	- -	18	46
5-Iodouracil	71	84	96
5-Nitrouracil*	430	65	90
Orotic acid (uracil-6-carboxylic acid)	- -	6	24
6-Propyl-2-thiouracil*	- -	3	24
Thymine	290	73	86
Uridine	- -	11	34
Uridine 5'-monophosphate disodium salt	- -	2	12
Alloxan	- -	14	37
5-Bromouridine	210	75	90
5-Iodouridine	43	89	98
5-Methylbarbituric acid	- -	3	5
5-Phenylbarbituric acid*	- -	0	11
Thymidine	180	77	89
Barbital	- -	− 1	4
cf. 5,6-Dihydrouracil	- -	1	5
3. NORHARMAN DERIVATIVES			
Norharman hydrochloride	4700	21	51
Harman hydrochloride	- -	17	- -
Harmine hydrochloride	< 3500	> 18	> 53
cf. Harmaline hydrochloride	- -	7	< − 50
cf. Reserpine**	790	> 66	- -
4. MISCELLANEOUS			
Chlorpromazine hydrochloride	- -	0	62
2,2'-Dipyridyl	- -	− 5	− 8
EDTA disodium salt	- -	14	41
Inosine	- -	9	30
5'-Inosinic acid sodium salt	- -	5	24
Nalidixic acid*	- -	7	37
Oxolinic acid*	- -	< − 2	> 26
1,10-Phenanthroline*	- -	− 102	− 210

Specificity of action is critical for enzyme inhibitors in application to *in vivo* studies. Table 5 shows a comparison of inhibitory effects of selected compounds on poly(ADP-ribose) synthetase and mono(ADP-ribosyl)transferase from hen heterophils, indicating that most of potent inhibitors of poly(ADP-ribose) synthetase act much more strongly on this enzyme than on arginine-specific mono(ADP-ribosyl)transferase. Although effects on other endogenous mono(ADP-ribosyl)transferases remain to be investigated, a big difference in the IC_{50} values for the two enzymes indicates that the use of these inhibitors *in vivo* at appropriate concentrations would clearly differentiate effects on poly(ADP-ribosyl)ation from mono(ADP-ribosyl)ation.

Table 5. Comparison of inhibitory potencies of various compounds on mono(ADP-ribosyl)transferase and poly(ADP-ribose) synthetase

Compound	IC_{50} (μM)		
	Mono-(ADP-ribosyl)-transferase (a)	Poly-(ADP-ribose) synthetase (b)	(a/b)
1,5-Dihydroxyisoquinoline	890	0.39	2282
6(*5H*)-Phenanthridinone*	> 500	0.30	> 1667
3-Hydroxybenzamide	9000	9.1	989
4-Amino-1,8-naphthalimide*	> 100	0.18	> 556
4-Hydroxyquinazoline	2600	9.5	274
2-Nitro-6(*5H*)-phenanthridinone*	83	0.35	237
1-Hydroxyisoquinoline	1500	7.0	214
Benzamide	4500	22	205
2-Methyl-4(*3H*)-quinazolinone*	1100	5.6	196
5-Iodouridine	7200	43	167
Chlorthenoxazin*	> 1200	8.5	> 141
2-Methylchromone	6300	45	140
1(*2H*)-Phthalazinone*	510	12	43
Benzoyleneurea*	200	8.1	25
1,8-Naphthalimide*	20	1.4	14
Flavone*	260	22	12
Oleic acid (C18:1, *cis*-9)*	200	82	2.4
Palmitoleic acid (C16:1, *cis*-9)*	200	95	2.1
Linoleic acid (C18:2, *cis*-9,12)*	90	48	1.9
γ-Linolenic acid (C18:3, *cis*-6,9,12)*	180	120	1.5
Arachidonic acid (C20:4, *cis*-5,8,11,14)*	66	44	1.5
Linolenic acid (C18:3, *cis*-9,12,15)*	110	110	1.0
Menadione sodium bisulfite	440	720	0.61
Vitamin K_3 (menadione)*	120	420	0.29
Novobiocin	280	2200	0.13

Acknowledgements. The authors would like to thank Prof. M. Shimoyama (Shimane Medical University) for kind donation of purified mono(ADP-ribosyl)transferase used in this study. This work was supported in part by a postdoctoral fellowship of the Japan Society for the Promotion of Science for Foreign Researchers in Japan.

References

1. Ueda K., Kawaichi, M., and Hayaishi, O. (1982) in ADP-Ribosylation Reactions: Biology and Medicine (Hayaishi, O., and Ueda, K., eds) pp. 117-155, Academic Press, New York
2. Ueda, K., and Hayaishi, O. (1985) Annu. Rev. Biochem. 54, 73-100
3. Althaus, F. R. (1987) in ADP-Ribosylation of Proteins (Althaus, F. R., and Richter, C., eds) pp. 24-29 Springer-Verlag, Berlin
4. Purnell, M. R., and Whish, W. J. D. (1980) Biochem. J. 185, 775-777
5. Purnell M. R., and Whish, W. J. D. (1980) Biochem. Soc. Trans. 8, 175-176
6. Sims, J. L., Sikorski, G. W., Catino, D. M., Berger, S. J., and Berger, N. A. (1982) Biochemistry 21, 1813-1821
7. Cantoni, O., Sestili, P., Spadoni, G., Balsamini, C., Cucchiarini, L., and Cattabeni, F. (1987) Biochem. Int. 15, 329-337
8. Rankin, P. W., Jacobson, E. L., Benjamin, R. C., Moss, J., and Jacobson, M. K. (1989) J. Biol. Chem. 264, 4312-4317
9. Sestili, P., Spadoni, G., Balsamini, C., Scovassi, I., Cattabeni, F., Duranti, E., Cantoni, O., Higgins, D., and Thomson, C. (1990) J. Cancer Res. Clin. Oncol. 116, 615-622
10. Banasik, M., Komura, H., Saito, I., Abed, N. A. N., and Ueda, K. (1989) in ADP-Ribose Transfer Reactions: Mechanisms and Biological Significance (Jacobson, M. K., and Jacobson, E. L., eds) pp.130-133, Springer-Verlag, New York
11. Banasik, M., Komura, H., and Ueda, K. (1990) FEBS Lett. 263, 222-224
12. Banasik, M., Komura, H., Shimoyama, M., and Ueda, K. (1991) submitted
13. Althaus, F. R. (1987) in ADP-Ribosylation of Proteins (Althaus, F. R., and Richter, C., eds) pp. 68-70, Sprinter-Verlag, Berlin
14. Cleaver, J. E., Milam, K. M., and Morgan, W. F. (1985) Radiat. Res. 101., 16-28
15. Tanigawa, Y., Tsuchiya, M., Imai, Y., and Shimoyama, M. (1984) J. Biol. Chem. 259, 2022-2029
16. Shimoyama, M. (1991) Seikagaku 63, 123-129 (in Japanese)

AUTOMODIFICATION AND NADASE ACTIVITY OF POLY(ADP-RIBOSE) POLYMERASE

Jean Lagueux, Yvan Desmarais, Luc Ménard, Sylvie Bourassa and
Guy G. Poirier

Laboratoire du Métabolisme du Poly(ADP-ribose), Endocrinologie Moléculaire,
Centre de Recherche du CHUL et Université Laval, Quebec G1V 4G2, Canada.

INTRODUCTION

Poly(ADP-ribose) polymerase is a nuclear enzyme which catalyses the hydrolysis of NAD^+ to ADP-ribose and nicotinamide (1-3). Most of this ADP-ribose is covalently attached to nuclear proteins as poly(ADP-ribose) (4-5). The polymerase can carry out branching of ADP-ribose residues on average at every 20 residues (6). In this paper, we are analyzing the kinetic properties of poly(ADP-ribose) polymerase as a function of automodification and also the distribution of these on the various domains of the polymerase.

RESULTS

We have analyzed the production of nicotinamide (total NADase) as a function of polymer synthesis. This data shows that the total NADase diminishes as the polymerase gets more modified (Fig. 1). We find that both

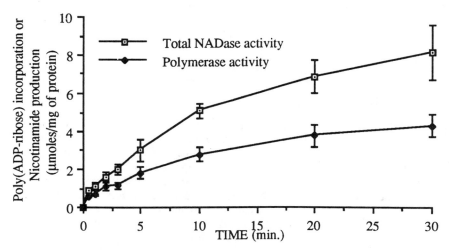

Figure 1 Determination of nicotinamide and poly(ADP-ribose) levels as a function of time. Poly(ADP-ribose) polymerase was added to the standard incubation media (60 μl) containing 0.2 μCi-1.7 μCi[^{32}P]NAD$^+$ or 5 μCi [^{3}H](nicotinamide) NAD$^+$. At different times, proteins were precipitated with 25% TCA, the polymer was evaluated by filtration on glass fiber filters and the level of nicotinamide was measured in the TCA soluble fraction by HPLC on SAX-Partisil column.(figure reproduce from ref 8)

the polymerase and the NADase activity drops as the polymerase modify 10 sites (Fig. 2). Furthermore, our data indicates that it takes over 30 minutes to label all the sites on polymerase, thus indicating that evaluating if the polymerase has a processive or distributive mode of elongating the polymer is very complicated since at all times, polymers are added to the enzyme.

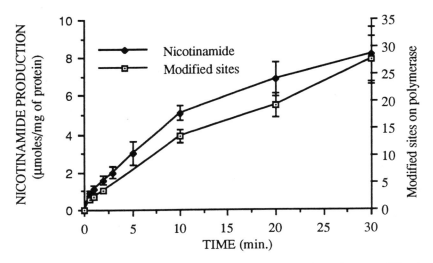

Fig 2.   Comparison between NADase activity and number of automodification sites as a function of time.(figure reproduce from ref 8)

Furthermore, Western blotting experiments indicated that all times, all the polymerase molecules were modified (data not shown). The NADase activity of the polymerase remains the same in the presence of glycohydrolase and a decrease in the number of labeled sites are observed (7). Finally, we have analyzed the distribution of sites on the polymerase when the polymerase is fully modified. This was done by removing the polymers with poly(ADP-ribose) glycohydrolase (8) and the sites were analyzed by selective proteolytic cleavage with chymotrypsin and papain. The distribution of the functional domains was then done by autoradiography of the domains and their identification with domain specific monoclonal antibody CII/10 (DNA binding domain) (Fig. 3) and CI/2 (NAD binding domain) and confirmed with a rabbit polyclonal antibody. It was observed that polymers were found to be attached to the DNA binding domain, automodification domain and NAD binding domain.

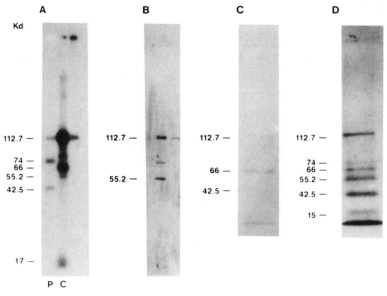

Fig. 3. Separation of [^{32}P] poly(ADP-ribosyl)ated and poly(ADP-ribose) polymerase domains by proteolytic digestion. Poly(ADP-ribose) polymerase was poly(ADP-ribosyl)ated at 200 μM NAD$^+$ for 30 min and was incubated with poly(ADP-ribose) glycohydrolase. After 5 min of hydrolysis with the glycohydrolase the PARP, was digested with papain and chymotrypsin. Panel A shows an autoradiography of glycohydrolase treated automodified poly(ADP-ribose) polymerase with chymotrypsin and papain. Panel B shows a western blot of the same sample which was revealed by alkaline phosphate method with monoclonal antibody CI/2 which recognizes the NAD$^+$ binding domain (8). Panel C shows a western blot of the same automodified polymer fragments shown in panels A and B but revealed with monoclonal antibody CII/10 for the DNA binding domain. Panel D shows a Western blot of proteolytic digest of automodified polymerase revealed with a polyclonal antibody. Gel electrophoresis on a gradient gel of 5 to 18% polyacrylamide was performed..(figure reproduce from ref 8)

DISCUSSION

In this paper, we have found that there is at the enzyme concentration that we have used a low level of alternate NADase activity. Furthermore, it is observed that after few sites are modified, on the polymerase, there is an inhibition of the polymerase both NADase activity and polymer adding activity. Finally, it is observed that there are modification sites outside the automodification possibly 7-8 in each domain. This confirms the data obtained by Simonin et al. (9) for the 54 kdA domain, where this domain could carry out autonomous polymer synthesis.

ACKNOWLEDGEMENTS

We want to thank Elaine Leclerc for her excellent secretarial assistance. This work was supported by The Natural Sciences and Engineering Research Council of Canada, The Medical Research Council of Canada and the programme pour la formation de chercheurs et d'aide à la recherche (Québec).

REFERENCES

1. Althaus, F.R. and Richter, C. (1987) In Molecular biology, biochemistry and biophysics, vol. 37. Edited by M. Solioz. Springer-Verlag, Berlin, Heidelberg, pp. 1-320.
2. de Murcia, G., Huletsky, A. and Poirier, G.G. (1988) Biochem. Cell. Biol. 66: 626-635.
3. Alvarez-Gonzalez, R. and Jacoboson, M.K. (1987) Biochemistry 26: 3218-3224.
4. Singh, N., Leduc, Y., Poirier, G.G. and Cerutti, P.A. (1985) Carcinogenesis (London) 6: 1489-1494.
5. Huletsky, A., de Murcia, G., Muller, S., Hengartner, M., Ménard, L., Lamarre, D. and Poirier, G.G. (1989) J. Biol. Chem. 264: 8878-8886.
6. Keith, G., Desgrès, J. and De Murcia, G. (1990) Anal. Biochem. 191: 309-313.
7. Ménard, L., Thibault, L. and Poirier, G.G. (1990) Biochim. Biophys. Acta 1049: 45-58.
8. Desmarais, Y., Ménard, L., Lagueux, J. and Poirier, G.G. (1991) Biochim. Biophys. Acta 1078: 179-186.
9. Simonin, F., Ménissier-de Murcia, J., Poch, O., Muller, S., Gradwohl, G., Molinette, M., Penning, C., Keith, G. and de Murcia, G. (1990) J. Biol. Chem. 265: 19249-19256.

The Poly(ADP-ribose)-Protein Shuttle of Chromatin

F.R. Althaus, S. Bachmann, S.A. Braun, M.A. Collinge, L. Höfferer, M. Malanga, P.L. Panzeter, C. Realini, M.C. Richard, S. Waser, B. Zweifel

University of Zürich-Tierspital, Institute of Pharmacology & Biochemistry, Winterthurerstrasse 260, CH-8057 Zürich, Switzerland

Poly ADP-ribosylation: A Posttranslational Protein Modification?

Since the discovery of poly(ADP-ribose) in Paul Mandel's laboratory in 1966 (1), poly ADP-ribosylation has been classified as a posttranslational protein modification (for review see 2-4), While this definition is formally correct, it may not precisely reflect the primary function of poly(ADP-ribose). The term "posttranslational modification" projects the idea that the modifying residue modifies protein function by *covalent* modification of the target acceptor. Over the past five years, we have focussed on the possibility that ADP-ribose polymers may primarily act on protein function(s) by *non-covalent* interactions with nuclear proteins. The working hypothesis envisions poly(ADP-ribose) metabolism as a protein shuttle mechanism in chromatin. The results show that poly ADP-ribosylation may selectively regulate DNA template accessibility for proteins involved in DNA processing functions. In this mechanism, poly(ADP-ribose) assumes the function of an alternative polynucleotide binding site for histones in chromatin. Testing this concept in models of various complexities, we conclude that protein shuttling by poly ADP-ribosylation may be involved in local unfolding of nucleosomes during DNA excision repair (5,6).

The Poly(ADP-ribose)-Protein Shuttle

The scheme in Fig. 1 summarizes essential features of the poly(ADP-ribose)-protein shuttle mechanism. Four principle reaction steps, numbered accordingly in Fig. 1, have been elucidated so far and will be discussed below. Briefly, the interaction of poly(ADP-ribose)polymerase with DNA is quite well understood. Two zinc-finger motifs in the N-terminal DNA-binding domain of this enzyme are involved in the differential recognition of single and double strand DNA breaks (7,8). The function of another DNA-binding motif, a helix-turn-helix motif is not known yet. Upon binding to DNA ss- or dsDNA breaks, poly(ADP-ribose)polymerase is activated and polymers are attached to the acceptors. The arrow (step 1 of Fig. 1) indicates that polymer addition to the acceptor protein is a strictly processive reaction. In addition, poly(ADP-ribose)polymerase is adaptive with regards to the polymer pattern formed. Step 2 symbolizes that a selected group of chromatin proteins, i.e. histones and protamine, may be taken off from the DNA template by the automodified polymerase. The mechanism involved is a preferential non-covalent binding of these proteins to the polymerase-bound ADP-ribose polymers. This state is readily reversed by the enzyme poly(ADP-ribose)glycohydrolase, which degrades ADP-ribose polymers

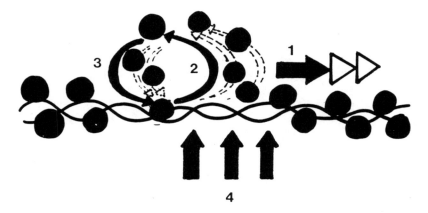

Figure 1: Scheme summarizing essential features of the protein shuttling mechanism. The numbers *1* to *4* denote individual reaction steps of the shuttle mechanism. *1* Processive mode of the poly(ADP-ribose) reaction; *2* binding of histones to polymerase-bound polymers; *3* reestablishment of histone binding to DNA following digestion of polymers by poly(ADP-ribose)glycohydrolase; *4* increased accessibility of DNA template to other proteins. For further discussion, see text.

and thereby releases polymer-bound histones for reassociation with DNA (step 3 of Fig. 1). This reversible shuttling of histones (and protamine) provides temporary DNA accessibility for other proteins (step 4), which may have a function in DNA processing, particularly in DNA repair. Below follows a step by step summary of the experimental evidence that led to the elucidation of the poly(ADP-ribose)-protein shuttle mechanism.

Step 1: Processivity of the Poly(ADP-ribose)polymerase Reaction; Adaptation of the Reaction Products in the Presence of Histones. We first studied the rules which govern the biosynthesis of specific numbers, size patterns, and branching frequencies of ADP-ribose polymers by poly(ADP-ribose)polymerase. For this purpose, we performed a step by step analysis of the molecular properties of ADP-ribose polymers formed in the auto-and heteromodification reaction of poly(ADP-ribose) polymerase in a reconstituted *in vitro* system (for details see 9). In parallel, we isolated and quantified the polymer carrying proteins. We found that i) the average polymer size remained constant throughout a reaction time af 45 min, when polymer synthesis came to a halt; ii) the relative frequency of individual polymer size classes was unchanged from 1 min to 45 min of polymer synthesis. An exception to this was the fraction of branched ADP-ribose polymers (10,11), which took up to 10 min to reach the final amount; iii) the number of modified acceptor molecules increased in parallel with increasing polymer numbers. From these and other data, we concluded that polymer addition to acceptor proteins is a strictly processive reaction (9). In fact, processivity of polymer addition occurred also under more complex reaction conditions, i.e. when purified polymerase was incubated in the presence of histones (12,13) in isolated nucleosomal core particles, and in isolated nuclei from rat liver (12). However, when the individual polymer molecules were analyzed, important differences were noted dependent on the microenvironment of the polymerase, and these patterns were highly conserved in each system once polymer synthesis had started. This suggested that nuclear factors were involved in the regulation of the polymer termination and branching reaction.

An *in vitro* test system was set up for the identification of these putative regulatory factors (for details see 12,13). We found that histone H1 and core histones were potent regulators of both the numbers and the sizes of ADP-ribose polymers. Each histone species induced the polymerase to synthesize a specific polymer size pattern. The number of branched polymers was also specifically adapted in the presence of histones. Various other basic and/or DNA binding proteins as well as other known stimulators of poly(ADP-ribose)polymerase (spermine, $MgCl_2$, nicked DNA) were ineffective as polymer size regulators. Testing specific proteolytic fragments of histone H1, the polymer number and polymer size modulating activity could be mapped to specific polypeptide domains. The relevance of these adaptations of poly(ADP-ribose)polymerase to protein ADP-ribosylation *in vivo* is not known at present. However, the pattern of ADP-ribose polymers produced in isolated nuclei was almost identical to the pattern found in nucleosomal core particles and to the composite pattern synthesized by purified poly(ADP-ribose)polymerase *in vitro* in the presence of all four core histones and histone H1 (12,13). We therefore concluded that histones are the predominant regulators of ADP-ribose polymer numbers, sizes, and branches. It remains to be seen whether and how the enzyme poly(ADP-ribose)glycohydrolase may also affect these parameters *in vivo.*

Step 2: Dissociation of Histone-Polymer Complexes; Non-covalent Interactions of Histones with ADP-ribose Polymers. Using mobility shift gel electrophoresis (14), we have determined the consequences of poly ADP-ribosylation on DNA-histone complexes. Purified poly(ADP-ribose)polymerase was reconstituted with 5'-end labeled core DNA fragments and histones (H1, H2A, H2B, H3, H4) purified by reversed phase high-pressure-liquid chromatography as described by Gurley (15). Following activation of the polymerase by addition of NAD, polymer formation was monitored and the binding of individual histone species to DNA and nascent ADP-ribose polymers was determined by quantifying free and histone-complexed DNA in the mobility shift gels (16,17). The results showed that all histone-DNA complexes are dissociated concomitant with poly(ADP-ribose) synthesis. Inhibition of polymer synthesis with benzamide or a-NAD prevented the complex dissociation (16,17). Addition of automodified poly(ADP-ribose)polymerase or free ADP-ribose polymers had a similar effect on the dissociation of histone-DNA complexes. Furthermore, these effects of ADP-ribose polymers could not be achieved with other polyanions even if they were added at excess negative charge equivalents (17). These and other data (17) demonstrated that ADP-ribose polymers affected DNA-histone interactions by a mechanism other than simple charge neutralization.

The non-covalent interactions of ADP-ribose polymers with histones and other basic and/or DNA binding proteins were examined using a novel assay which allows for the quantification of protein-binding to individual polymer molecules under high stringency conditions (for details see 11). Surprisingly, of 25 basic and/or DNA-binding proteins tested, only histones H1, H2A, H2B, H3, H4, and protamine, which replaces histones in the final stages of spermatogenesis, bound to poly(ADP-ribose) under these conditions. Branched polymers were a highly preferred target of this binding, and different histone species differentially bound to mixed polymer populations (for details see 11). - Thus, non-covalent interactions of ADP-ribose polymers may dramatically alter the DNA-binding function of a specific group of chromatin proteins, i.e. histones and protamine. In fact, a properly structured polymer of about 40 ADP-ribose residues is sufficient to dissociate the DNA of a single nucleosome (11).

Step 3: Poly(ADP-ribose)glycohydrolase Catalyzes the Conversion of Histone-Poly(ADP-ribose)-Complexes into Histone-DNA Complexes. The reassembly of histone-DNA complexes could be achieved by degradation of ADP-ribose polymers in the histone-polymer complexes with purified poly(ADP-ribose)glycohydrolase (17). Several aspects of this reaction are noteworthy. The reassembly of histone-DNA complexes was a two step reaction, the first one leading to the formation of a DNA-histone complex which, by mobility shift gel electrophoresis, was indistinguishable from the complex prior to dissociation (step 2). This step was complete after a subset of the ADP-ribose polymers in the histone-polymer complexes had been degraded by poly(ADP-ribose)glycohydrolase. Secondly, the DNA in the newly formed DNA-histone complex was partially sensitive to micrococcal nuclease digestion. The establishment of the original micrococcal nuclease resistance of histone-DNA complexes required degradation of residual ADP-ribose polymers by poly(ADP-ribose)glycohydrolase (17, Realini and Althaus, manuscript in preparation). Another notable observation was that poly(ADP-ribose)glycohydrolase degraded polymers in an exoglycosidic and endoglycosidic manner (18,19). The kinetics of these reactions were highly dependent on the structure and size of the polymer substrate as has been observed by others (20). Furthermore, the tight association of histones with branched sites (cf. step 2, 11) caused significant protection from poly(ADP-ribose) degradation. - Thus, step 3 of the shuttle mechanism, i.e. the degradation of polymers and subsequent reassembly of histone-DNA complexes emerges as a complex but highly ordered process. We propose that differential degradation of ADP-ribose polymers may determine the order of reassembly of histone-DNA complexes.

Step 4: Shuttling of Histones Establishes Local Accessibility of DNA Domains for Other Proteins. The working model outlined in Fig. 1 predicts in step 4 that reversible shuttling of histones on the DNA templates generates DNA domains accessible to other proteins. This possibility was tested as follows. Histone-DNA complexes were reconstituted with poly(ADP-ribose)polymerase and polymer formation was allowed to proceed as described above (step 2). Prior to loading the reaction mixture onto mobility shift gels, DNA accessibility was probed with micrococcal nuclease or DNase I (17). As expected, DNA became fully accessible to these enzymatic probes. In fact, the sensitivities to micrococcal nuclease and DNase I were similar to that of uncomplexed DNA (17). Extending from these observations, we determined whether template accessibility could also be established for an enzyme known to operate in DNA excision repair. Purified DNA helicase from calf thymus (donated to us by Drs. Thömmes and Hübscher from our institute) was incubated with ssM13 DNA annealed with a short radiolabeled primer, and DNA helicase activity was monitored in a primer release assay (21). Interestingly, in the presence of histones, DNA helicase was inactive in this system. Sequestration of histones by ADP-ribose polymers allowed the helicase to perform the strand separation reaction (17). - These results suggested that the poly(ADP-ribose) shuttle mechanism may indeed establish DNA template accessibility for various proteins.

Characteristics of the Poly(ADP-ribose)-Protein Shuttle Mechanism
The results obtained using reconstituted *in vitro* systems revealed several interesting insights into the molecular function of the poly ADP-ribosylation system of chromatin. The two major components of this system, poly(ADP-ribose)polymerase and poly(ADP-ribose)glycohydrolase, may cooperate to shuttle histones off and back onto DNA templates. This system exhibits several remarkable features, such as targetability to specific sites in chromatin (ss- or dsDNA breaks, step 1), processivity (step 1), adaptability to histone species (step 2), selectivity of target handling (step 2), and reversibility (step 3). Finally, it establishes local and temporary access to specific sites in DNA (step 4).

Relevance of the Poly(ADP-ribose)-Shuttle Mechanism to the Role of Poly ADP-ribosylation in DNA Excision Repair
The repair of DNA damage in eukaryotic cells is closely coupled with local changes of chromatin structure such that newly synthesized repair patches transiently appear in free DNA domains with increased accessibility to enzymatic and chemical probes (22). This phenomenon has been attributed to local unfolding of nucleosomes in chromatin (22). We have noted that the onset of nucleosomal unfolding, which starts at the incision step of repair, coincides with a drastic stimulation of the turnover of ADP-ribose residues in poly(ADP-ribose). In fact, the number of ADP-ribose residues processed by poly ADP-ribosylation system may increase several thousand fold dependent on the amount DNA damage (23). In addition, elevated ADP-ribose turnover returns to the low constitutive levels observed in undamaged cells with a half life very similar to the half life of the rapid phase of nucleosomal refolding (22). In view of these results, we speculated that a histone shuttle mechanism driven by the poly ADP-ribosylation system may be involved in the unfolding/refolding of nucleosomes during DNA excision repair (24). Therefore, we wanted to study nucleosomal unfolding in mammalian cells which were completely depleted of all chromatin-associated ADP-ribose polymers. In addition, several repair reactions were monitored in unfolded and folded nucleosomal domains of chromatin (6, 24). The results showed that excision of bulky DNA adducts occurred primarily in unfolded domains (6) and was closely linked with the appearance of newly synthesized repair patches. In poly(ADP-ribose)-depleted cells, the repositioning of repair patches into these domains was completely blocked, although overall repair patch synthesis was unaltered. Concomitantly, adduct excision was completely blocked, and adducts tended to accumulate in free DNA domains (6). These results suggested a tight coupling of the excision step with nucleosomal unfolding by a mechanism involving dynamic turnover of poly(ADP-ribose).

Conclusion

The current status of our results suggests that the eukaryotic poly ADP-ribosylation system may operate as a histone-shuttle mechanism in chromatin which is involved in the nucleosomal unfolding of chromatin during excision repair. This mechanism assigns a specific role to the automodification reaction of poly(ADP-ribose)polymerase and provides a rational explanation for the formation of specific and highly regulated polymer patterns. Moreover, our results emphasize the importance of non-covalent interactions of ADP-ribose polymers with a selected group of chromatin proteins. The specificity of these interactions is a very surprising observation (cf. 11). Using the approaches described in this report, it will be possible to determine the role of heteromodifications of proteins in this mechanism. Also, it will be possible to define the family of nuclear proteins which depends on the poly ADP-ribosylation system to perform DNA processing reactions.

Acknowledgments

We thank our colleagues Drs. Pia Thömmes and Ulrich Hübscher for providing us with an aliquot of purified DNA helicase from calf thymus, and Dr. Myron K. Jacobson for helping us with the purification of poly(ADP-ribose)glycohydrolase. This work was supported by grants (to F.R.A.) from the Swiss National Foundation for Scientific Research (31-31203.91), the Krebsliga des Kantons Zürich, and the Jubiläumsstiftung des Kantons Zürich. M.A.C. and M.C.R. were supported by fellowships from the Commission of the European Communities.

References

1. Chambon, P., Weill, J.D., Doly, J., Strosser, M.T., Mandel, P. (1966) *Biochem. Biophys. Res. Commun. 25*, 638-643.
2. Ueda, K. (1986) In *Pyridine Nucleotide Coenzymes* (Dolphin, D., Poulson, R. & Avramovic, O., eds.) Part B, pp. 549-597, John Wiley & Sons, New York.
3. Althaus, F.R. & Richter, C. (1987) *ADP-ribosylation of Proteins: Enzymology and Biological Significance*, Springer Berlin.
4. Jacobson, M.K. & Jacobson, E.L. (1989) *ADP-ribose Transfer Reactions: Mechanism and Biological Significance*, Springer Berlin.
5. Althaus, F.R., Collinge, M., Loetscher, P., Mathis, G., Naegeli, H., Panzeter, P., and Realini, C. (1989) In: *Chromosomal Aberrations* (Obe, G. & Natarajan, A.T., eds.) Springer Berlin, pp. 22-30.
6. Mathis, G., and Althaus, F.R. (1990) *Carcinogenesis 11*, 1237-1239.
7. Gradwohl, G., Ménissier-de Murcia, J., Molinete, M., Simonin, F., Koken, M., Hoejmakers, J.H.J., and de Murcia, G. (1990) *Proc. Natl. Acad. Sci. USA 87*, 2990-2994.
8. Ikejima, M., Noguchi, S., Yamashita, R., Ogura, T., Sugimura, T., Gill, D.M., and Miwa, M. (1990) *J. Biol. Chem. 265*, 21907-21913.
9. Naegeli, H., Loetscher, P., and Althaus, F.R. (1989) *J. Biol. Chem. 264*, 14382-14385.
10. Panzeter, P.L. and Althaus, F.R. (1990) *Nucleic Acids Res. 18*, 2194.
11. Panzeter, P.L., Realini, C., and Althaus, F.R. (1991) This volume.
12. Naegeli, H. and Althaus, F.R. (1991) *J. Biol. Chem. 266, in press.*
13. Naegeli, H. and Althaus, F.R. (1991) This volume.
14. Mathis, G., Althaus, F.R. (1987) *Biochem. Biophys. Res. Commun. 143*, 1049-1054.

15.Gurley, L.R., Valdez, J.G., Prentice, D.A. and Spall, W.D: (1983) *Anal. Biochem.* *129*, 132-144.

16.Althaus, F.R., Loetscher, P., Mathis, G., Naegeli, H., Panzeter, P., and Realini, C. (1991) In: *Light in Biology and Medicine* (Douglas, R.J., Moan, J., and Ronto, G., eds.) Vol 2, *in press.*

17.Realini, C. (1991) *Poly ADP-ribosylation: A protein shuttle mechanism in chromatin* Ph.D. Thesis, University of Zürich.

18.Ikejima, M. and Gill, D.M. (1988) *J. Biol. Chem. 263*, 11037-11040.

19.Braun, S.A. (1991) *Molecular Mechanisms of Poly(ADP-ribose)glycohydrolase action* Ph.D. Thesis, University of Zürich.

20.Hatakeyama, K., Nemoto, Z., Ueda, K., and Hayaishi, O. (1986) *J. Biol. Chem. 261*, 14902-14911.

21.Thömmes, P. and Hübscher, U. (1990) *J. Biol. Chem. 265*, 14347-14354.

22.Smerdon, M. (1990) In: *DNA Repair Mechanisms and their Biological Significance* (Lambert, M.W. & Laval, J., eds) Plenum Press New York, pp. 271-294.

23.Alvarez-Gonzalez, R. and Althaus, F.R. (1989) *Mutat. Res. 218*, 67-74.

24.Mathis, G. and Althaus, F.R. (1986) *J. Biol. Chem. 261*, 5758-5765.

25.Mathis, G. and Althaus, F.R. (1990) *Cell Biol. Toxicol. 6*, 35-45.

Part 4
Mono ADP-Ribosylation

Endogenous Mono-ADP-Ribosylation of Wild-Type and Mutant Elongation Factor 2 in Eukaryotic Cells

W. J. Iglewski, J. L. Fendrick, J. M. Moehring and T. J. Moehring.

1. INTRODUCTION

Eukaryotic ADP-ribosyltransferases transfer the adenosine 5'-diphosphate ribosyl (ADPR) portion of nicotinamide adenine dinucleotide (NAD^+) to elongation factor 2 (EF-2) *in vitro*. These transferases have been isolated from beef liver and polyoma virus-transformed baby hamster kidney (PyBHK) cells [1, 2]. Similar enzymes have been isolated from rabbit reticulocytes and rat liver [3, 4]. These eukaryotic enzymes are of interest since they covalently modify the same EF-2 tryptic peptide as does diphtheria and *Pseudomonas* A toxins. ADP-ribosylation of EF-2 by these two bacterial toxins occurs at an unique amino acid on the molecule termed 2-[3-carboxylamido-3-(trimethylammonio)propyl]histidine or commonly known as diphthamide [5]. Diphtheria toxin is able to reverse the forward reaction catalyzed by the eukaryotic ADP-ribosyltransferases which indicates that the products of the two different reactions are identical [1, 2]. Both bacterial toxins inactivate the cellular EF-2 which causes an inhibition of protein synthesis and subsequent cell death. It is therefore assumed that the eukaryotic ADP-ribosyltransferases have a role in the regulation of protein synthesis at the step of nascent peptide chain elongation [6].

Further investigations have demonstrated that the *in vivo* product, ADP-ribosylated EF-2 (ADPR-EF-2), of the eukaryotic ADP-ribosyltransferase reaction can be isolated by immunoprecipitation from PyBHK cells [7]. The endogenous eukaryotic transferase is different from the bacterial transferases in that the former only ADP-ribosylates a fraction of the total cellular EF-2 and that the ADPR portion can be removed from the EF-2 within the cell [3]. This finding is consistent with the idea that the reversible reaction of endogenous EF-2 ADP-ribosylation has a regulatory role in the cell. Physiological conditions such as serum concentration of the growth medium influence cellular EF-2 ADP-ribosylation. A 30 to 35% fraction of the total EF-2 of cells cultured in 2% serum is ADP-ribosylated [7]. Through the technique of immunoprecipitation using antiserum made against ADPR-EF-2 a wide variety of cell types have been probed for the presence of endogenous ADPR-EF-2. These studies have been extended to include a set of somatic mutant Chinese hamster ovary (CHO) cells that are resistant to EF-2 ADP-ribosylation by diphtheria toxin [8]. When the different cell types were cultured in the presence of ^{32}P-orthophosphate a ^{32}P-labeled protein was isolated that had the same MW as EF-2 which was identified as endogenous ADPR-EF-2 in CHO-K1 cells. The mutant CHO cell lines, of which some lack diphthamide, were found to contain endogenous ADP-ribosylated mutant EF-2.

2. IMMUNOPRECIPITATION OF ADP-RIBOSYLATED EF-2

Cells from primary cell cultures and continuous cell lines derived from various tissues of human (MRC-5, HEp-2), monkey (VERO), canine (MDCK), hamster (PyBHK, PyBHKR, CHO-K1), mouse (P3X63-AG8.653, 45.6TG1.7, L929) and chicken (chick embryonic firbroblasts) origin were used to provide presumptive evidence for the ubiquitous nature of ADP-ribosylated EF-2. Cells were labeled with 32P-orthophosphate and the endogenous ADPR-EF-2 immunoprecipitated from the cell lysates for analysis by SDS-PAGE and autoradiography as previously described [7, 9]. ^{32}P-labeled EF-2 was immunoprecipitated from all of the cell lysates suggesting that ADP-ribosylation of EF-2 is neither species or tissue specific. It is also pointed out that PyBHKR is a mutant cell line whose EF-2 is resistant to ADP-ribosylation by diphtheria toxin but not by the endogenous ADP-ribosyltransferase from wild-type PyBHK cells [10].

The kinetics of endogenous EF-2 ADP-ribosylation in PyBHK and CHO-K1 cells was then examined in the presence of 2% or 10% serum. ^{32}P-labeled ADPR-EF-2 was first detected by 2 hr and reached a maximum level by 6 to 8 hr for both cell lines. Reduced serum concentration only slightly enhanced the ^{32}P-labeling of EF-2. This indicated that the activity of the endogenous ADP-ribosyltransferase is similar in both the PyBHK and CHO-K1 cells.

Since it was observed that endogenous ADP-ribosylation occurred in the PyBHKR diphtheria toxin resistant mutant cells, other toxin resistant mutant cell lines were examined. The mutant cell lines tested were derived from the parental CHO-K1 line and have been extensively characterized. Three of the mutants designated RPE-3b, P1R2.D17 and RPE.33d have defects in the biosynthesis of diphthamide and represent complementation groups 1, 2 and 3 respectively [11]. RE1.22c and PyBHKR are cells containing mutations in the EF-2 structural gene [10, 11]. Wild-type CHO-K1 and PyBHK cells plus their respective mutants were examined for their ability to incorporate ^{32}P-label into EF-2 in the presence or absence of diphtheria toxin [7]. Similar intensities of ^{32}P-labeling of endogenously ADPR-EF-2 in wild-type and mutant cell lines were observed. However, only the two wild-type cell lines showed increased intensities of ^{32}P-labeling of the ADPR-EF-2 when treated with diphtheria toxin. All mutant cells exposed to the toxin had the same intensities of labeled ADPR-EF-2 as mutants not treated with toxin. This further emphasizes that the EF-2 structural gene and diphthamide mutations result in EF-2 toxin resistance but does not abolish the ability of mutant EF-2 to be endogenously ADP-ribosylated. It also demonstrates that the endogenous ADP-ribosyltransferase does not require the presence of the diphthamide side chain for its action.

3. CHARACTERIZATION OF ^{32}P-LABELED EF-2

The ^{32}P-labeled ADPR-EF-2 bands of the wild-type CHO-K1 and PyBHK cells and their respective mutants that were isolated by immunoprecipitation and SDS-PAGE followed by autoradiographic detection were eluted from the gels and digested with snake venom phosphodiesterase (SVP) [9]. SVP digestion of ADPR-EF-2 results in the production of AMP which can be separated from ADPR and orthophosphate by HPLC on an Ultrasil SAX column. SVP digestion of the ^{32}P-labeled ADPR-EF-2 isolated from the wild-type CHO-K1 and PyBHK cells both gave a single peak of radioactivity by HPLC analysis that chromatographed as AMP. Analysis of the two EF-2 structural mutants and the three diphthamide mutants all showed that SVP digestion of their ADPR-EF-2 resulted in formation of AMP.

Tryptic peptide digestion was performed on the eluted ^{32}P-labeled ADPR-EF-2 isolated from all of the wild-type and mutant cells and analyzed by HPLC analysis [9]. Elution profiles demonstrated that the same tryptic peptide of EF-2 was labeled in the wild-type as the mutant cells by the endogenous ADP-ribosyltransferase. Analysis of purified EF-2 ADP-ribosylated by diphtheria toxin in the presence of radioactive NAD^+ labeled the same EF-2 tryptic peptide as the endogenous ADP-ribosyltransferase when analyzed by HPLC.

4. REVERSAL OF ENDOGENOUS ADP-RIBOSYLTRANSFERASE

ADPR-EF-2 that has been catalyzed by eukaryotic ADP-ribosyltransferase can be reversed in vitro [2]. Reversal requires diphtheria toxin fragment A, excess nicotinamide and an acidic pH. Therefore, if a portion of purified EF-2 from the PyBHK cells is ADP-ribosylated by the endogenous transferase it would be possible to reverse the reaction in the presence of ^{14}C-carbonyl labeled nicotinamide and produce ^{14}C-labeled NAD^+.

When purified EF-2 was combined with 5 μg fragment A and 3 mM [^{14}C-carbonyl]-nicotinamide at a pH of 6.5 for the reversal reaction ^{14}C-NAD^+ was generated which cochromatographed with NAD^+. Identical results were observed for the reversal reaction when purified EF-2 was first ADP-ribosylated in the presence of fragment A and unlabeled NAD^+ [9]. This adds further evidence that a fraction of the EF-2 purified from eukaryotic cells is ADP-ribosylated in vivo and is identical to that catalyzed by diphtheria toxin.

5. SUMMARY

It has been shown that the endogenous ADP-ribosyltransferase exists in a variety of eukaryotic cells and not only ADP-ribosylates EF-2 *in vivo* but does it in a similar manner as diphtheria toxin and *Pseudomonas* toxin A. The endogenous ribosylated product is identical to that made by the diphtheria toxin ADP-ribosylated EF-2. However the diphthamide mutant cell lines indicate that the eukaryotic transferase recognizes a different determinant in the EF-2 ADPR acceptor site than the toxins. Endogenous ADP-ribosylation of EF-2 may represent a normal cellular function. Thus a mutation conferring resistance to the eukaryotic ADP-ribosyltransferase may be lethal.

Acknowledgments. We thank Patrick Muller and Sammy Ndive for purification of EF-2. This research was supported by Grants GM 41644 and AI 09100 from the National Institutes of Health.

6. REFERENCES

1. Iglewski WJ, Lee H, Muller P (1984) ADP-ribosyltransferase from beef liver which ADP-ribosylates elongation factor-2. FEBS Lett 173:113-118

2. Lee H, Iglewski WJ (1984) Cellular ADP-ribosyltrasferase with the same mechanism of action as diphtheria toxin and *Pseudomonas* toxin A. Proc Natl Acad Sci USA 81:2703-2707

3. Sitkov AS, Davydova EK, Ovchinnikov (1984) Endogenous ADP-ribosylation of elongation factor 2 in polyribosome fraction of rabbit reticulocytes. FEBS Lett 176:251263

4. Sayhan O, Ozdemirle M, Nurten R, Bermek E (1986) On the nature of cellular ADP-ribosyltransferase from rat liver specific for elongation factor 2. Biochem Biophys Res Com 139:1210-1214

5. VanNess BG, Howard JB, Bodley JW (1980) ADP-ribosylation of elongation factor 2 by diphtheria toxin. NMR spectra and proposed structures of ribosyl-diphthamide and its hydrolysis products. J Biol Chem 225:10710-10716

6. Iglewski WJ, Fendrick JL (1990) ADP ribosylation of elongation factor 2 in animal cells, p. 511-524. *In* J Moss, M Vaughan (eds) ADP-ribosylating toxins and G proteins. Insights into signal transduction. Amer Soc Mocrobiol, Washington, D.C.

7. Fendrick JL, Iglewski WJ (1989) Endogenous ADP-ribosylation of elongation factor 2 in polyoma virus-transformed baby hamster kidney cells. Proc Natl Acad Sci USA 86:554-557

8. Moehring TJ, Danley DE, Moehring JM (1979) Codominant translation mutants of Chinese hamster ovary cells selected with diphtheria toxin. Somatic Cell Genet 5:469-480

9. Fendrick JL, Iglewski WJ, Moehring JM, Moehring TJ (1991) Characterization of the endogenous ADP-ribosylation of wild-type and mutant elongation factor-2 in eukaryotic cells. (in preparation)

10. Iglewski WJ, Lee H (1983) Purification and properties of an altered form of elongation factor 2 from mutant cells resistant to intoxication by diphtheria toxin. Eur J Biochem 134:237-240

11. Moehring JM, Moehring, TJ (1988) The post-translational trimethylation of diphthamide studied *in vitro*. J Biol Chem 263:3840-3844

HETEROGENEITY OF ADPribosylation REACTION IN Sulfolobus solfataricus.

*M.Rosaria Faraone-Mennella, *Filomena De Lucia, *Piera Quesada,**Mario De Rosa,+Agata Gambacorta,+Barbara Nicolaus & *Benedetta Farina

*Dip.Chimica Org.& Biol., Facoltà di Scienze,**Ist. Biochim. Macromol., I Fac. Medicina, Università di Napoli; +Ist.M.I.B., CNR, Arco Felice, Napoli (Italy)

The post-translational modification of proteins by ADPribose is widely distributed among living organisms (1). In prokaryotes, a single ADPR unit is transferred to acceptor proteins by mono-ADPR transferases (1). Many bacterial toxins are mono-ADPR transferases that modify host-cell proteins (1). Only three bacterial ADPribosylating systems, in *Rodospirillum rubrum, Escherichia coli* and *Pseudomonas maltophilia*, have been described, characterized by endogenous enzyme and substrate proteins (1, 2).

Exogenous ADPribosylation of proteins has also been demonstrated in the extreme thermophile *S. solfataricus* (S.s.)(3). This microorganism belongs to the Chraenarcheota domain of Archaea, a kingdom which, in the evolutionary scale, is well distinct from Eukarya and Bacteria (4, 5).

In our laboratory, an ADPribosylating system which functions at high temperatures (80°C) has already been detected in S.s. (6).

The studies described here were undertaken in order to characterize ADPribosylation in this archaebacterium and to determine if it occurs by means of NAD glycohydrolase(s) and/or ADPribosylating enzyme(s).

MATERIALS AND METHODS

Cell culture, homogenate preparation and ammonium sulphate fractionation. Growth conditions of S.s. (ATCC 49155) were set up according to (5). Cells collected from the stationary phase of growth were washed with isotonic saline solution, lysed by freeze-thawing and suspended in 20mM Tris-HCl buffer, pH 7.5 (1:3, w/v). The crude homogenate was heat-treated (85°C) for 5 min and centrifuged at 11,000xg for 30min. Ammonium sulphate was added to the supernatant to 40% final concentration and, after centrifugation, the pellet (40% AS) was dialysed and used for assays.

Standard assay. 40% AS (200μg protein) was incubated with 0.1M Tris-HCl buffer, pH 7.5 and 2mM NaF, in the presence of 0.64mM [adenosine-^{14}C(U)]NAD (Amersham; 10,000 cpm/nmol) or 0.64mM [adenosine-^{14}C(U)]ADPR (10,000 cpm/nmol; prepared according to (7)) at 80°C in a final volume of 125μl for 10min, unless otherwise stated. All compounds used resulted stable under these conditions. ^{14}C incorporation in the 20% trichloroacetic acid-insoluble fraction (TCA-fraction) was measured by liquid scintillation.

The blank for each assay was obtained by incubating 40% AS, heated at 110°C for 48h, with either [^{14}C]NAD or [^{14}C]ADPR under standard conditions.

Reaction product analysis. The labelled TCA fraction, after column centrifugation (7), was incubated in 0.1M HCl, 3.0M NH$_2$OH at 37°C; or 0.3M

NaOH at 56°C. At given times, 125μl aliquots were withdrawn and protein-bound radioactivity was determined after TCA precipitation and filtration (6).

The TCA fraction obtained by incubating (20min) 40% AS (3mg protein) with 2.4μmoles [^{14}C]NAD (10,000 cpm/nmol) in 3.75ml final volume under standard conditions was alkali-treated and chromatographed on boronate resin according to (7). The radioactive fraction eluted with 10mM HCl was analyzed using RP-HPLC.

RESULTS AND DISCUSSION

Characterization of ADPribosylation in S.s. was performed using the 40% AS from the bacterial crude homogenate incubated in the presence of [^{14}C]NAD under standard conditions. The uptake of ^{14}C in the TCA fraction was time-, temperature- and NAD-dependent. The optimum temperature was between 80°C-100°C and the labelling saturation was reached at 0.64mM [^{14}C]NAD.

The high thermostability of the examined system is shown in TABLE 1. At 100°C, radioactivity levels were reduced to 50% after 1.5h preincubation of 40% AS, while the radioactivity was totally lost after 3h. As opposed to this, 85% and 68% labelling was still present after 24h preincubation at 60°C and 80°C respectively (TABLE 1).

Table 1 ^{14}C incorporation into proteins after exposure of S.s. 40% AS to high temperatures.

Temperature (°C)	Time (h)[*]	Decrease of ^{14}C label (%)
60	24.0	15
80	24.0	32
100	1.5	50

*Time of preincubation in the absence of NAD, followed by standard assay.

The chemical stability of the ADPR-protein adducts was determined after column centrifugation of the labelled TCA fraction (TABLE 2). In 0.33M NaOH at 56°C, 68% protein-bound labelling was solubilized in 2h. 3M NH$_2$OH treatment at 37°C for 2h released 45% radioactivity; this value did not vary after 17h (data not shown). High stability (87%) was determined in the presence of 0.1M HCl at 37°C over 2h.

Table 2 Chemical stability of protein-bound ADPribose in S.s. labelled TCA fraction.

Treatment	ADPribose released [*] (%)
0.1M HCl	13
3.0M NH$_2$OH[**]	45
0.3M NaOH	68

* 2h incubation
** pH 7.5 at this pH, in the absence of NH$_2$OH, radioactivity release was about 20%

Analysis of the alkali-digested labelled TCA fraction produced mainly 5'-AMP (Figure 1). This result indicates that in S.s. mono-ADPribosylation is prevalent.

Fig.1- RP-HPLC analysis of the alkali-digested TCA fraction after chromatography on boronate resin (2,800 cpm).
Elution from C-18 column (cm 0.46x25): A= 1% Methanol in 7mM NH_4-formate buffer, pH 5.8; B= 100% methanol. Star-ting buffer: A (3min), followed by a gradient to 10% B (1min). At 7min, another gradient (1min) to 20% B was applied and elution went on isocratically for up to 15min. Standards were eluted as shown by arrows.

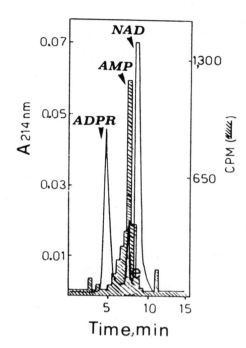

However, it is not sufficient to prove that binding of mono-ADPR to proteins occurs unequivocally due to the action of ADPribosyl transferase. In fact, among the processes of mono-ADPribosylation in eukaryotes, some authors (8, 9) showed that the mitochondrial one was quite peculiar. It is the result of a two-step mechanism, where the enzymatic hydrolysis of NAD carried out by a NAD glycohydrolase is followed by the non-enzymatic attachment of ADPR to acceptors.

In order to stress which mechanism(s) occurs in *S. solfataricus*, the assays either with [14C]ADPR or [14C]NAD were carried out at different pHs and incubation times, with and without 10mM isonicotinic acid hydrazide (INH), inhibitor of eukaryotic NAD glycohydrolase(s) (TABLE 3). After 10 min incubation and at pH 3.6, radioactivity levels in proteins were much higher in the presence of [14C]ADPR than of [14C]NAD. 14C uptake was completely inhibited when INH was added to the incubation mixture containing [14C]NAD.

On the other hand, after 10 min and at alkaline pH (7.5) [14C]NAD appeared to be a better substrate than [14C]ADPR and inhibition due to INH was reduced (TABLE 3).

Table 3. ^{14}C uptake from [^{14}C]NAD and [^{14}C]ADPR at different pHs in S.s..

pH	Protein-bound radioactivity (cpm)	
	[^{14}C]ADPR	[^{14}C]NAD
3.6	6,700	1,500
3.6[a]	ND[b]	0
7.5	375	600
7.5[a]	ND	162

a. In the presence of 10mM INH
b. Not determined

After 40 min incubation and pH 7.5, ^{14}C labelling of proteins from [^{14}C]ADPR was again prevalent (data not shown).

These results strongly suggest that in S.s. both ADPribosylation mechanisms (through the action of NAD glycohydrolase and ADPribosylating enzyme(s)) are present. Incubation time and pH thus determine the prevalence of one or the other system.

The co-existence of both enzymatic and "non-enzymatic" ADPribosylation in S.s. represents a very interesting finding in terms of evolution, since these organisms and their enzymes have developed a quite unusual resistance to high temperatures. Furthermore, the evidence that endogenous ADPribosylation is present in S.s. too, is in line with the discovery of the same process in other bacteria (1, 2) and indicates that it can have a well defined significance within prokaryotes.

Acknowledgements
We wish to thank the Servizio di Fermentazioni of MIB (V.Calandrelli, E.Esposito, I.Romano) for technical assistance. Financial support from CNR (Target Project in Biotechnology and bioinstrumentation).

REFERENCES
1. "ADPribosylation of Proteins" (1987) (Althaus, F.R., Richter, C. eds.), Springer-Verlag, Berlin.
2. Edmonds, C., Griffin, G.E., Johnstone, A.P. (1989) Biochem J 261: 113-118
3. Kessel, M., Klink, F. (1980) Nature 287: 250
4. Woese, C.R., Kandler, O., Wheelis, M.L. (1990) Proc Natl Acad Sci USA 87: 4576-4579
5. De Rosa, M., Gambacorta, A., Nicolaus, B., Giardina, P., Poerio, E., Buonocore, V. (1984) Biochem J 224: 407-414
6. Quesada, P., Faraone-Mennella, M.R., De Rosa, M., Gambacorta, A., Nicolaus, B., Farina, B. (1989) in "ADPribose Transfer Reactions", (Jacobson, E.L. & Jacobson, M.K., eds.) Springer-Verlag, pp.101-104
7. Payne, D.M., Jacobson, E.L., Moss, J., Jacobson, M.K. (1985) Biochemistry 24: 7540-7549
8. Hilz, H., Kock, R., Fanick, W., Klapporth, K., Adamietz, P. (1984) Proc Natl Acad Sci USA 81: 3929-3933
9. Frei, B., Richter, C. (1988) Biochemistry 27: 529-535

A novel C3-like ADP-ribosyltransferase produced by Clostridium limosum

Ingo Just, Gisela Schallehn[1] and Klaus Aktories

Institut für Pharmakologie und Toxikologie der Universität des Saarlandes, D-6650 Homburg-Saar, Germany

Introduction

Various strains of *Clostridium botulinum* type C and D produce the exoenzyme C3 (1, 2, 3) which ADP-ribosylates the small GTP-binding proteins rho (4, 5) and rac (6). C3 is encoded by phages which also encode the structural gene for botulinum neurotoxins (7, 8). However, it is now generally accepted that C3 is structurally, immunologically and functionally distinct from the neurotoxins (9). Here we report on the purification and characterization of a novel C3-like ADP-ribosyltransferase produced by *Clostridium limosum*, which modifies small GTP-binding proteins.

Results

When various strains of *Clostridium limosum* were tested for ADP-ribosyltransferase activity, one strain which was isolated from a human lung abscess, selectively labelled human platelet membrane proteins (Mr 20,000 - 25,000) in the presence of 0.1 μM [^{32}P]NAD (Fig.1). Labelling of membrane proteins was inhibited after NADase treatment of [^{32}P]NAD and in the presence of high concentrations of unlabelled NAD (100 μM). In order to characterize the ADP-ribosyltransferase activity, the enzyme was purified from the supernatant of a 48 h culture of *C.limosum* by ammonium sulfate precipitation, Sephadex G-25 gelfiltration and DEAE A-50 Sephadex anion exchange chromatography. The ADP-ribosyltransferase was further purified to near homogeneity by preparative gel electrophoresis. The purified enzyme migrated on SDS-PAGE with a Mr of about 25,000. In immunoblot analysis anti-C3 antibody cross-reacted with *C.limosum* exoenzyme. For inhibition of activity of the *C. limosum* exoenzyme an about 50-fold higher concentration of the C3 antibody was required than for inhibition of C3 ADP-ribosyltransferase activity. Determination of the pI revealed very basic isoelectric points (C3: 10.6; *C.limosum* exoenzyme: 10.3).

---

[1]Institut für Medizinische Mikrobiologie der Universität, 5300 Bonn, Germany

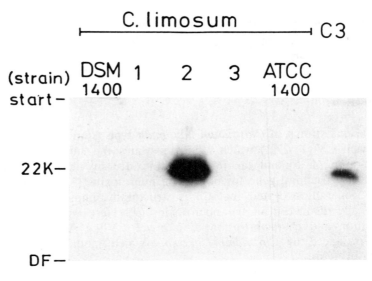

Fig. 1 Platelet membranes were incubated with the culture supernatants of the indicated strains of C.limosum in the presence of [^{32}P]NAD.
Samples were subjected to 12.5% SDS-PAGE. The autoradiography of
the SDS-PAGE is shown.

The partial proteolytic peptide analysis by HPLC exhibited no close
relationship of both enzymes. However, amino acid sequence analysis of
six peptide fragments (representing 69 of the 211 amino acid residues)
revealed a homology of 74% between both ADP-ribosyltransferases.
Despite the high sequence homology only *C.limosum* exoenzyme showed
a significant auto-ADP-ribosylation in the presence of 0.01% sodium
dodecyl sulfate. The auto-ADP-ribosylated exoenzyme shared the same
high stability of the ADP-ribose bond towards neutral hydroxylamine (0.5
M, 2h) as it was described for the C3-catalyzed ADP-ribose-protein
linkage in rho/rac (10). This finding suggests that asparagine is also the
acceptor amino acid for the auto-ADP-ribosylation of the *C.limosum*
exoenzyme. *C.limosum* exoenzyme and C3 revealed the same Km-value
for NAD (0.3 μM) and almost the same specific enzyme activity (C3: 6.4
nmol ADP-ribose/min/mg; *C.limosum* exoenzyme: 3.1 nmol/min/mg). Both
enzymes apparently modified the same substrate proteins. When human
neutrophil and platelet membranes were ADP-ribosylated by C3, the
subsequent ADP-ribosylation by *C.limosum* exoenzyme in the presence of
[^{32}P]NAD did not cause any further incorporation of [^{32}P]ADP-ribose.

The same was true when the first ADP-ribosylation was catalyzed by *C.limosum* exoenzyme and the second by C3. In order to study the substrate specificity in more detail, we employed the method of de-ADP-ribosylation. For this purpose, rho/rac proteins were ADP-ribosylated by C3. Thereafter, unreacted NAD was destroyed by NADase treatment and C3 was removed by immunoprecipitation. The addition of *C.limosum* exoenzyme in the presence of 30 mM nicotinamide induced the cleavage of the ADP-ribose. Also rho/rac proteins previously ADP-ribosylated by *C.limosum* exoenzyme were de-ADP-ribosylated by C3 indicating that most likely both enzymes modify the substrate proteins at the identical amino acid (asparagine-41 in rho) (11).

Treatment of NIH3T3 cells and FAO cells with the supernatant of *C.limosum* or with the purified exoenzyme caused rounding up of the cells. Identical morphological changes were observed after C3 treatment.

Discussion

Here we report on a novel ADP-ribosyltransferase produced by *C.limosum*. The M_r of the *C.limosum* exoenzyme is about 25,000 and the pI 10.3. The transferase activity and the K_m for NAD are almost identical for C3 and the *C.limosum* exoenzyme. As reported for C3, the *C.limosum* exoenzyme ADP-ribosylates the small GTP-binding proteins rho and rac, most likely at the identical amino acid (asparagine). The close relationship between C3 and *C.limosum* exoenzyme was corroborated by sequence analysis of proteolytic peptides showing an about 70% homology of both enzymes. Thus, the data indicate that modification of rho/rac is not unique to C3. It appears that a class of related enzymes exists which modifies this family of small GTP-binding proteins. *C.limosum* exoenzyme is definitely produced in the absence of botulinum neurotoxins, therefore the novel ADP-ribosyltransferase is free of any contamination with these very potent agents. This facilitates enzyme handling and the interpretation of in-vivo experiments.

References

1. Aktories, K., Weller, U., Chhatwal, G.S. (1987) FEBS Lett 212: 109-113
2. Aktories, K., Rösener, S., Blaschke, U., Chhatwal, G.S.(1988) Eur J Biochem 172: 445-450
3. Rubin, E.J., Gill, D.M., Boquet, P., Popoff, M.R. (1988) Mol Cell Biol 8: 418-426
4. Braun, U., Habermann, B., Just, I., Aktories, K., Vandekerchhove, (1989) FEBS Lett 243: 70-76

5. Narumiya, S., Sekine, A., Fujiwara, M. (1988) J Biol Chem 263: 17255-17257

6. Didsbury, J., Weber, R.F., Bokoch, G.M., Evans, T., Snyderman, R. (1989)
    J Biol Chem 264: 16378-16382

7. Popoff, M.R., Boquet, P., Gill, D.M., Eklund, M.W. (1990) Nucl Acids Res 18: 1291

8. Eklund, M.W., Poysky, F.T., Reed, S.M., Smith, C.A. (1971) Science 172: 480-482

9. Rösener, S., Chhatwal, G.S., Aktories, K. (1987) FEBS Lett 224: 38-42

10. Aktories, K., Just, I., Rosenthal, W. (1988) Biochem Biophys Res Commun 156: 361-367

11. Sekine, A., Fujiwara, M., Narumiya, S. (1989) J Biol Chem 264: 8602-8605

THE 52 kDa ADP-RIBOSYLATED PROTEIN IN THE RAT HEART PLASMA MEMBRANE: IS IT G_{sa}?

Kathryn K. McMahon and Kristien J. Piron

G proteins[1] are a family of heterotrimeric molecules that are implicated in a variety of signal transduction processes (for review see 1). ADP-ribosylation is a covalent modification which transfers the ADP-ribose moiety of NAD to cellular acceptor proteins. Several bacterial toxins posses this ADP-ribosyltransferase activity and modify several G proteins. For example, cholera toxin can ADP-ribosylate the α subunits of the G proteins G_s and transducin (1). These toxins may mimic endogenous processes and G proteins may be endogenously ADP-ribosylated. Indeed several authors have demonstrated that $G_{s\alpha}$ can be ADP-ribosylated by endogenous ADP-ribosyltransferases (2-6). In liver membranes, a 55 kDa protein which was speculated to be $G_{s\alpha}$ is ADP-ribosylated in both the absence and presence of cholera toxin. In the presence of isoproterenol ADP-ribosylation of this protein is increased (6). In NG108-15 hybrid cells (5) and in platelets (4) ADP-ribosylation of $G_{s\alpha}$ has been implicated in heterologous desensitization to prostacyclins. Recently, it was also shown that in canine sarcolemma, adenylyl cyclase might be regulated by endogenous, reversible ADP-ribosylation of $G_{s\alpha}$ (7).

Previously, we identified a 52 kDa ADP-ribosylated protein in the rat heart plasma membrane which is mono-ADP-ribosylated on an arginine residue (8). This ADP-ribosylated 52 kDa product and ADP-ribosylated cholera toxin substrate comigrated on SDS gels (8). Therefore the objectives were to investigate whether isoproterenol would affect the ADP-ribosylation of the 52 kDa product and whether this endogenously ADP-ribosylated protein was $G_{s\alpha}$.

METHODS

The subcellular fractionation was performed on hearts of Sprague-Dawley rats as previously described (8). Fractions F2 through F4 have previously been shown to contain the plasma membrane fraction (8) and were therefore used in all experiments. The $[^{32}P]$ADP-ribosylation was as described previously (8) using $[^{32}P]$NAD (New England Nuclear, Boston, MA) in the presence of 0.1 mM NADP. The reaction was incubated for 30 minutes at $30°$ C and was terminated by the addition of SDS-sample buffer. Proteins in the reaction then were separated by 10% SDS-PAGE. For some experiments the radioactive region was identified by autoradiography, removed from the gel and counted in a liquid scintillation counter. For other

---

[1] G proteins, guanine nucleotide binding proteins; G_{α}, alpha subunit of the stimulatory guanine nucleotide binding protein.

experiments, the separated proteins were transferred to Immobilon P (Millipore, Bedford, MA) and Western blots were performed according to Harris et al. (9), for the 584 antibody (kindly provided by Dr. S. Mumby, University of Texas Health Sciences Center at Dallas). Purified $G_{s\alpha}$ (large and small forms) was kindly provided by Dr. P. Casey (Duke University) and was run as an internal standard for Western blots. The 584 antibody was used at a 1:1000 dilution. The secondary antibody was a horseradish peroxidase labeled donkey anti-rabbit antibody and was used at 1:5000 for 1 hour. The final detection was by an enhanced chemiluminescence system (Amersham, Arlington Heights, IL). Thereafter the blots were subjected to autoradiography for detection of the $[^{32}P]$ADP-ribosylated products.

RESULTS AND DISCUSSION

Isoproterenol ($1\mu M$) was previously shown to enhance the ADP-ribosylation of a 55 kDa protein in the plasma membrane of cultured differentiated RL-PR-C hepatocytes (6). The possibility existed that this was the same protein that was present and ADP-ribosylated in the rat heart plasma membrane and it was investigated whether isoproterenol affected the ADP-ribosylation of the cardiac protein in a similar manner. The ADP-ribosylation assay was performed in the presence of $1\mu M$, $10\mu M$ and $100\mu M$ isoproterenol. Table 1 demonstrates that the presence of isoproterenol did not affect the level of ADP-ribosylation of the 52 kDa protein. Therefore, it could be concluded that isoproterenol (i.e., β-adrenergic receptor stimulation) does not activate the arginine-specific ADP-ribosyltransferase or inhibit the ADP-ribosylarginine hydrolase and does not affect the ADP-ribosylation of the 52 kDa protein.

$G_{s\alpha}$ and the 52 kDa ADP-ribosylated protein have several common characteristics. 1) They are both localized in the plasma membrane. 2) When $G_{s\alpha}$ is ADP-ribosylated by cholera toxin, it comigrates with the 52 kDa endogenous ADP-ribosylation product (8). 3) Cholera toxin ADP-ribosylated $G_{s\alpha}$ (10) and the 52 kDa ADP-ribosylation product (11) both display isoelectric point heterogeneity when run on two dimensional gels and their pI values are within the same range of 5.5 to 6. Therefore it was investigated whether the 52 kDa protein was $G_{s\alpha}$. For this purpose Western blotting with the 584 anti-$G_{s\alpha}$ antibody was performed. This antibody is directed against amino acids 325 to 339 of $G_{s\alpha}$ (12). Fig. 1 (lanes 1 and 2) shows the Western

Table I: Effect of isoproterenol on the ADP-ribosylation of the 52 kDa product.

	%	n
Basal	100	7
Isoproterenol, $1\mu M$	96 ± 12	6
Isoproterenol, $10\mu M$	100 ± 15	7
Isoproterenol, $100\mu M$	107 ± 14	7

Heart membranes were ADP-ribosylated in the presence of the indicated concentrations of isoproterenol. After SDS-PAGE the amount of radioactivity incorporated in the 52 kDa region was counted and expressed as percent of basal. Values are the mean $\pm$ S.E.M.

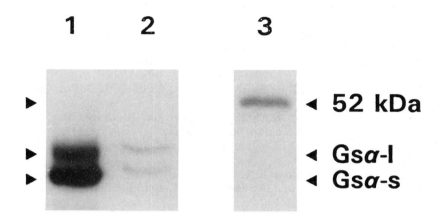

Fig. 1: Western blot with the anti-$G_{s\alpha}$ 584 antibody against 1 µg of purified $G_{s\alpha}$ (lane 1) or 75 µg of the [^{32}P]ADP-ribosylated plasma membrane fraction (lane 2). Lane 3 is an autoradiogram of lane 2.

blot with this anti-$G_{s\alpha}$ antibody 584. Lane 1 contains the purified $G_{s\alpha}$ while lane 2 contains the ADP-ribosylated plasma membrane fraction. In both lanes the two forms of $G_{s\alpha}$ could be detected with the 584 antibody. Lane 3 is an autoradiogram of lane 2 and demonstrates that neither of the bands detected with anti-$G_{s\alpha}$ coincided with the ADP-ribosylated protein. Therefore it was concluded that the 52 kDa ADP-ribosylation product that is observed in the rat heart plasma membrane fraction was not $G_{s\alpha}$. This is in contrast with Quist et al. (7) who observed the ADP-ribosylation of $G_{s\alpha}$ in canine sarcolemma. However, they detected multiple ADP-ribosylated bands on SDS-gels in their sarcolemmal preparation, while in our membranes only one major band is being ADP-ribosylated. Therefore, it cannot be excluded that either species differences or variations in the membrane preparation or the ADP-ribosylation assay are responsible for these different results.

In conclusion, the 52 kDa ADP-ribosylated protein in the adult rat heart plasma membrane is not $G_{s\alpha}$ and the ADP-ribosylation of this protein is not affected by isoproterenol.

ACKNOWLEDGEMENT

We want to thank Dr. S. Mumby for her gift of the anti-$G_{s\alpha}$ 584 antibody and Dr. P. Casey for his gift of purified $G_{s\alpha}$. The authors gratefully acknowledge the secretarial assistance of Mrs. Josie Aleman and Mrs. Mary Alba. This work was supported by the Seed Grant Program of TTHSC (Institute for Biomedical Research).

REFERENCES

1. Freissmuth, M., Casey, P.J., and Gilman, A.G. (1989) FASEB J. 3, 2125-2131.

2. Jacquemin, C., Thibout, H., Lambert, B. and Correze, C. (1986) Nature 323, 182-184.

3. Clark, J.A., Terwilliger, R.Z., Nestler, E.J. and Duman, R.S. (1989) Society for Neuroscience Abstracts Vol. 15, 435.

4. Molina y Vedia, L., Nolan, R.D. and Lapetina, E.G. (1989) Biochem. J. 261, 841-845.

5. Donnelly, L.E., Boyd, R.S., and MacDermot, J. (1991) Br. J. Pharmacol. 102, 34P.

6. Reilly, T.M., Beckner, S., McHugh, E.M. and Blechner, M. (1981) Biochem. Biophys. Res. Commun. 98, 1115-1120.

7. Quist, E.E., Coyle, D.L., Aboul-Ela, N. and Jacobson, M.K. (1991) FASEB J. 5, A1506.

8. Piron, K.J. and McMahon, K.K. (1990) Biochem. J. 270, 591-597.

9. Harris, B.A., Robishow, J.D., Mumby, S.M. and Gilman, A.G. (1985) Science 229, 1274-1277.

10. Graziano, M. P., Freissmuth, M. and Gilman, A.G. (1989) J. Biol. Chem. 264, 409-418.

11. Piron, K.J. and McMahon, K.K. (1991) FASEB J. 5, A1177.

12. Mumby, S.M. and Gilman, A.G. (1991) Methods of Enzymology, 195, 215-233.

Biochemical and Developmental Characterization of ADP-ribosylation Factors, A Family of 20 kDa Guanine Nucleotide-binding Proteins.

S.-C. Tsai, S. R. Price, M. Tsuchiya, C. F. Welsh, R. Adamik, J. Moss, and M. Vaughan

The cholera toxin A subunit (CT-A) is an ADP-ribosyltransferase; its principal cellular substrates are the α subunits of G_s, the stimulatory G protein of adenylyl cyclase. The activity of CT-A is enhanced by soluble and membrane proteins termed the ADP-ribosylation factors (ARFs), a family of $\sim$20 kDa guanine nucleotide-binding proteins (Bobak et al., 1990; Kahn and Gilman, 1984; 1986; Tsai et al., 1987; 1988). In the presence of GTP or non-hydrolyzable GTP analogues, but not GDP, its analogues, or adenine nucleotides, bovine brain soluble and membrane ARFs enhanced ADP-ribosylation of $G_{s\alpha}$, simple guanidino compounds, and proteins unrelated to the cyclase system (Tsai et al., 1987; 1988). Kinetic data were consistent with the conclusion that ARF acts as an allosteric activator of the cholera toxin catalytic unit (Noda et al., 1990).

Characterization of an ARF-Toxin Complex

In the presence of SDS and GTPγS, but not GDPβS or adenine nucleotides, ARF and CTA formed a complex, which behaved on Ultrogel AcA 44 column chromatography as a >100 kDa aggregate containing similar amounts of CTA and ARF (Tsai et al., 1991a). Samples of fractions containing the complex auto-ADP-ribosylated CTA_1, but did not modify $G_{s\alpha}$ or the low molecular weight protein standards. Monomeric ARF and CTA emerged after carbonic anhydrase (30 kDa). These fractions ADP-ribosylated $G_{s\alpha}$, bovine serum albumin and phosphorylase b, but no auto-ADP-ribosylation of CTA_1 was observed. When monomeric fractions were pooled, concentrated and again incubated with GTPγS and SDS, no additional complex was detected. Thus, only a small fraction of ARF and CTA apparently is capable of forming a relatively stable complex which is very active in auto-ADP-ribosylation of CTA_1 but not $G_{s\alpha}$ whereas the monomeric proteins exhibit a preference for $G_{s\alpha}$ and certain other proteins over auto-ADP-ribosylation of CTA_1.

```
              10        20        30        40        50        60
hARF1   MGNIFANLFKGLFGKKEMRILMVGLDAAGKTTILYKLKLGEIVTTIPTIGFNVETVEYKNI
hARF3   *****G**L*S*I************************************************
bARF2   ***V*EK***S*************************************************
hARF4   **LTISS**SR****Q*******************************************
hARF5   **LTVSA**SRI****Q******************************************
hARF6   **KVLSK----I**N*******L****************QS******V*******T***V
yARF    **LFASK**SN***N**********G*****V********VI************Q****

              70        80        90        100       110       120
hARF1   SFTVWDVGGQDKIRPLWRHYFQNTQGLIFVVDSNDRERVNEAREELMRMLAEDELRDAVLL
hARF3   ***********************************************************
bARF2   ***********************************************T***********
hARF4   C***********R*****K*******************IQ*VAD**QK**LV********
hARF5   C*************************************Q*SAD**QK**Q*********
hARF6   K*N****************YTG*********CA**D*ID***Q**H*IINDR*M***II*
yARF    ***********R**S****YR**E*V********S*IG****VMQ***N*****N*AW*

              130       140       150       160       170       180
hARF1   VFANKQDLPNAMNAAEITDKLGLHSLRHRNWYIQATCATSGDGLYEGLDWLSNQLRNQK
hARF3   ****************************************************A***K*K*
bARF2   **V***********************Q******************************K***
hARF4   L***********AIS*M*****Q***N*T**V*****Q*T**********E*-SKR
hARF5   *******M****PVS*L******QH**S*T**V*****Q*T***D*****HE*-SKR
hARF6   I********D**KPH**QE****TRI*D****V*PS***********T**TSNYKS--
yARF    *********E**S****E*****I*N*P*F********E******E****S*K*ST
```

Fig. 1. The deduced amino acid sequences of reported human, bovine, and yeast ARFs were aligned using CLUSTAL (Higgins and Sharp, 1989). Human ARFs 1, and 3-6 are indicated by hARF 1, and hARF 3 (Bobak *et al.*, 1989), hARF 4 (Monaco *et al.*, 1990), hARF 5, and hARF 6 (Tsuchiya *et al.*, 1991), respectively; bARF 2 (Price *et al.*, 1988) and yARF (Sewell and Kahn, 1988) represent bovine ARF 2 and yeast ARF, respectively. Asterisks indicate amino acids identical with ARF 1. Gaps introduced for optimal alignment are indicated as hyphens.

Expression of mammalian ARFs

Based on amino acid sequences deduced from six cDNAs, ARF proteins can be grouped into three classes (Fig. 1): class 1 containing human ARFs 1 and 3 and bovine ARF 2; class 2 with human ARFs 4 and 5, and class 3 with human ARF 6. The six mammalian ARFs exhibit >64% amino acid identity and contain the consensus sequences for guanine nucleotide binding and hydrolysis. To investigate requirements of ARFs for activation of CTA, recombinant ARF 2 and ARF 6 proteins were expressed in E. coli and purified; both were active. Half maximal enhancement of transferase activity by rARF 6 required ~0.04 μM GTP whereas 3 μM was required by rARF 2, which, unlike ARF 6, also required DMPC/cholate. Based on auto-ADP-ribosylation of CTA_1, activities were sARF II > rARF 6 > rARF 5 >> rARF 2.

Developmental Regulation of ARF

Rabbit anti-bovine sARF II antibodies reacted with two ARF bands on immunoblots from brain cytosol of bovine, chicken, rat and frog, only a single band (the faster moving ARF I) was detected in liver, heart, kidney, lung, and spleen. Neuronal tissue had the highest content of both sARF I and sARF II. With these antibodies, we detected an increase of immunoreactive ARF protein in rat brain cytosol during postnatal development (Tsai et al., 1991b). Amounts of ARF were maximal at 27-60 days and correlated with changes in ARF activity assessed by enhancement of toxin ADP-ribosyltransferase activity. Among six known ARFs, only ARF 3 mRNAs (3.7 and 1.2 kb), increased with development. ARF 2 and 4 mRNAs decreased and ARF 1, 5, and 6 mRNAs were unchanged. As partial amino acid sequence of sARF II is identical to deduced amino acid sequences from ARF 1 and ARF 3 cDNAs, it appears that sARF II is a product of the ARF 3 gene, possibly involved in neuronal maturation during postnatal development of rat brain.

References

Bobak, DA, Nightingale, MS, Murtagh, JJ, Price, SR, Moss, J, Vaughan, M (1989) Molecular cloning, characterization, and expression of human ADP-ribosylation factors: Two guanine nucleotide-dependent activators of cholera toxin. Proc Natl Acad Sci USA 86:6101-6105

Bobak, DA, Bliziotes, MM, Noda, M, Tsai, S-C, Adamik, R, Moss, J (1990) Mechanism of activation of cholera toxin by ADP-ribosylation factor (ARF): Both low- and high-affinity interactions of ARF with guanine nucleotides promote toxin activation. Biochemistry 29:855-861

Higgins, DG, Sharp PM (1989) Fast and sensitive multiple sequence alignments on a microcomputer. CABIOS 5:151-153

Kahn, RA, Gilman, AG (1984) Purification of a protein cofactor required for ADP-ribosylation of the stimulatory regulatory component of adenylate cyclase by cholera toxin. J Biol Chem 259:6228-6234

Kahn, RA, Gilman AG (1986) The protein cofactor necessary for ADP-ribosylation of G_s by cholera toxin is itself a GTP-binding protein. J Biol Chem 261:7906-7911

Monaco, L, Murtagh, JJ, Newman, KB, Tsai, S-C, Moss, J, Vaughan, M (1990) Selective amplification of an mRNA and related pseudogene for a human ADP-ribosylation factor, a guanine nucleotide-dependent protein activator of cholera toxin. Proc Natl Acad Sci USA 87:2206-2210

Noda, M, Tsai, S-C, Adamik, R, Moss, J, Vaughan, M (1990) Mechanism of cholera toxin activation by a guanine nucleotide-dependent 19 kDa protein. Biochem Biophys Acta 1034:195-199

Price, SR, Nightingale, M, Tsai, S-C, Williamson, KC, Adamik, R, Chen, H-C, Moss, J, Vaughan, M (1988) Guanine nucleotide-binding proteins that enhance choleragen ADP-ribosyltransferase activity: nucleotide and deduced amino acid sequence of an ADP-ribosylation factor cDNA. Proc Natl Acad Sci USA 85:5488-5491

Sewell, JL, Kahn, RA (1988) Sequences of the bovine and yeast ADP-ribosylation factor and comparison to other GTP-binding proteins. Proc Natl Acad Sci USA 85:4620-4624

Tsai, S-C, Adamik, R, Moss, J, Vaughan, M (1991a) Guanine nucleotide dependent formation of a complex between choleragen (cholera toxin) A subunit and bovine brain ADP-ribosylation factor. Biochemistry 30:3647-3703

Tsai, S-C, Adamik, R, Tsuchiya, M, Chang, PP, Moss, J, Vaughan, M (1991b) Differential expression during development of ADP-ribosylation factors, 20 kDa guanine nucleotide-binding protein activators of cholera toxin. J Biol Chem, in press

Tsai, S-C, Noda, M, Adamik, R, Moss, J, Vaughan, M (1987) Enhancement of choleragen ADP-ribosyltransferase activities by guanyl nucleotides and a 19-kDa membrane protein. Proc Natl Acad Sci USA 84:5139-5142

Tsai, S-C, Noda, M, Adamik, R, Chang, P, Chen, H-C, Moss, J, Vaughan, M (1988) Stimulation of choleragen enzymatic activities by GTP and two soluble proteins purified from bovine brain. J Biol Chem 263:1768-1772

Tsuchiya, M, Price, SR, Tsai, S-C, Moss, J, Vaughan, M (1991) Molecular identification of ADP-ribosylation factor mRNAs and their expression in mammalian cells. J Biol Chem 266:2772-2777

Nonenzymatic ADP-ribosylation of Cysteine

L.J. McDonald, L. Wainschel, N.J. Oppenheimer, and J. Moss

Introduction

Mono-ADP-ribosylation is a post-translational modification of proteins catalyzed by bacterial toxins and eukaryotic amino acid-specific ADP-ribosyltransferases (Moss and Vaughan, 1988; Williamson and Moss, 1990). Several of the known acceptor proteins for mono-ADP-ribosylation by the bacterial toxins are heterotrimeric G-proteins that are involved in signal transduction (Moss and Vaughan, 1988). Cysteine residues in the α-subunits of G-proteins are ADP-ribosylated by pertussis toxin and by a human erythrocyte enzyme (West et al., 1985; Tanuma et al., 1988). NAD:cysteine ADP-ribosyltransferases have been characterized with free cysteine or other sulfhydryl compounds as model acceptors (Tanuma et al., 1988; Lobban and Van Heyningen, 1988).

All ADP-ribosyltransferases, such as pertussis toxin, are also NAD glycohydrolases that generate free ADP-ribose (Moss and Vaughan, 1988). ADP-ribose is recognized as a reactive molecule able to combine nonenzymatically with protein amino and carboxyl groups, and possibly with cysteine sulfhydryl groups (Frei and Richter, 1988; Kun et al., 1976). Free cysteine is also reactive, forming thiazolidine carboxylic acids when incubated with aldehydes (Schubert, 1936, 1939; Ratner and Clarke, 1947). We hypothesized that formation of ADP-ribosylcysteine could occur from an analogous nonenzymatic reaction of cysteine with ADP-ribose(aldose). In this paper, formation of such a compound is described. The compound was tentatively identified as ADP-ribosethiazolidine carboxylic acid based on incorporation of radioactivity from precursors, reaction specificity for cysteine with a free amino group, products of its chemical degradation, and NMR structural data. This compound was distinguished from the ADP-ribosylcysteine produced in protein by pertussis toxin by nature of its sensitivity to hydroxylamine.

Results and Discussion

During analysis of reaction mechanisms for cysteine-specific ADP-ribosyltransferase activities, free ADP-ribose was found to combine nonenzymatically with cysteine. The product, which incorporated [adenine-^{14}C]ADP-ribose and [^{14}C]cysteine in 1:1 ratio, was separated from its precursors by ion-exchange HPLC (Moss et al., 1983). With ~ 0.1 mM ADP-ribose and ~ 50 mM L-cysteine the reaction proceeded to completion within 1 - 2 h at 30°C. Under identical conditions, several amino acids were tested for reactivity with ADP-ribose. Cysteine was the only amino acid that reacted with ADP-ribose, while serine, alanine, lysine and arginine did not, strongly suggesting the cysteine sulfhydryl group was involved in the reaction. Several sulfhydryl compounds were subsequently tested, and further specificity was observed: L- and D-cysteine, cysteine methyl ester and cysteamine reacted with ADP-ribose, but N-acetylcysteine, 2-mercaptoethanol, dithiothreitol, or glutathione did not. The reactive compounds all contained an intact amino group in addition to the sulfhydryl.

The nature of the bond formed in ADP-ribosylcysteine was examined by assessing the stability of the compound in the presence of hydroxylamine and other reagents. The ADP-ribosylcysteine linkage produced in guanine nucleotide-binding proteins by pertussis toxin-catalyzed ADP-ribosylation is hydroxylamine-stable but sensitive to mercuric ion (Hsia et al., 1985; Meyer et al., 1988). In contrast, the nonenzymatically-formed ADP-ribosylcysteine was sensitive to both hydroxylamine (0.5 M, pH 7) and mercuric ion. In the presence of 0.5 M hydroxylamine, ADP-ribosylcysteine was degraded with a half-life of 13.4 ± 0.3 min (n=3). Degradation of ADP-ribosylcysteine in hydroxylamine released free cysteine and ADP-ribose hydroxamate, while degradation by mercuric ion produced ADP-ribose and a cysteine-mercuric ion adduct. These results indicated that the two starting compounds were present intact in the product.

A sample of nonenzymatically-formed ADP-ribosylcysteine was analyzed by proton NMR. The ^{1}H NMR spectrum of the compound revealed the presence of two diastereomeric forms in nearly 1:1 ratio, and the vicinal coupling constants were consistent with a noncyclic sugar in the ribosylcysteine moiety. Based on the specificity of the reaction for a sulfhydryl compound with a free amine, and on the structural data obtained, we propose that the nonenzymatic ADP-ribosylcysteine is formed by the reaction scheme shown in Fig. 1. This reaction is analogous to the previously described formation of thiazolidine carboxylic acids by the combination of aldehydes with cysteine (Horton and Hutson, 1963; Schubert, 1936, 1939; Ratner and Clarke, 1947). In this scheme, ADP-ribose is in equilibrium with its aldose form, which then reacts with free cysteine to form ADP-ribosethiazolidine carboxylic acid. The thiazolidine ring contains the ribose anomeric carbon along with the nitrogen, sulfur, and the two intervening carbons from cysteine.

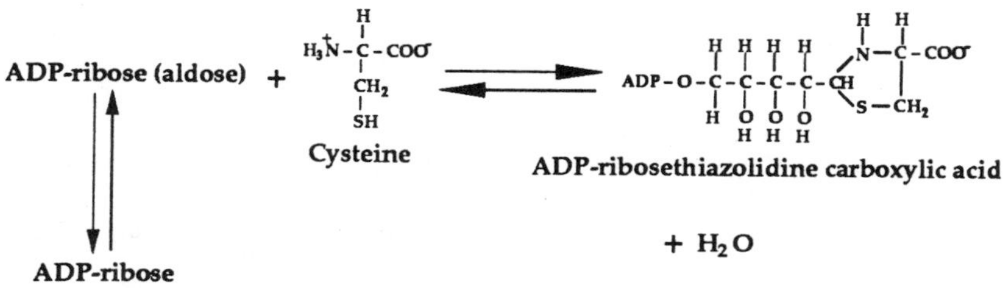

FIGURE 1. Proposed Reaction for the Formation of ADP-ribosethiazolidine Carboxylic Acid from ADP-ribose and Cysteine.

The thiazolidine linkage joining ADP-ribose and cysteine is quite different from the bond formed by the pertussis toxin-catalyzed ADP-ribosylation of proteins on cysteine residues. Pertussis toxin links the cysteine sulfhydryl to the ADP-ribose anomeric carbon in a thioglycoside bond (Meyer et al., 1988). Since the thiazolidine is a ring structure joining the cysteine and ADP-ribose groups, both of the bonds joining the precursors (C–S and C–N bonds) must be broken to recover cysteine and ADP-ribose. The chemical agents could attack the ring structure directly, or alternatively the agents may trap the thiazolidine precursors when in their free forms due to the active equilibrium process of thiazolidines formation (Schubert, 1936; Ratner and Clarke, 1947).

Both of the known examples of NAD:cysteine ADP-ribosyltransferases, pertussis toxin and the human erythrocyte enzyme, are proposed to ADP-ribosylate free sulfhydryl compounds such as cysteine, cysteine methyl ester, or dithiothreitol (Tanuma et al., 1988; Lobban and Van Heyningen, 1988). Following reported reaction conditions (Lobban and Van Heyningen, 1988), in incubations with $[^{14}C]$NAD and 250 mM cysteine, pertussis toxin produced the same hydroxylamine-sensitive ADP-ribosylcysteine that was formed nonenzymatically from ADP-ribose and cysteine. Dithiothreitol was also not ADP-ribosylated by pertussis toxin. These results indicated that under these reaction conditions, neither sulfhydryl compound was directly ADP-ribosylated by pertussis toxin, reactions that would have created a hydroxylamine-stable thioglycoside linkage. Instead, pertussis toxin through its NAD glycohydrolase activity generated ADP-ribose that then reacted nonenzymatically with cysteine.

In summary, a thiazolidine-type of ADP-ribosylcysteine was formed under conditions expected to be encountered in the analysis of cysteine-specific ADP-ribosyltransferases, i.e., high mM concentration of cysteine, sub-mM NAD, and a source of NADase activity. The use of an alternate acceptor thiol compound not readily reactive with ADP-ribose, such as glutathione or N-acetylcysteine may obviate the complications imposed by the ready formation of the thiazolidine-type of ADP-ribosylcysteine linkage under conditions where NADase activity is present.

Supported in part by GM-22982.

References

Frei B, Richter C (1988) Mono(ADP-ribosylation) in rat liver mitochondria. Biochemistry 27:529-535

Horton D, Hutson DH (1963) Developments in the chemistry of thio sugars. Advan Carbohydrate Chem 18:123-199

Hsia JA, Tsai S-C, Adamik R, Yost DA, Hewlett EL, Moss J (1985) Amino acid-specific ADP-ribosylation: sensitivity to hydroxylamine of [cysteine(ADP-ribose)]protein and [arginine(ADP-ribose)]protein linkages. J Biol Chem 260:16187-16191

Kun E, Chang ACY, Sharma ML, Ferro AM, Nitecki D (1976) Covalent modification of proteins by metabolites of NAD^+. Proc Natl Acad Sci USA 73:3131-3135

Lobban MD, Van Heyningen S (1988) Thiol reagents are substrates for the ADP-ribosyltransferase activity of pertussis toxin. FEBS Lett 233:229-232

Meyer T, Koch R, Fanick W, Hilz H (1988) ADP-ribosyl proteins formed by pertussis toxin are specifically cleaved by mercury ions. Biol Chem Hoppe-Seyler 369:579-583

Moss J, Vaughan M (1988) ADP-ribosylation of guanyl nucleotide-binding regulatory proteins by bacterial toxins. Adv Enzymol 61:303-379

Moss J, Yost DA, Stanley SJ (1983) Amino acid-specific ADP-ribosylation: stability of the reaction products of an NAD:arginine ADP-ribosyltransferase to hydroxylamine and hydroxide. J Biol Chem 258:6466-6470

Ratner S, Clarke HT (1947) The action of formaldehyde upon cysteine. J Am Chem Soc 59:200-206

Schubert MP (1936) Compounds of thiol acids with aldehydes. J Biol Chem 114:341-350

Schubert MP (1939) The combination of cysteine with sugars. J Biol Chem 130:601-603

Tanuma S, Kawashima K, Endo H (1988) Eukaryotic mono(ADP-ribosyl)transferase that ADP-ribosylates GTP-binding regulatory G_i protein. J Biol Chem 263:5485-5489

West RE, Moss J, Vaughan M, Liu T, Liu T-Y (1985) Pertussis toxin-catalyzed ADP-ribosylation of transducin: cysteine 347 is the ADP-ribose acceptor site. J Biol Chem 260:14428-14430

Williamson KC, Moss J (1990) Mono-ADP-ribosyltransferases and ADP-ribosylarginine hydrolases: a mono-ADP-ribosylation cycle in animal cells. In: Moss J and Vaughan M (eds) ADP-Ribosylating Toxins and G-Proteins: Insights into Signal Transduction. ASM, Washington, pp 493-510

Molecular and Immunological Characterization of ADP-ribosylarginine Hydrolases

J. Moss, S.J. Stanley, M.S. Nightingale, J.J. Murtagh, Jr., L. Monaco, K. Mishima, H.-C. Chen, M.M. Bliziotes, and S.-C. Tsai

Introduction

Mono-ADP-ribosylation of arginine residues appears to be a reversible modification of proteins in animal and bacterial cells (Williamson and Moss, 1990). NAD:arginine ADP-ribosyltransferases catalyze the forward reaction, whereas ADP-ribosylarginine hydrolases release ADP-ribose, regenerating the (arginine)protein acceptor (Williamson and Moss, 1990). ADP-ribosyltransferases have been examined in animal species and appear to be a family of ~28-39 kDa proteins, that differ in intracellular localization, and in kinetic, physical and regulatory properties (Moss et al., 1980; Yost and Moss, 1983; Tanigawa et al., 1984; Soman et al., 1984; West and Moss, 1986; Peterson et al., 1990). The reaction is stereospecific; the transferase catalyzes the formation of α-ADP-ribosylarginine from β-NAD (Moss et al., 1979). In agreement with the hypothesis that the ADP-ribosylarginine hydrolases catalyze the reversal of the transferase-catalyzed reaction is the observation that the enzyme prefers α-ADP-ribosylarginine as substrate (Moss et al., 1986).

A turkey erythrocyte ADP-ribosylarginine hydrolase was previously purified and characterized (Moss et al., 1985; 1986; 1988). It was a protein of ~39 kDa by gel permeation chromatography (Moss et al., 1988). Hydrolase activity was stimulated in the presence of either Mg^{2+} and dithiothreitol, with optimal activity in the presence of both (Moss et al., 1985). Mg^{2+} stabilized the hydrolase, whereas dithiothreitol accelerated thermal denaturation (Moss et al., 1986). The putative product of the hydrolase reaction, arginine, was modified in the presence of NAD by the turkey erythrocyte transferase, consistent with the conclusion that removal of the ADP-ribose did not chemically alter the guanidino moiety (Moss et al., 1985). In addition to ADP-ribosylarginine, the hydrolase also cleaved 2'-phospho-ADP-ribosylarginine and ADP-ribosylguanidine (Moss et al., 1986). In contrast to the arginine component of the substrate, the ADP-ribose moiety was critical for substrate recognition; both phosphoribosylarginine and ribosylarginine were poor substrates (Moss et al., 1986). ADP-ribose was also a potent inhibitor of the hydrolase, significantly more active than ADP, AMP or adenosine (Moss *et al.*, 1986).

Characterization of ADP-ribosylarginine Hydrolases in Animal Tissues

ADP-ribosylarginine hydrolases were characterized in a variety of animal tissues and species. Hydrolase activity in brain was higher in rat and mouse than it was in pig, dog, sheep, guinea pig, rabbit and calf. In rat tissues, activity was elevated in brain, spleen and testis relative to lung, heart, liver, muscle and kidney. Hydrolase activity was primarily recovered in the soluble fraction.

The hydrolases were partially purified from turkey erythrocytes and mouse, rat, and calf brain. They behaved as ~38 kDa proteins by gel filtration under non-denaturing conditions; the highly purified rat brain hydrolase had a mobility on sodium dodecyl sulfate-polyacrylamide gel electrophoresis (SDS-PAGE) consistent with a protein of 39 kDa. The hydrolases would thus appear to be active in their monomeric state. The partially purified proteins required Mg^{2+} (e.g., calf and guinea pig brain) and in some instances (e.g., mouse and rat brain, turkey erythrocytes), dithiothreitol, for optimal activity.

To define the molecular properties of these ubiquitous enzymes, ADP-ribosylarginine hydrolase was purified ~20,000-fold from rat brain using successive chromatography on DE-52, phenyl Sepharose, hydroxylapatite, organomercurial agarose, and Ultrogel AcA 54. The purified hydrolase represented a major protein at 39 kDa on SDS-polyacrylamide gels. On native, non-denaturing polyacrylamide gels and on isoelectric focusing gels, hydrolase activity appeared to coincide with protein. Rabbit anti-rat hydrolase polyclonal antibodies were immunoaffinity-purified by adsorption on hydrolase linked to nitrocellulose. These antibodies reacted with 39 kDa proteins on immunoblots of partially purified hydrolases from rat, mouse, and calf brains and turkey erythrocytes. These immunological studies are consistent with partial conservation of hydrolase structure across animal species.

Cloning of the ADP-ribosylarginine Hydrolase

To obtain further information regarding the structure of the ADP-ribosylarginine hydrolase, the cDNA was cloned from a rat brain library. The purified rat brain enzyme appeared not to be post-translationally modified at the amino terminus. Based on amino acid sequence obtained from the amino terminus of the intact rat brain hydrolase and from high performance liquid chromatography (HPLC)-purified tryptic peptides, degenerate oligonucleotide and polymerase chain reaction (PCR)-generated cDNA probes were prepared and used to screen a rat brain Lambda ZAP library. Two independent clones were isolated which contained overlapping nucleotide sequences. A composite sequence included a 1086-bp open reading frame. The deduced amino terminal sequence, following the presumed initiator methionine, was identical to the amino terminal sequence of the intact hydrolase, suggesting that the purified protein was neither blocked nor proteolyzed; the amino acid sequence obtained from a tryptic peptide was located just upstream of the TAG codon.

The identity of the cDNA and of the purified sequenced protein was confirmed by expression of the gene in *E. coli* as a glutathione transferase-linked fusion protein. The glutathione transferase-hydrolase fusion protein, purified by chromatography on glutathione beads, exhibited ADP-ribosylarginine hydrolase activity that, like the purified rat brain enzyme, was dependent on Mg^{2+} and dithiothreitol for maximal activity. On immunoblot, the fusion protein reacted with immunoaffinity-purified rabbit anti-rat brain hydrolase polyclonal antibodies. Recombinant hydrolase, released from the fusion protein by thrombin digestion, exhibited a mobility of ~39 kDa on SDS-polyacrylamide gels. A PCR-generated coding region cDNA hybridized to a 1.7 kb band on Northern analysis of poly(A)$^+$ RNA from rat and mouse brain and PC12 cells, but not from bovine, rabbit, and chicken brains or cultured cells of human origin (e.g., IMR32, HL60). Total RNA from rat tissues contained a 1.7 kb mRNA that reacted on Northern analysis with a hydrolase-specific oligonucleotide.

Summary

ADP-ribosylarginine hydrolases were expressed ubiquitously in animal species; in rat tissues, activities were relatively higher in spleen, brain, and testis than in lung, liver, kidney and skeletal muscle. The rat brain hydrolase was purified ~20,000-fold and exhibited one major band having a mobility on SDS-PAGE consistent with a protein of 39 kDa. Affinity-purified rabbit anti-rat brain hydrolase polyclonal antibodies reacted with 39 kDa proteins on immunoblots of partially purified hydrolase preparations from turkey erythrocytes and mouse, rat and calf brains. The hydrolase was encoded by a 1086 bp open reading frame in cDNA cloned from a rat brain library. A PCR-generated coding region cDNA hybridized to a 1.7 kb band on Northern analysis of rat and mouse poly(A)$^+$ RNA. Both the immunological and molecular biological data are consistent with partial conservation of hydrolase structure across animal species.

References

Moss J, Jacobson MK, Stanley SJ (1985) Reversibility of arginine-specific mono(ADP-ribosyl)ation: identification in erythrocytes of an ADP-ribose-L-arginine cleavage enzyme. Proc Natl Acad Sci USA 82:5603-5607

Moss J, Oppenheimer NJ, West Jr. RE, Stanley SJ (1986) Amino acid specific ADP-ribosylation: substrate specificity of an ADP-ribosylarginine hydrolase from turkey erythrocytes. Biochemistry 25:5408-5414

Moss J, Stanley SJ, Oppenheimer NJ (1979) Substrate specificity and partial purification of a stereospecific NAD- and guanidine-dependent ADP-ribosyltransferase from avian erythrocytes. J Biol Chem 254:8891-8894

Moss J, Stanley SJ, Watkins PA (1980) Isolation and properties of an NAD- and guanidine-dependent ADP-ribosyltransferase from turkey erythrocytes. J Biol Chem 255:5838-5840

Moss J, Tsai S-C, Adamik R, Chen H-C, Stanley SJ (1988) Purification and characterization of ADP-ribosylarginine hydrolase from turkey erythrocytes. Biochemistry 27:5819-5823

Peterson JE, Larew JS-A, Graves DJ (1990) Purification and partial characterization of arginine-specific ADP-ribosyltransferase from skeletal muscle microsomal membranes. J Biol Chem 265:17062-17069

Soman G, Mickelson JR, Louis CF, Graves DJ (1984) NAD:guanidino group specific mono ADP-ribosyltransferase activity in skeletal muscle. Biochem Biophys Res Commun 120:973-980

Tanigawa Y, Tsuchiya M, Imai Y, Shimoyama M (1984) ADP-ribosyl-transferase from hen liver nuclei. J Biol Chem 259:2022-2029

West Jr. RE, Moss J (1986) Amino acid specific ADP-ribosylation: specific NAD:arginine mono-ADP-ribosyltransferases associated with turkey erythrocyte nuclei and plasma membranes. Biochemistry 25:8057-8062

Williamson KC, Moss J (1990) Mono-ADP-ribosyltransferases and ADP-ribosylarginine hydrolases: a mono-ADP-ribosylation cycle in animal cells. In: Moss, J and Vaughan, M (eds) ADP-ribosylating toxins and G proteins: Insights into signal transduction. American Society for Microbiology, Washington, D.C., pp 493-510

Yost DA, Moss J (1983) Amino acid-specific ADP-ribosylation. J Biol Chem 258:4926-4929

Characterization of the Family of Mammalian Genes Encoding ADP-ribosylation Factors

Randy S. Haun, Inez M. Serventi, Su-Chen Tsai, Chii-Ming Lee, Eleanor Cavanaugh, Joel Moss, and Martha Vaughan

ADP-ribosylation factors (ARFs) are ~20-kDa guanine nucleotide-binding proteins that stimulate the *in vitro* cholera toxin-catalyzed ADP-ribosylation of the α subunit of G_s (the stimulatory GTP-binding protein of the adenylyl cyclase system) (Kahn and Gilman, 1984, 1986; Bobak *et al.*, 1990). Membrane (mARF) and soluble (sARF I and sARF II) ADP-ribosylation factors that enhance cholera toxin-catalyzed ADP-ribosylation of $G_{s\alpha}$ and simple guanidino compounds were identified in bovine brain (Tsai *et al.*, 1987, 1988). Subsequent molecular cloning of the cDNAs that encode the ARFs revealed the existence of a larger family of related guanine nucleotide-binding proteins. To date, six mammalian ARF cDNAs (ARFs 1 to 6) have been isolated from bovine and/or human libraries (Sewell and Kahn, 1988; Price *et al.*, 1988; Bobak *et al.*, 1989; Monaco *et al.*, 1990; Tsuchiya *et al.*, 1991). Based on the deduced amino acid sequences, mammalian ARFs represent a family of 20-kDa guanine nucleotide-binding proteins clearly different from members of the *ras* and *ras*-like (20-25 kDa) superfamily. In particular, the signature sequences of the regions believed to be involved in guanine nucleotide binding and GTP hydrolysis in the ARF proteins more closely resemble those in the heterotrimeric G protein α subunits than those found in the *ras, rho, rac, rap, ral*, and *rab* families (Price *et al.*, 1990). The GTP-hydrolysis domain GXXXXGK is completely conserved across the mammalian ARFs as GLDAAGK. The sequence DVGG which forms the DXXG consensus sequence that has been proposed to coordinate binding to Mg^{2+} and the β phosphate of GDP and the sequence NKQD (the NKXD consensus sequence) which is thought to contribute to the specificity of interaction with the purine ring of GTP, are found in all the deduced ARF sequences (Price *et al.*, 1990).

Sequence similarities among the ARFs extend beyond these guanine nucleotide-binding and GTP-hydrolysis domains and combined with the putative protein sizes can be used to divide the ARF family into three classes (Tsuchiya *et al.*, 1991). Class I is composed of ARFs 1 to 3, 181 amino acids

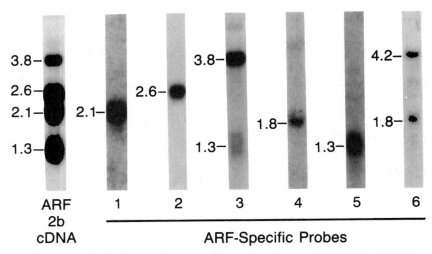

Figure 1. Hybridization of rat brain poly(A)+ RNA with ARF2 cDNA and ARF-specific oligonucleotides.

in length; ARF 4 and 5 each contain 180 amino acids and belong to class II; and ARF 6 with 175 amino acids forms class III. The class I ARFs are 95-96% identical in deduced amino acid sequences and have differences essentially only near the amino and carboxyl termini. ARF 4 and 5 (class II) are 90% identical. They share 79-80% sequence identity with the class I proteins and differ from them within the body of the coding region as well as near the amino and carboxyl termini. In contrast, ARF 6 differs considerably in deduced amino acid sequence from either the class I (68-69%) or the class II (64%) ARFs.

Despite the similarities in deduced amino acid sequences and sizes of ARF proteins, Northern and Southern analysis and chromosomal localization using ARF-specific oligonucleotides revealed that ARF mRNAs are the products of six different genes and their sizes are conserved across mammalian species (Tsuchiya et al., 1989, 1991). As depicted in Fig. 1, single major mRNAs were detected for ARFs 1, 2, 4, and 5 of 2.1, 2.6, 1.8, and 1.3 kb, respectively; whereas, ARF 3- and 6-specific oligonucleotides detected two mRNAs of 3.8 and 1.3 kb or 4.2 and 1.8 kb, respectively.

Expression of the ARF gene products differs in different tissues and times of development and is widespread in mammalian species. mRNAs corresponding to ARFs 1, 3, 4, 5, and 6 were detected in poly(A)$^{+}$ RNA from human, bovine, rabbit, rat, and mouse; ARF 2 mRNA has been detected in all these species except human (Bobak et al., 1989; Monaco et al., 1990; Tsuchiya et al., 1990, 1991). Based on the fact that highly specific ARF coding region probes hybridize across mammalian species, it appears that there is considerable conservation in nucleotide sequence.

To understand the molecular mechanisms responsible for the expression and regulation of this family of guanine nucleotide-binding proteins, the mammalian genes were isolated and characterized. Sequence analysis of the class I genes (human ARF 1 and 3 and bovine ARF 2) indicates that these genes share a common linear arrangement. Each is composed of five exons and four introns. The organization of the genes partitions the functional domains of the ARF proteins into separate exons, similar to other GTP-binding protein genes (e.g., $G_{s\alpha}$, N-*ras*). The first exon contains only untranslated sequences. Translation initiates within exon 2 which also encodes the GTP-hydrolysis domain, GXXXXGK. The sequence DVGG is encoded within exon 3; whereas, in contrast to other GTP-binding protein genes, intron 4 divides the sequence NKXD between exons 4 and 5. Finally, exon 5 encodes the carboxy-terminal portion of the class I ARFs and contains a long 3'-untranslated region.

In each class I ARF, introns are located at the same position within the coding sequence. Preliminary characterization of the human ARF 4 and 5 genes indicates that the coding regions for these class II genes are disrupted at identical locations that differ from those of the class I ARF genes. Therefore, the division of the ARFs into classes based upon similarities in protein structure is consistent with the structure of the respective genes.

Analysis of the ARF 3 gene also identified the mechanism responsible for the appearance of two ARF 3 mRNA species. The cryptic polyadenylation signal AACAAA is found at position 1091 (relative to the site of transcription initiation) and is contained within the fifth exon corresponding to the 3'-untranslated region of the mRNA. A polypyrimidine tract (which is normally adjacent to authentic polyadenylation signals) is found 26 bp downstream from this sequence and may explain why this signal is used to generate the 1.2 kb ARF 3 mRNA. Isolation and characterization of the ARF 6 gene will be required to determine whether it also uses differential polyadenylation to produce the two ARF 6 mRNAs or whether a different mechanism (e.g., alternative splicing) is utilized.

Initial sequence analysis indicates that the class I genes do not possess typical promoter elements found in many polymerase II-transcribed genes (e.g., TATA and CAAT boxes). Similar to other genes lacking these features, the 5'-flanking regions are GC-rich and contain many sequence elements (CCGCCC) that may bind the transcription factor Sp1. Studies have been initiated to analyze the elements responsible for the expression and regulation of the ARF genes. An analysis of the 5'-flanking region of these genes may help elucidate the mechanisms that are utilized for their tissue-preferential and developmentally-regulated expression.

References

Bobak DA, Nightingale MS, Murtagh JJ, Price SR, Moss J, Vaughan M (1989) Molecular cloning, characterization, and expression of human ADP-ribosylation factors: Two guanine nucleotide-dependent activators of cholera toxin. Proc Natl Acad Sci USA 86:6101-6105

Bobak DA, Tsai S-C, Moss J, Vaughan M (1990) Enhancement of cholera toxin ADP-ribosyltransferase activity by guanine nucleotide-dependent ADP-ribosylation factors. In: Moss J and Vaughan M (eds) ADP-ribosylating toxins and G proteins: Insights into signal transduction. American Society for Microbiology, Washington, DC, pp 439-456

Kahn RA, Gilman AG (1984) Purification of a protein cofactor required for ADP-ribosylation of the stimulatory regulatory component of adenylate cyclase by cholera toxin. J Biol Chem 259:6228-6234

Kahn RA, Gilman AG (1986) The protein cofactor necessary for ADP-ribosylation of G_s by cholera toxin is itself a GTP-binding protein. J Biol Chem 261:7906-7911

Monaco L, Murtagh JJ, Newman KB, Tsai S-C, Moss J, Vaughan M (1990) Selective amplification of an mRNA and related pseudogene for a human ADP-ribosylation factor, a guanine nucleotide-dependent protein activator of cholera toxin. Proc Natl Acad Sci USA 87:2206-2210

Price SR, Nightingale M, Tsai S-C, Williamson KC, Adamik R, Chen H-C, Moss J, Vaughan M (1988) Guanine nucleotide-binding proteins that enhance choleragen ADP-ribosyltransferase activity: Nucleotide and deduced amino acid sequence of an ADP-ribosylation factor cDNA. Proc Natl Acad Sci USA 85:5488-5491

Price SR, Barber A, Moss J (1990) Structure-function relationships of guanine nucleotide-binding proteins. In: Moss J and Vaughan M (eds) ADP-ribosylating toxins and G proteins: Insights into signal transduction. American Society for Microbiology, Washington, DC, pp 397-424

Sewell JL, Kahn RA (1988) Sequences of the bovine and yeast ADP-ribosylation factor and comparison to other GTP-binding proteins. Proc Natl Acad Sci USA 85:4620-4624

Tsai S-C, Noda M, Adamik R, Chang P, Chen H-C, Moss J, Vaughan M (1988) Stimulation of choleragen enzymatic activities by GTP and two soluble proteins purified from bovine brain. J Biol Chem 263:1768-1772

Tsai S-C, Noda M, Adamik R, Moss J, Vaughan M (1987) Enhancement of choleragen ADP-ribosyltransferase activities by guanyl nucleotides and a 19-kDa membrane protein. Proc Natl Acad Sci USA 84:5139-5142

Tsuchiya M, Price SR, Nightingale MS, Moss J, Vaughan M (1989) Tissue and species distribution of mRNA encoding two ADP-ribosylation factors, 20-kDa guanine nucleotide binding proteins. Biochemistry 28:9668-9673

Tsuchiya M, Price SR, Tsai S-C, Moss J, Vaughan M (1991) Molecular identification of ADP-ribosylation factor mRNAs and their expression in mammalian cells. J Biol Chem 266:2772-2777

Purification and characterisation of NAD:Arginine mono ADP-Ribosyl Transferase from chicken erythrocytes; identification of some enzyme inhibitors.

Jamal Sabir, Manoochehr Tavassoli and Sydney Shall.

Cell and Molecular Biology Laboratory, School of Biological Sciences, University of Sussex, Brighton, East Sussex, BN1 9QG, ENGLAND.

Introduction

We are beginning a systematic study of the molecular physiology of the mono-ADP-Ribosyl Transferase enzymes; there are known to be at least three classes of such enzymes, specific for arginine, cysteine and histidine residues, respectively. We have begun by purifying and characterising an arginine specific mono-ADP-Ribosyl transferase from chicken erythrocytes.

Purification.

As a first step we have isolated and purified an arginine mono-ADP-Ribosyl Transferase from chicken erythrocytes. Mono ADP-Ribosyl Transferases have been purified from a number of sources previously. We have used slightly different procedures from those described in the literature. A seven step purification procedure included a high speed centrifugation of a cell homogenate, sequential chromatography through hydrophobic phenyl-sepharose, blue-sepharose, and green-agarose, followed by a gel filtration step and finally a separation on anion exchange chromatography on an FPLC mono Q system. The overall degree of purification was in excess of 800,000 fold, and there was greater than 100% recovery of enzyme activity, which probably indicates that there were inhibitory compounds in the original cell extract. The highest specific enzyme activity that we obtained was 15.5 umol/min/mg. The enzyme shows a heavy main band of 26.5 kDa, on separation with a "Phast" gel system. On G-75 Sephadex, the relative molecular mass appears to be 27.5 kDa.

Enzymic properties.

The enzyme activity is routinely assayed with poly-arginine and NAD as substrates, and the acid-insoluble radio-activity is estimated. The enzyme reaction is linear for about 20 to 30 minutes. There is a fairly sharp pH optimum at around 8.0, and a temperature optimum of about 30'C. We have determined the Km for NAD under these optimal conditions; and using poly-arginine as the second substrate, the Km is estimated to be 130.0+-3.2 uM. Using a different assay system, we have estimated the Km for agmatine; the Km is estimated to be 2.982+-0.009 mM. Similarly, we have estimated the Km for arginine methyl ester; the value estimated is 25.63+-0.07 mM.

The product of the reaction with NAD and poly-arginine, was analyzed by treating the enzymic product after reaction with snake venom phosphodiesterase; the radioactive products were separated and identified on PEI-Cellulose thin layer chromatography. The only product observed was AMP; no phosphoribosyl AMP was detected.

The effect of several salts on the enzyme activity was determined. It was observed that 1.0 and 10.0 mM sodium, potassium and ammonium chloride all stimulated the enzyme activity when added to an enzyme reaction performed in Tris-Cl buffer (pH 7.0, 50 mM). All three salts however, inhibited the enzyme activity at

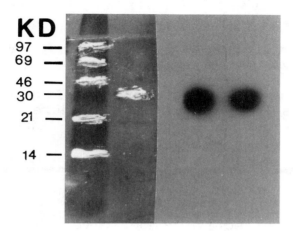

Figure. 1. Enzyme assay after SDS polyacrylamide gel electrophoresis; an "activity" gel assay. The left-hand column shows the sizes of the molecular weight markers, which appear in the second column. The third column marks the position of the major protein stained band; and the last two columns show the "activity" gel assay after autoradiography.

higher concentrations (100 and 500 mM). Similar results were observed with calcium and magnesium chloride. The inhibition by salt seems to be independent of the specific salt, and would therefore appear to be due simply to the ionic strength of the solution.

Stability of the purified enzyme.

The stability of the purified enzyme was measured at 4'C in the presence and absence of 50% (v/v) propylene glycol. The glycol clearly stabilised the enzyme somewhat, but under both conditions the enzyme was relatively stable; the rate of loss of enzyme activity at 4'C was 0.0160 %/hour, and 0.0133%/hour, in the presence and absence of propylene glycol, respectively. At higher temperatures the enzyme was notably less stable, although here too the propylene glycol increased the stability.

Enzyme assay after SDS PAGE; an "activity" gel assay.

We have developed an activity gel assay for this enzyme. This procedure shows a relative molecular mass for the enzyme activity of 28 KDa (Fig. 1).

Table 1

Some effective competitive inhibitors of chicken arginine mono-ADP-Ribosyl transferase.

Compound	K_i (uM)
1,4 Naphthoquinone	4.6
5,8 dihydroxy-1,4-naphthoquinone	29.0
4-amino-1-naphthol hydrochloride	32.0
1,2 naphthoquinone	49.0

Enzyme inhibitors.

Enzyme inhibitors are often useful reagents in analyzing both the enzyme mechanism as well as the molecular physiology of the enzyme in intact cells. We have begun a survey of likely inhibitors, and have established that naphthoquinones are significant inhibitors of this enzyme; 1,4 naphthoquinone is a competitive inhibitor with respect to NAD, and shows a K_i of about 5.0 uM.

Several other naphthoquinones are also enzyme
inhibitors, although they are all weaker than
1,4 naphthoquinone. All four naphthoquinones
tested were competitive inhibitors (Table 1).

A number of other compounds were found
to be weaker inhibitors. It is of interest
that formyl-luminol, 3-formyl-aminobenzamide,
and 3-ureidobenzamide were weaker inhibitors.
We found that 3-guanidino-benzamide, 3-iodo-
benzyl-guanidine (MIBG) and 4-formyl-
aminobenzamide were extremely weak inhibitors.
Indeed, in our hands, 3-methoxy-benzamide and
3-amino-benzamide were not at all inhibitory
at 1.0 mM.

Amino acid sequence.

The N-terminal 13 amino acids have been
sequenced in a gas-phase sequencer.

Target Proteins for Arginine-Specific ADP-Ribosyltransferase in Chickens

Makoto Shimoyama, Koichi Mishima, Seiji Obara, Kazuo Yamada, Masaharu Terashima and Mikako Tsuchiya

Eukaryotic arginine-specific ADP-ribosyltransferases are present in various types of tissues and cells. Although some of the related enzymes have been purified and characterized (Moss et al., 1980, Tanigawa et al., 1984, Peterson et al., 1990, Larew et al., 1991) the physiological role of the enzymes remains obscure, presumably because much less is known of the endogenous acceptor protein. Identification and characterization of target proteins for endogenous arginine-specific ADP-ribosylation provide clues to the physiological function of ADP-ribosylation.

Direct Evidence for Endogenous GTP-dependent ADP-Ribosylation of Gsα Concomitant with Increase in the Basal Adenylyl Cyclase Activity

Incubation of chicken spleen cell membranes with [adenylate-^{32}P]NAD in the presence of GTP and subsequent analysis of the membrane with SDS-PAGE followed by autoradiography showed the incorporation of ^{32}P into a 45 kDa protein. (Obara et al., 1991). This incorporation was greatly diminished by adding arginine during the incubation. The radioactive compound released by glycine-NaOH treatment from the area corresponding to the 45 kDa protein on the gels was identified as ADP-ribose by reverse-phase HPLC. Treatment of the membrane with or without cholera toxin in the presence of GTP and labelled NAD showed that the major labelled band was the same and corresponded to the 45 kDa protein, except that the toxin enhanced modification of the 45 kDa protein. When the membranes were preincubated with unlabelled NAD, under conditions used for endogenous ADP-ribosylation, and then were incubated with cholera toxin and labelled NAD, labelling of the 45 kDa protein greatly diminished. Densitometric scanning of the autoradiograph revealed increase in the Gsα modification by 0.1, 1, 10, and 100 μM GTP to be 1.4-, 1.8-, 1.9-, and

2.2-fold over the control value, respectively. These results indicate that chicken spleen cell membrane contain the short form of Gsα and that the Gsα can be modified GTP-dependently by membrane-intrinsic ADP-ribosyltransferase (Obara et al., 1991).

The GTP-dependent ADP-ribosylation of Gsα in the membrane led to subsequent activation of adenylyl cyclase. As shown in Fig. 1B, basal adenylyl cyclase activity increased with increase in the preincubation time for ADP-ribosylation. In the presence of 1 mM novobiocin, a potent inhibitor of arginine-specific ADP-ribosyltransferase (Banasik et al., 1989), during the preincubation, activation of adenylyl cyclase was completely blocked. The time course of endogenous ADP-ribosylation of Gsα increased with increase in the incubating time, while a complete block occurred with 1 mM novobiocin (Fig. 1A).

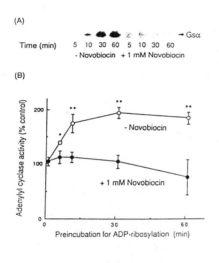

Fig.1. ADP-ribosylation of Gsα of chicken spleen cell membrane concomitant with an increase in adenylyl cyclase activity. (A) The membranes (100 μg) were incubated for the indicated time in the presence of 0.1 mM labelled NAD plus 0.1 mM GTP, with or without 1 mM novobiocin and analyzed by SDS-PAGE followed by autoradiography. (B) The membranes were preincubated for the indicated time with unlabelled NAD plus GTP in the presence or absence of novobiocin and subjected to adenylyl cyclase assay (Salomon et al., 1974). The figure in B represents the mean±S.D. of four experiments. *P<0.05; compared to control, **P<0.01; compared to control.

Heretofore no direct evidence was reported for the endogenous ADP-ribosylation of Gsα. We demonstrated evidence for endogenous GTP-dependent ADP-ribosylation of Gsα concomitant with increase in adenylyl cyclase activity with chicken spleen cell membrane (Obara et al., 1991). ADP-ribosylation of Gsα by cholera toxin has several consequences. Modification of Gs promotes dissociation of the α and β plus γ subunits (Kahn and Gilman, 1984) and this dissociation of regulatory proteins may play an important role in activating adenylyl cyclase (Cerione et al., 1985). All these results taken together strongly indicate that the endogenous arginine-specific ADP-ribosyltransferase regulates the basal activity of adenylyl cyclase by altering structure of the Gs subunit.

Co-Localization of p33 and Arginine-Specific ADP-Ribosyltransferase in the Granules of Chicken Peripheral Polymorphonuclear-Pseudoeosinophilic Granulocytes (Heterophils)

Among various chicken tissues and peripheral blood cells tested, peripheral heterophils showed the highest activity of ADP-ribosyltransferase (Mishima et al., 1991). In man and in such other mammals as the dog, these leukocytes possess neutral-staining granules (neutrophils) (Sturkie, 1986). Percoll density gradient centrifugation of the postnuclear fraction of the heterophils showed the co-localization of ADP-ribosyltransferase and 33 kDa protein (p33), a major acceptor protein for the enzyme, in the granule fraction (Fig. 2).

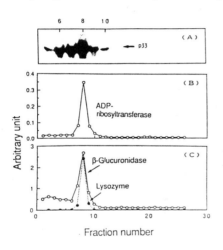

Fig. 2. Percoll density gradient centrifugation of heterophil postnuclear fraction. Heterophil postnuclear fraction (2 mg protein) was fractionated on Percoll gradients. After removal of Percoll from the pooled fraction by centrifugation, individual gradient fractions were subjected to the ADP-ribosylation system containing purified ADP-ribosyltransferase and labelled NAD, followed by SDS-PAGE and autoradiography (A), and to assays for ADP-ribosyltransferase (B) and for granule marker enzymes (C). Numbers above and below are related to the fractions on the Percoll gradient.

The enzyme and p33 were purified from the postnuclear pellet fraction to apparent homogeneity, respectively (Mishima et al., 1991). The molecular mass of the heterophil enzyme was estimated by SDS-PAGE to be 27.5 kDa. Kinetic properties of the heterophil enzyme were similar to these of liver enzyme (Tanigawa et al., 1984). SDS-PAGE analysis after limited trypsin proteolysis of p33s, purified from chicken heterophils and liver, showed much the same pattern. Thus, it appears that ADP-ribosyltransferase and p33 present in chicken peripheral heterophils are identical to those in the liver, respectively.

Heterophils were collected from peritoneal exudate induced in the chicken, permeabilized with saponin, and incubated with labelled NAD. A predominant incorporation of ^{32}P into p33 was detected (Mishima et al., 1991).

In the avian species, the liver serves as a predominant extramedullary hematopoietic organ (Romanoff, 1960). With hematoxyllin-eosin staining, we confirmed histologically the presence of numerous heterophils in interlobular connective tissue of chicken liver. These results together with the data described above suggest that the p33 and the enzyme originally found in the liver (Tanigawa et al., 1984, Mishima et al., 1989) derive from heterophils present in interlobular connective tissue of the liver.

References

Banasik, M., Abed, N. A. N., Saito, I., Komura, H., Ueda, K. 7th International Conference on Cyclic Nucleotides, Calcium and Protein Phosphorylation (Kobe) Abst.534; 1989.

Cerione, R. A., Staniszewski, C., Caron, M. G., Lefkowitz, R. J., Codina, J., Brown, A. M., Birnbaumer, L. A role for Ni in the hormonal stimulation of adenylate cyclase. Nature. 318: 293-295; 1985.

Kahn, R. A., Gilman, A. G. ADP-ribosylation of Gs promotes the dissociation of its α and β subunits. J. Biol. Chem. 259: 6235-6240; 1984.

Larew, J. S-A., Peterson, J. E., Graves, D. J. Determination of the kinetic mechanism of arginine-specific ADP-ribosyltransferase using a high performance liquid chromatographic assay. J. Biol. Chem. 266:52-57; 1991.

Mishima, K., Tsuchiya, M., Tanigawa, Y., Shimoyama, M. DNA-dependent mono(ADP-ribosyl)ation of p33, an acceptor protein in hen liver nuclei. Eur. J. Biochem. 179: 267-273; 1989.

Mishima, K., Terashima, M., Obara, S., Yamada, K., Shimoyama, M. Arginine-specific ADP-ribosyltransferase and its acceptor protein p33 in chicken polymorphonuclear cells. J. Biochem.. (in press); 1991.

Obara, S., Yamada, K., Yoshimura, Y., Shimoyama, M. Evidence for endogenous GTP-dependent ADP-ribosylation of stimulatory guanine nucleotide-binding protein Gsα. Eur. J. Biochem. (in press) ; 1991.

Peterson, J. E., Larew, J. S.-A., Graves, D. J. Purification and partial characterization of arginine-specific ADP-ribosyltransferase from skeletal muscle microsomal membranes. J. Biol. Chem. 265: 17062-17069; 1990.

Romanoff, A.L; Peterson J. E., ed. The Avian Embryo: Structural and Functional Development, The Macmillan Co.,New York, 1960: p.571-1225.

Salomon, Y., Londos, C., Rodbell, M. A highly sensitive adenylate cyclase assay. Anal Biochem 58: 541-548; 1974.

Sturky, P. D. Body fluids: Blood. In: Sturky, P. D., ed. Avian physiology. New York: Springer-Verlag; 1986:p. 102-129.

Tanigawa, Y., Tsuchiya, M., Imai, Y., Shimoyama, M. ADP-ribosyltransferase from hen liver nuclei. J. Biol. Chem. 259: 2022-2029; 1984.

NAD+ BIOSYNTHESIS IN HUMAN PLACENTA : CHARACTERIZATION OF HOMOGENEOUS NMN ADENYLYLTRANSFERASE

Emanuelli M. , Raffaelli N. , Ruggieri S. , Balducci E. , Natalini P. , Magni G.

INTRODUCTION

The nuclear location of NMN adenylyltransferase (NMNAT), the last enzyme in the main biosynthetic pathway of NAD+, prompted us to investigate about a possible involvement of this enzyme in the regulation of the cellular activity by hypothesizing its eventual relationship with poly(ADP-ribose) polymerase (ADPRP), another chromatin bound enzyme. This hypothesis was based on the discovery of an inhibitory effect exerted by NMNAT on ADPRP in vitro in reconstituted systems, composed by enzymes purified both from different sources (heterologous system) (1) and from identical sources (homologous system) (2). In order to verify such a phenomenon in man and to better elucidate its functional significance we carried out the purification of NMNAT from human placenta, where the system involved in NAD+ comsumption has been fully characterized (3).

METHODS

Assay: Enzyme activity was routinely tested by a continuous spectrophotometric coupled enzyme assay (4). The standard reaction mixture consisted of 60 mM Hepes buffer, pH 7.6, 1.18 mM NMN, 1.47 mM ATP, 20.7 mM $MgCl_2$, 35 mM semicarbazide hydrochloride, 0.45% (v/v) ethanol, 7.8 units of yeast alcohol dehydrogenase, 0.59 mg/ml bovine serum albumin and an appropriate amount of sample to be assayed, in a final volume of 0.85 ml.The reaction was started by the addition of NMN and continuously followed at 340 nm in a Beckman DU-70 spectrophotometer. The temperature was maintained at 37°C. A different assay procedure, based on HPLC techniques (5), was adopted when the kinetic of the reaction was studied in the backward direction. One enzyme unit is defined as the amount of enzyme which catalyzes the synthesis of 1 μmol of NAD+ per minute at 37°C.

Purification: All operations were performed at 4°C. Term human placentae were obtained immediately after delivery and immersed in 0.129 M sodium citrate. The tissue was trimmed free of adherent membranes, washed, cut into small pieces, blotted on paper towels, suspended in 4 volumes of 100 mM buffer A (potassium phosphate buffer, pH 7.4, containing 1 mM DTT, 1 mM MgCl2, 0.5 mM EDTA and 1mM PMSF) and homogenized in a Waring Blender. The homogenate was centrifuged at 13,000 x g for 20 min and the supernatant was referred as the crude extract. To the crude extract was added dropwise 1 M acetic acid until pH 5.0 was reached, the pellet was then collected by centrifugation at 25,000 x g for 10 min and dissolved in a small volume of 100 mM buffer A containing 3 M KCl. The suspension was adjusted to pH 7.4, stirred 30 min and centrifuged at 25,000 x g for 20 min. The pellet was then resuspended and centrifuged at 25,000 x g for 20 min. The resulting supernatants were pooled and submitted to the purification procedure illustrated in Table I.

TABLE I

STEP	TOTAL PROTEIN (mg)	TOTAL ACTIVITY (Units)	PURIFICATION (x-fold)	YIELD (%)	SPEC.ACT. (U/mg)
crude extract	11696	4867	--	100	0.00042
pH 5.0 fraction	1707	4554	6.4	94	0.0027
Phenyl Sepharose[a]	144	4560	76	94	0.0317
Green A[b]	6.4	3664	1364	75	0.573
Hydroxyapatite[c]	1.6	2198	3338	45	1.402
TSK Phenyl5PW[d]	0.116	1099	22548	23	9.47

[a] equilibrated with 100 mM buffer A, 3 M KCl; eluted with a linear gradient from 2 to 0 M KCl in 100 mM buffer A.
[b] equilibrated with 100 mM buffer A; eluted with a linear gradient from 0 to 1.2 M KCl in 10 mM buffer A.
[c] equilibrated with 5 mM buffer A; eluted with a linear gradient from 5 to 50 mM potassium phosphate pH 7.4, containing 2 M KCl.
[d] equilibrated with 50 mM buffer A, 2 M KCl; eluted with a discontinuous gradient from 2 to 0 M KCl in 50 mM buffer A.

Polyacrylamide Gel Electrophoresis: The homogeneity of purified enzyme was established by analytical polyacrylamide gel electrophoresis carried out in homogeneous buffer system (0.015 M sodium borate, pH 9.0). After electrophoresis gels were either stained for protein or sliced into 2 mm sections to correlate the stained band with the enzyme activity. Gel electrophoresis under denaturating conditions was conducted essentially as described by Laemmli in 15% polyacrylamide gels.

Isoelectrofocusing: Isoelectrofocusing was conducted on a 110-ml LKB 8100 ampholine column filled with a linear, 0-65 % (w/v), gradient of glycerol containing 1 % (w/v), pH 3.5-10, ampholine carrier ampholytes.

Molecular weight determinations: The molecular weight of the purified enzyme was determined by gel filtration on a Superose 12 FPLC column by elution with 50 mM buffer A 0.5 M KCl, pH 7.2, at a flow rate of 0.5 ml/min.

Amino Acid Analysis: Samples of protein were hydrolyzed for 45 min at 155°C in 6 N HCl, in sealed evacuated tubes, as described by Hare. Cysteine was determined as cysteic acid after performing acid oxidation. The analysis was performed on a Chromakon 500 (Kontron Instruments) amino acid analyzer and the amino acids were detected after post-column reaction with o-phthalaldehyde.

Kinetic analysis: Kinetic parameters were determined on homogeneous preparation of NMNAT. The effect of pH on maxi mum velocity was studied at 37°C by using a 50 mM Bis-Tris and Tris-HCl buffering mixture. The activity was determined according to (5).

RESULTS AND DISCUSSION

The purified NMNAT, in contrast with the yeast enzyme (4), was stable in 50 mM buffer A 1 M KCl, pH 6.8, for several months, but it was very sensitive to freezing and thawing. The maintenance of activity is, in any storage condition, strictly dependent on the presence of high KCl concentrations, whereas the stability exhibited by the yeast enzyme seems to be independent upon the KCl presence. The final preparation, submitted to polyacrylamide gel electrophoresis under non denaturating conditions, showed one diffuse band with an Rm of 0.25. Determination of NMNAT activity in a parallel gel showed a perfect correspondence with the stained diffuse band. The molecular weight of the native enzyme, estimated as described under Methods, was 132,000 ± 10,000, in agreement with the value found for the enzyme purified to homogeneity from bull testis (unpublished results). Both in the presence and in the absence of 2-mercaptoethanol a single band was observed upon SDS PAGE of Mr 33,000 ± 1,000, suggesting that the enzyme is an oligomer composed of four apparently identical subunits, without interchain disulfide bridges. The amino acid composition of native enzyme (Table II) showed a relative excess of the acidic amino acid residues. Isoelectrofocusing experiments in the absence of urea gave

Table II

Amino acid	%[**]	Amino acid	%[**]
Aspartic and asparagine	11.1	Tyrosine	3.6
Glutamic acid and glutamine	12.7	Methionine	1.6
Serine	6.3	Valine	7.9
Histidine	3.0	Phenylalanine	3.3
Glycine	6.4	Isoleucine	4.6
Threonine	4.8	Leucine	10.3
Arginine	6.4	Lysine	9.1
Alanine	6.6	Half-Cystine	2.2

[**]Moles of amino acid in 100 moles of all amino acids. Values are the average of four determinations

multiple pI's values in the acidic range, namely, at pH 4.7, 5.7 and 6.6. This behavior is in agreement with that observed both for yeast (4) and chicken enzyme (6) and can be ascribed to aggregation of the enzyme molecules, as observed for other nuclear proteins (7). The behavior of the enzyme activity at different H^+ concentrations was consistent with a plateau ranging from pH 6.0 to pH 9.0. The results of the kinetic studies obtained by analyzing the reaction in both directions are summarized in Table III. NMNAT exhibits linear kinetic with respect to NMN, ATP, NAD^+ and PPi; our Km values are one order of magnitude lower compared with those determined for NMNAT purified from other sources (4,8). The product inhibition studies (not shown) diagnosed a mechanism of the Ordered Bi-Bi type, according to Cleland nomenclature. All available data seem to indicate NMN as the first substrate to bind and PPi as the last product to be released. A variety of compounds including pyrimidine and

Table III

Substrate	Michaelis Constant (mM)
ATP	0.023
NMN	0.038
NaMN	0.180
NAD⁺	0.067
PPi	0.125

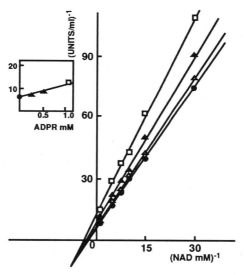

Fig.-1- Inhibition by ADP-ribose at various fixed concentrations:(●) none, (△) 0.25 mM, (▲) 0.5 mM, (▢) 1,0 mM. Inset: secondary plot of intercept vs. ADP-ribose.

purine bases, oxy- and deoxynucleosides, oxy- and deoxynucleotides, cyclic AMP, several halogenated and methylated nucleobases, oxy- and deoxyribose, oxy- and deoxyribose 1-phosphate, glucose, glucose 1-phosphate, glucose 6-phosphate, glucose 1,6-diphosphate, fructose, fructose 1,6-diphosphate, were examined for their ability to cause variations of enzyme activity. Among them only ß-NMNH and ADP-ribose resulted effective on depressing the enzyme activity. The inhibition exerted by ADP-ribose toward NAD⁺ was noncompetitive showing a Ki=950 µM (Figure 1). The level of ADP-ribose necessary for substantial inhibition of NMNAT appears higher than the levels reported for mammalian cells in vivo. However, during particular cellular conditions, probably sufficient elevations in ADP-ribose content are achieved, thus explaining its possible action as transient regulator of NMNAT.

REFERENCES

1. Ruggieri,S., Gregori,L., Natalini, P., Vita,A., Emanuelli,M., Raffaelli,N., and Magni,G.(1990) Biochemistry 29: 2501-2506.
2. Raffaelli,N., Emanuelli,M., Magni,G., Natalini,P., Ruggieri,S.(1990) Ital. Biochem. Soc. Trans. (IBST) 1: 295.
3. Ushiro,H., Yokoyama,Y., and Shizuta,Y.(1987) J. Biol.Chem. 262: 2352-2357.
4. Natalini,P., Ruggieri,S., Raffaelli,N., and Magni,G.(1986) Biochemistry 25: 3725-3729.
5. Stocchi,V., Cucchiarini,L., Magnani,M., Chiarantini,L.,Palma,P., and Crescentini,G.(1985) Anal. Biochem. 146: 118-124.
6. Cantarow,W, and Stollar,B.D.(1977) Arch. Biochem. Biophys. 180: 26-33.
7. Adamietz,P., Klapproth,K., and Hilz,H.(1979) Biochem. Biophys. Res. Commun. 91: 1232-1238.
8. Ferro,A.M., and Kuehl,L.(1975) Biochim. Biophys. Acta 410: 285-298.